Artificial Intelligence for Intelligent Systems

The aim of this book is to highlight the most promising lines of research, using new enabling technologies and methods based on AI/ML techniques to solve issues and challenges related to intelligent and computing systems. Intelligent computing easily collects data using smart technological applications like IoT-based wireless networks, digital health care, transportation, blockchain, 5.0 industry, and deep learning for better decision-making. AI-enabled networks will be integrated in the smart cities' concept for interconnectivity. Wireless networks will play an important role. The digital era of computational intelligence will change the dynamics and lifestyle of human beings. Future networks will be introduced with the help of AI technology to implement cognition in real-world applications. Cyber threats are dangerous to encode information from networks. Therefore, AI intrusion detection systems need to be designed for identification of unwanted data traffic.

This book:

- Provides a better understanding of artificial intelligence–based applications for future smart cities
- Presents a detailed understanding of artificial intelligence tools for intelligent technologies
- Showcases intelligent computing technologies in obtaining optimal solutions using artificial intelligence
- Discusses energy-efficient routing protocols using artificial intelligence for flying ad hoc networks (FANETs)
- Covers machine learning–based intrusion detection system (IDS) for smart grid

It is primarily written for senior undergraduate and graduate students and academic researchers in the fields of electrical engineering, electronics and communication engineering, and computer engineering.

Intelligent Data-Driven Systems and Artificial Intelligence

Series Editor: Harish Garg

Modelling of Virtual Worlds using the Internet of Things
Simar Preet Singh and Arun Solanki

Data-Driven Technologies and Artificial Intelligence in Supply Chain
Tools and Techniques
Mahesh Chand, Vineet Jain and Puneeta Ajmera

Artificial Intelligence for Intelligent Systems
Fundamentals, Challenges, and Applications
Inam Ullah Khan, Mariya Ouaissa, Mariyam Ouaissa, Muhammad Fayaz, and Rehmat Ullah

For more information about this series, please visit: www.routledge.com/Intelligent-Data-Driven-Systems-and-Artificial-Intelligence/book-series/CRCIDDSAAI

Artificial Intelligence for Intelligent Systems

Fundamentals, Challenges, and Applications

Edited by Inam Ullah Khan, Mariya Ouaissa,
Mariyam Ouaissa, Muhammad Fayaz,
and Rehmat Ullah

CRC Press
Taylor & Francis Group
Boca Raton London New York

CRC Press is an imprint of the
Taylor & Francis Group, an **informa** business

Designed cover image: shutterstock

First edition published 2025
by CRC Press
2385 NW Executive Center Drive, Suite 320, Boca Raton FL 33431

and by CRC Press
4 Park Square, Milton Park, Abingdon, Oxon, OX14 4RN

CRC Press is an imprint of Taylor & Francis Group, LLC

Library of Congress Cataloging-in-Publication Data
Names: Khan, Inam Ullah, editor.
Title: Artificial intelligence for intelligent systems : fundamentals, challenges, and applications / edited by Inam Ullah Khan [and 4 others].
Description: First edition. | Boca Raton, FL : CRC Press, 2024. | Series: Intelligent data-driven systems and artificial intelligence | Includes bibliographical references and index.
Identifiers: LCCN 2024004095 (print) | LCCN 2024004096 (ebook) | ISBN 9781032603179 (hardback) | ISBN 9781032803180 (paperback) | ISBN 9781003496410 (ebook)
Subjects: LCSH: Artificial intelligence. | Computational intelligence.
Classification: LCC Q335 .A78734 2024 (print) | LCC Q335 (ebook) | DDC 006.3—dc23/eng/20240521
LC record available at https://lccn.loc.gov/2024004095
LC ebook record available at https://lccn.loc.gov/2024004096

ISBN: 9781032603179 (hbk)
ISBN: 9781032803180 (pbk)
ISBN: 9781003496410 (ebk)

DOI: 10.1201/9781003496410

Typeset in Times New Roman
by Apex CoVantage, LLC

Contents

PART II
Secure artificial intelligence in computing systems

Preface

Recent technological infrastructure developments have diverse applications. Smart efforts from academia and industry have already been put to research on artificial intelligence (AI) for intelligent computation. Although, AI is designed to optimize 5G, 6G, the internet of things (IoT), cloud computing, edge, ad hoc networks, and intrusion detection systems, the AI-based intelligent system, network security, and data analytics need to be used for better communication. As the overall topological structure has advanced with the passage of time, there exist several issues that need to be addressed. AI is deployed in many real-time applications, including agriculture, transportation, health care, and industry. Nowadays, machine learning (ML) models have the capabilities of intelligent predication. Cognitive computing techniques schedule resources that provide better communication channels, and novel algorithms have been made possible using the ML approach to give low-cost solutions.

The aim of this book is to highlight the most promising lines of research, using new enabling technologies and methods based on AI/ML techniques to solve issues and challenges related to intelligent and computing systems. Intelligent computing easily collects data using smart technological applications like IoT-based wireless networks, digital health care, transportation, blockchain, 5.0 industry, and deep learning for better decision-making. AI-enabled networks will be integrated in the smart cities' concept for interconnectivity. Wireless networks will play an important role. The digital era of computational intelligence will change the dynamics and lifestyles of human beings. Future networks will be introduced with the help of AI technology to implement cognition in real-world applications. Cyber threats are dangerous to encode information from networks. Therefore, AI intrusion detection systems need to be designed for identification of unwanted data traffic.

This book is divided into three parts. Part I introduces the recent trends and challenges of AI. Part II goes in depth to highlight the security of intelligent systems using AI, and part III explores the role of big data analytics in current applications.

Part I: Recent trends and challenges of artificial intelligence

Here, in the opening section, we set the stage. Chapter 1 examines the interdisciplinary uses of AI and how it is revolutionizing several sectors, by introducing the concept of AI and the significance of a multidisciplinary approach in harnessing its potential. Chapter 2 introduces deep neural networks and their structures, methodologies, properties, and limits. Also discussed are the significant distinctions between deep neural networks and standard ML, as well as the big obstacles ahead. Chapter 3 focuses on providing a comprehensive comparative analysis of AI, deep learning (DL), and ML. It briefly overviews the state-of-the-art developments in these fields, highlighting key trends, challenges, and applications. Chapter 4 sheds light on the challenges and possibilities offered by these technologies by addressing the fundamental principles of big data mining and distributed processing. Also discussed are the distributed processing frameworks' roles in the administration of large datasets and the extraction of valuable intelligence. In Chapter 5, we embark on an exploration of quantum AI—an innovative domain, where quantum computing and AI converge to reshape our technological landscape.

Part II: Secure artificial intelligence in computing systems

Our journey then takes a deep dive into the security of intelligent systems-based AI. Chapter 6 investigates the importance of AI in the realm of cybersecurity as well as its potential influence on the protection of computerized information and systems. This chapter also summarizes the most recent research on the applications and implications of ML and DL techniques in cybersecurity. Chapter 7 discusses IoT security issues and threats, reviews current countermeasures, and emphasizes the promise of blockchain technology. It explores the security aspects of blockchain, highlights its main uses, and compares implementations. Chapter 8 focuses on data traffic management in AI-IoT networks to reduce congestion. It explores the underlying causes of congestion in AI-IoT networks and presents a comprehensive overview of existing congestion control mechanisms. Chapter 9 is mainly based on AI-based anomaly IDS. Also, AI-based anomaly IDS's real-time applications, future trends, and solutions are properly incorporated. In addition, technical issues regarding IDS are well explained, which will help AI/ML engineers, researchers, and practitioners in the near future.

Part III: Big data analytics applying in current applications

Big data analytics in current applications emerges as a focal point. Chapter 10 focuses on particular objectives and the application of certain methods, the many subfields of AI research, and big data–based health care. Some of the common objectives of AI research include the ability to argue; communicate knowledge; plan, learn, and understand natural language; sense, move, and control objects. Chapter 11 investigates how big data can be used in health care. It looks at where the data comes from, how it is analyzed, how it is used, current trends and issues, and future possibilities in this field. Chapter 12 aims to improve the diagnosis of hepatitis C by comprehensively analyzing several ML algorithms in this study. In Chapter 13, the authors propose an approach of profound learning as a method for combining algorithms for the classification of Alzheimer's disease. Chapter 14 develops a prediction model that successfully separates individuals with liver disorders from those who do not, to investigate the use of ML algorithms in the early diagnosis of liver diseases. Chapter 15 provides an in-depth introduction to the world of intelligent transportation channels for intelligent cities, exploring the fundamental principles, technologies, and potential benefits associated with their implementation. Chapter 16 covers an in-depth discussion of digital twin (DT) technology and the way Industry 4.0 and 5.0 are making use of it. It explores the key concept of a DT and its vital components while reviewing a variety of DTs now being utilized in the industry, including product, process, and system ones.

About the editors

Dr. Inam Ullah Khan is the founder of the Internet of Flying Vehicles-Lab at AI-EYS. Currently, he is working as global mentor/guest lecturer at Impact Xcelerator, IE School of Science and Technology, Madrid, Spain. Previously, he was working as visiting researcher at King's College London, United Kingdom. Also, he was a faculty member at different universities in Pakistan, which include the Center for Emerging Sciences, Engineering and Technology (CESET), Islamabad; Abdul Wali Khan University, Garden Campus and Timergara Campus; University of Swat; and Shaheed Zulfikar Ali Bhutto Institute of Science and Technology (SZABIST), Islamabad Campus. He completed his PhD in Electronics Engineering from the Department of Electronics Engineering, School of Engineering & Applied Sciences (SEAS), Isra University, Islamabad Campus. Also, he completed his MS degree in Electronics Engineering at the Department of Electronics Engineering, School of Engineering & Applied Sciences (SEAS), Isra University, Islamabad Campus. He completed his Bachelor of Computer Science degree at Abdul Wali Khan University, Mardan, Pakistan. He has authored/coauthored more than 50 research articles in reputable journals, conferences, and book chapters. His research interest includes network system security, intrusion detection, intrusion prevention, cryptography, optimization techniques, WSN, IoT, mobile ad hoc networks (MANETS), flying ad hoc networks, and machine learning. He served in many international conferences as session chair/technical program committee member. Also, he served as guest editor with many prestigious international journals. Apart from this, he was the general chair at the International Conference on Trends and Innovations in Smart Technologies (ICTIST '22), virtually from London, United Kingdom. In addition, he also served as the editor of several books. More interestingly, he was invited as a technology expert many times on Pakistan National Television and other media outlets.

Dr. Mariya Ouaissa is currently a professor in cybersecurity and networks at the Faculty of Sciences Semlalia, Cadi Ayyad University, Marrakech, Morocco. She is a PhD, graduated in 2019, in Computer Science and Networks, from Laboratory of Modelisation of Mathematics and Computer Science, ENSAM-Moulay Ismail University, Meknes, Morocco. She is a networks and telecoms engineer, graduated in 2013, from the National School of Applied Sciences, Khouribga, Morocco. She is a cofounder and IT consultant at the IT Support and Consulting Center, Meknes, Morocco. She was working for the School of Technology, Meknes, Morocco, as a visiting professor from 2013 to 2021. She is a member of the International Association of Engineers and International Association of Online Engineering, and from 2021, she is also an ACM professional member. She is an expert reviewer with the Academic Exchange Information Centre (AEIC) and a brand ambassador with Bentham Science. She has served and continues to serve on technical program and organizer committees of several conferences and events and has organized many symposiums/workshops/conferences as a general chair and is also a reviewer of numerous international journals. Dr. Ouaissa has made contributions in the fields of information security and privacy, internet of things (IoT) security, and wireless and constrained networks security. Her main research topics are IoT, M2M, D2D, WSN, cellular networks, and vehicular networks. She has published over 40 papers (in book chapters, international journals, and conferences/workshops), 12 edited books, and 8 special issues as guest editor.

Dr. Mariyam Ouaissa is currently an assistant professor in Networks and Systems at ENSA, Chouaib Doukkali University, El Jadida, Morocco. She received her PhD degree in 2019 from National Graduate School of Arts and Crafts, Meknes, Morocco, and her engineering degree in 2013 from the National School of Applied Sciences, Khouribga, Morocco. She is a communication and networking researcher and practitioner with industry and academic experience. Dr. Ouaissa's research is multidisciplinary and focuses on the internet of things, M2M, WSN, vehicular communications and cellular networks, security networks, the congestion overload problem, and resource allocation management and access control. She is serving as a reviewer for international journals and conferences, including *IEEE Access* and *Wireless Communications and Mobile Computing*. Since 2020, she has been a member of the International Association of Engineers (IAENG) and International Association of Online Engineering; since 2021, she is also an ACM professional member. She has published more than 30 research papers (this includes book chapters, peer-reviewed journal articles, and peer-reviewed conference manuscripts), 10 edited books, and 6 special issues as guest editor. She has served on program committees and organizing committees of several conferences and events and has organized many symposiums/workshops/conferences as a general chair.

Dr. Muhammad Fayaz is an assistant professor of Computer Science at UCA's School of Arts and Sciences. Before joining UCA, he worked as a visiting lecturer and teaching assistant at the University of Malakand in Chakdara (Pakistan) for two years. Throughout his academic career, he has published over 100 research papers in the reputed Web of Science and Scopus-indexed international journals and conferences. His research interests include combinatorial optimization problems, machine learning, image processing, internet of things (IoT), fuzzy inference systems, and other related areas. He earned his PhD in computer engineering from Jeju National University (JNU) in Jeju, South Korea, in 2019. During his studies, he worked at a Mobile Computing Lab at JNU. His responsibilities included preparing project proposals, designing IoT-based smart solutions, the application and implementation of artificial intelligence algorithms, and writing project reports. During his stay in South Korea, he also worked on several projects on the underground system and the SAR-based big data classification and analysis of disaster/damage type, funded by the Electronics and Telecommunication Research Institute and Korean Aerospace Research Institute, respectively.

Dr. Rehmat Ullah earned his PhD degree in electronics and computer engineering from Hongik University, South Korea. He currently works as an assistant professor at the school of technologies, Cardiff Metropolitan University, UK. Previously, he worked as a research fellow (postdoctorate) at the University of St Andrews, UK, and Queen's University Belfast, UK, and as an assistant professor at Gachon University, South Korea. Dr. Rehmat has been a member of the technical program committees of several flagship conferences such as ACM ICN, ACM IMC, ACM MobiHoc, IEEE VTC, and IEEE ICC, and he is an invited reviewer for numerous high-impact journals such as IEEE/Elsevier/ACM/Wiley. His work has appeared in more than 45 research papers in peer-reviewed conferences and top journals such as *IEEE Communications Magazine*, *IEEE Internet of Things Journal*, *IEEE Transactions on Network Science and Engineering*, *IEEE Wireless Communications Magazine*, *Journal of Network and Computer Applications*, and *Future Generation Computer Systems*. He currently holds four patents, and two US patents are under review. His research focuses on the broader areas of network and distributed systems, particularly the development of architectures, algorithms, and protocols for emerging paradigms, such as edge computing, internet of things, information centric networking, and 5G evolution and beyond, with a recent focus on federated learning for edge computing systems. This includes the design, measurement studies, prototyping, test bed development, and performance evaluations.

Contributors

A. M. Abdullahi
Faculty of Computing and Informatics
Albukhary International University
Kedah, Malaysia

Deepak Adhikari
University of Electronic Science and Technology of China
Chengdu, China

Abdullah Akbar
Department of Electrical Engineering
National University of Computer and Emerging Sciences, Peshawar Campus
Peshawar, Pakistan

S. Akhtar
Department of Computer Science
National University of Computer and Emerging Sciences, Chiniot-Faisalabad Campus
Chiniot, Pakistan

Md. Alif Sheakh
Daffodil International University
Dhaka, Bangladesh

H. Arshad
Department of Computer Science
National University of Computer and Emerging Sciences, Chiniot-Faisalabad Campus
Chiniot, Pakistan

Faiza Ayyob
University of Okara
Okara, Pakistan

Muhammad Azeem
University of Salford
Manchester, UK

Indu Bala
SEEE
Lovely Professional University
Punjab, India

Sabitha Banu A.
Sri Ramakrishna College of Arts and Science for Women
Coimbatore, India

Inam Bari
Department of Systems Engineering
Military Technological College
Muscat, Oman

Dipen Bepari
National Institute of Technology
Raipur, Chhattisgarh, India

M. Bilal
Department of Computing and Information Systems, School of Engineering and Technology
Sunway University
Petaling Jaya, Selangor, Malaysia

Muhammad Bux
Communication Engineering, Engineering and Architecture
University of Parma
Parma, Italy

Gagandeep
Chitkara University
Punjab, India

K. Gurucharan
ECE Department
Lendi Institute of Engineering & Technology
Vizianagaram, India

Muhammad Imad
Ulster University
Belfast, UK

Taminul Islam
Daffodil International University
Dhaka, Bangladesh

Rishalatun Jannat Lima
Daffodil International University
Dhaka, Bangladesh

Sanaa Jeehan
Department of Computer Science
National University of Computer and Emerging Sciences, Peshawar Campus
Peshawar, Pakistan

Abinaya Keerthana S.
Sri Ramakrishna College of Arts and Science for Women
Coimbatore, India

Habib Khan
Department of Computer Science
Islamia College University
Peshawar, Pakistan

Ijaz Khan
School of Electronics and Information Engineering
Harbin Institute of Technology
Harbin, China

S. S. Kiran
ECE Department
Lendi Institute of Engineering & Technology
Vizianagaram, India

Hanane Lamaazi
College of Information Technology
UAE University
Al Ain, UAE

Shahzad Latif
Department of Computer Science
SZABIST University, Islamabad Campus
Islamabad, Pakistan

Elezabeth Mathew
College of Information Technology
UAE University
Al Ain, UAE

Soumen Mondal
National Institute of Technology
Durgapur, West Bengal, India

Zaid Muhammad
Abdul Wali Khan University, Pabbi Campus
Pakistan

Maqsood Muhammad Khan
Department of Electrical Engineering
National University of Computer and Emerging Sciences, Peshawar Campus
Peshawar, Pakistan

Md. Musfiqur Rahman Foysal
Daffodil International University
Dhaka, Bangladesh

Md. Muzahidul Islam
Daffodil International University
Dhaka, Bangladesh

Farhood Nishat
Department of Information Security, Institute of System Engineering
Riphah International University, Islamabad Campus
Islamabad, Pakistan

Nafeya Rahman Heya
Daffodil International University
Dhaka, Bangladesh

Sajidur Rahman Sajid
Daffodil International University
Dhaka, Bangladesh

Mallavarapu Rajanbabu
ECE Department
Lendi Institute of Engineering & Technology
Vizianagaram, India

Muhammad Rouf
University of Engineering and Technology
Peshawar, Pakistan

Mst. Sazia Tahosin
Daffodil International University
Dhaka, Bangladesh

Anwar Shah
National University of Computer and Emerging Sciences, Faisalabad Campus
Islamabad, Pakistan

P. Srujana
ECE Department
Lendi Institute of Engineering & Technology
Vizianagaram, India

Hamed Taherdoost
Department of Arts, Communications and Social Sciences
University Canada West
Vancouver, BC, Canada

M. Tayyab
School of Computer Science
Taylor's University, Lakeside Campus
Subang Jaya, Malaysia

Inam Ullah
Department of Computer Engineering
Gachon University
Seongnam, Sujeong-gu, Republic of Korea

Najeeb Ullah
Department of Computer Science
National University of Science and Technology
Quetta, Pakistan

Rahat Ullah
University of Wah
Wah Cantonment, Pakistan

Ubaid Ullah
University of Wah
Wah Cantonment, Pakistan

Muhammad Usama
University of Wah
Wah Cantonment, Pakistan

Omar Usman Khan
Department of Computer Science
National University of Computer and Emerging Sciences, Peshawar Campus
Peshawar, Pakistan

Jyoti Verma
Chitkara University
Punjab, India

Fazal Wahab
College of Computer Science and Technology
Northeastern University
Shenyang, China
School of Electronics & Computer Science
University of Southampton Malaysia
Iskandar Puteri, Malaysia

Part I

Recent trends and challenges of artificial intelligence

Chapter 1

Unleashing the power of artificial intelligence

Exploring multidisciplinary frontiers for innovation and impact

Gagandeep and Jyoti Verma

1.1 INTRODUCTION

1.1.1 Brief overview of artificial intelligence and its capabilities

Artificial intelligence (AI) is the emulation of human intellect in computers that are designed to carry out tasks like learning, decision-making, and problem-solving and that normally need human intelligence. It's a vast field that includes a number of subfields, including robotics, computer vision, natural language processing, and machine learning.

Figure 1.1 visually illustrates the relationships between AI subfields and their specific applications, providing a clear overview of the diverse capabilities of AI. The graph structure and node styles are used to represent the hierarchy and associations within the AI field, aiding in understanding the various domains and applications encompassed by AI. Recent years have witnessed tremendous progress in AI thanks to increases in processing power, the availability of big datasets, and innovations in algorithmic techniques. Machine learning, which enables systems to learn from data and improve performance over time, is one of the core skills of AI. Large datasets can be analyzed by machine learning algorithms to find trends, forecast outcomes, and reveal new information. Deep learning, a branch of machine learning that uses artificial neural networks to teach them to recognize hierarchical data representations, is one such example. In domains like speech recognition, picture identification, and natural language processing, deep learning has shown impressive promise [1]. Another important AI feature that focuses on the communication between computers and human languages is natural language processing, or NLP. Machines can now comprehend, interpret, and produce human language thanks to NLP approaches. NLP is used in a variety of contexts, including sentiment analysis tools, language translation systems, and virtual assistants like Siri and Alexa [2]. AI's field of computer vision focuses on giving robots the ability to recognize and comprehend visual data. By leveraging image and video analysis techniques, computer vision enables systems to identify objects, recognize faces, and understand scenes. Applications of computer vision include

DOI: 10.1201/9781003496410-2

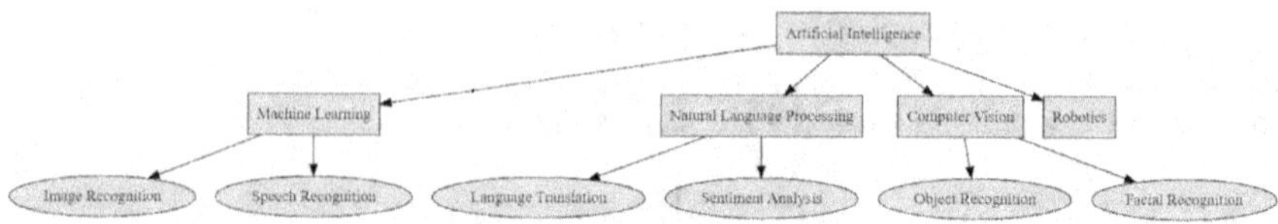

Figure 1.1 AI subfields and capabilities.

autonomous vehicles, facial recognition systems, and medical image analysis [3]. AI also plays a significant role in robotics, where intelligent machines are designed to interact with the physical world. AI-capable robots can carry out hazardous, repetitive, or very precise tasks. They are employed in a number of sectors, such as exploration, manufacturing, and health care [4]. The capabilities of AI extend beyond individual domains. AI systems can exhibit general intelligence, where they possess the ability to understand and perform tasks across different domains. However, achieving human-level general intelligence remains an ongoing challenge in the field [5].

1.1.2 Importance of multidisciplinary approach in AI applications

The multidisciplinary approach is crucial in AI applications, as it brings together expertise from various fields to tackle complex challenges and drive innovation. AI researchers and practitioners can create more reliable and efficient solutions by combining knowledge and methods from several fields, including computer science, mathematics, cognitive science, and domain-specific fields as indicated in Figure 1.2.

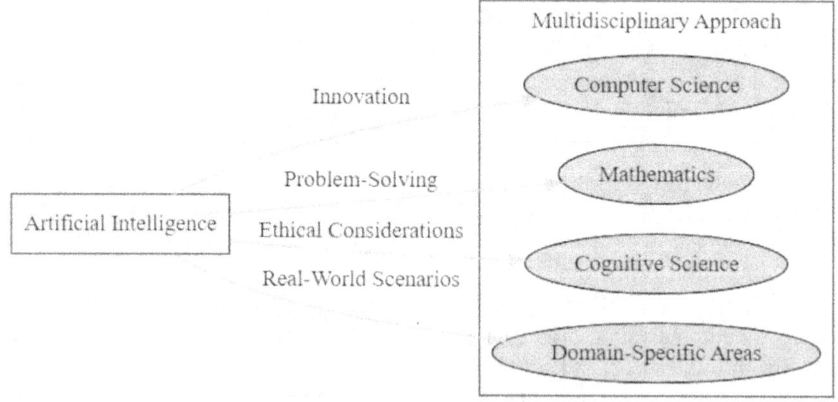

Figure 1.2 The multidisciplinary approach in AI.

The capacity to use domain expertise to improve AI algorithms and systems is one of the fundamental advantages of the multidisciplinary approach to AI. Domain-specific expertise helps in understanding the nuances of the problem, designing appropriate models, and interpreting the results in meaningful ways. As highlighted by Domingos [6], "In many cases, the biggest challenge to building a successful AI system is not the AI techniques, but the ability to access, capture, and represent the right domain knowledge." Additionally, collaboration between experts from different disciplines promotes cross-pollination of ideas and fosters innovation. By combining insights from diverse perspectives, researchers can develop novel approaches that push the boundaries of AI. A study by Brynjolfsson and McAfee [7] emphasized the importance of interdisciplinary collaboration in driving AI advancements and highlighted that breakthroughs often occur at the intersection of multiple fields. Furthermore, the multidisciplinary approach helps address ethical and societal implications associated with AI applications. It enables a more comprehensive analysis of the ethical, legal, and social aspects of AI, which promotes the creation of accountable and human-centered AI systems. A report by the AI100 Standing Committee and Study Panel [8] stressed the need for multidisciplinary engagement to address issues such as bias, fairness, accountability, and transparency in AI systems. Moreover, the multidisciplinary approach enables AI to tackle complex real-world problems that require a comprehensive understanding of the context. For instance, AI applications in health care often involve collaboration between AI experts, medical professionals, and biomedical researchers to develop solutions that improve patient outcomes and assist in diagnosis [9].

1.2 HEALTH CARE

1.2.1 AI in medical diagnosis and treatment

AI has become a potent diagnostic and therapeutic tool that is transforming the medical field by enhancing human potential and patient outcomes. AI systems are able to help with precise diagnosis, individualized treatment planning, and effective health care delivery through the analysis of large volumes of medical data and the use of sophisticated algorithms. Medical image processing is one field where AI has demonstrated great promise. Deep learning algorithms have proven to be quite effective at identifying anomalies and supporting radiologists in the interpretation of medical pictures. For instance, a study by Esteva et al. [9] showed that by examining pictures of skin lesions, a deep neural network could classify skin cancer at the dermatologist level. The algorithm's performance was shown to be on par with dermatologists with board certifications, demonstrating the potential of AI to improve diagnostic precision. AI is also essential for early disease diagnosis and detection. Scientists have created AI-based systems that can examine patient data, including genetic data and electronic health records, to find trends and risk factors linked to different diseases. For instance, in a study published in *Nature Medicine*, Choi et al. [10] developed an AI model that accurately predicted the development of cardiovascular disease by analyzing electronic health records. Furthermore, AI algorithms have been employed in treatment planning and decision-making. By analyzing patient-specific data, such as genetic information, medical history, and treatment outcomes, AI systems can provide personalized treatment recommendations. A notable example is the use of AI in precision oncology, where algorithms help identify optimal treatment strategies based on tumor characteristics and genetic profiles [11]. AI also supports health care providers in improving patient care and optimizing workflows. Chatbots and virtual assistants powered by AI can interact with patients, provide information, and assist in triaging cases based on symptom analysis. This can help alleviate the burden on health care professionals and ensure timely access to appropriate care. Moreover, AI-enabled predictive analytics models can aid in hospital resource management, predicting patient admissions, optimizing bed utilization, and improving operational efficiency [12].

1.2.1.1 Machine learning algorithms for disease detection

Machine learning algorithms have demonstrated remarkable capabilities in disease detection, aiding health care professionals in accurate and timely diagnosis. These algorithms analyze large datasets and learn patterns to identify disease indicators, contributing to early detection and improved patient outcomes. Numerous researches have demonstrated how well machine learning works in a variety of medical fields to identify diseases. Deep learning algorithms have demonstrated remarkable effectiveness in the field of radiology when it comes to identifying anomalies in medical images.

Figure 1.3 provides a graphic representation of how AI technologies, including convolutional neural networks (CNNs) and deep learning, support medical image processing by helping radiologists identify abnormalities and aid in the diagnostic process. For example, a CNN was used in a study by Gulshan et al. [13] to identify diabetic retinopathy in retinal fundus photos. The algorithm showed promise for automated illness screening by achieving high sensitivity and specificity, which are comparable to the performance of human ophthalmologists. Neurological problems have also been diagnosed using machine learning techniques. A support vector machine (SVM) algorithm was applied in a study by Fakoor, Ladhak, Nazi, and

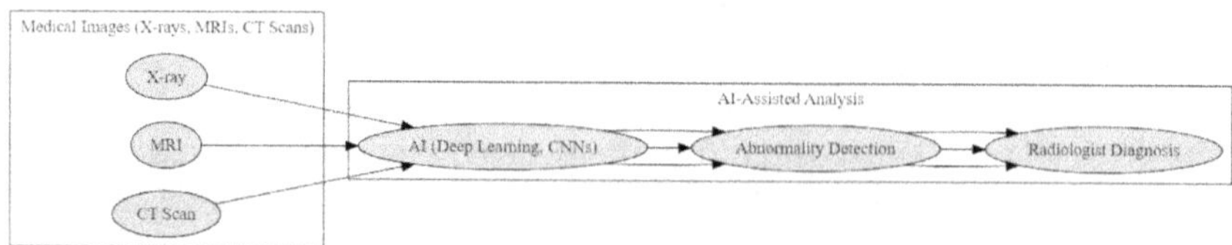

Figure 1.3 AI in medical imaging analysis.

Huber [14] to categorize Alzheimer's disease and moderate cognitive impairment based on structural magnetic resonance imaging (MRI) data. The SVM algorithm proved to be a useful tool for early identification, with great accuracy in differentiating between individuals with cognitive impairment and those in good health. Machine learning algorithms have demonstrated potential in the field of cardiovascular disease diagnosis and prognosis. For instance, a deep learning model was created in a work by Attia et al. [15] to identify frequent heart arrhythmia atrial fibrillation by analyzing ECG data. The algorithm's promise as an atrial fibrillation screening tool was demonstrated by its high sensitivity and specificity. Algorithms for machine learning have also been used to detect cancer. One noteworthy instance is the application of machine learning to mammography-based breast cancer diagnosis. In Kooi et al. [16], radiologists' and a deep learning algorithm's effectiveness in identifying breast cancer from mammography pictures were evaluated. The deep learning algorithm demonstrated accuracy on par with radiologists, suggesting that it could be used as a screening tool to help detect breast cancer.

1.2.1.2 AI-assisted surgical procedures

AI has attracted much attention lately because it has the potential to improve patient outcomes, efficiency, and precision in surgical procedures. Surgical operations are being improved and optimized by the incorporation of AI technology, such as computer vision, machine learning, and robotics. AI has been applied in a variety of surgical fields, as demonstrated by numerous researches that highlight its potential to completely transform health care procedures. Robot-assisted surgery is one field where AI has showed potential. AI has been added to robotic surgical systems, such as the da Vinci surgical system, to help surgeons perform difficult surgeries. For instance, Ahmad, Kulkarni, Wang, Catto, and Aboumarzouk [17] looked into the application of AI for robotically assisted radical prostatectomy in the field of urology. The authors demonstrated that AI algorithms could enhance surgical performance by providing real-time feedback and guidance to surgeons, leading to improved surgical outcomes. AI has also been utilized in image-guided surgical interventions. In the domain of neurosurgery, AI algorithms have been developed to assist in the segmentation and analysis of brain structures from preoperative imaging data. An example of this is the work by Charron, Lalys, and Jannin [18], where a deep learning model was employed to automate the segmentation of brain tumors in MRI scans. The AI system achieved high accuracy in tumor detection, aiding neurosurgeons in treatment planning and tumor resection. Furthermore, AI has shown potential in surgical decision support systems. To help surgeons make well-informed decisions, these systems use machine learning algorithms to assess patient data, including imaging results, clinical outcomes, and medical history. A machine learning algorithm was created by Jiang et al. [19] to forecast the likelihood of postoperative problems in patients undergoing heart surgery. The model utilized patient characteristics and preoperative variables to provide personalized risk assessment, supporting surgeons in optimizing patient care and outcomes.

1.2.2 AI for drug discovery and personalized medicine

AI has become a potent instrument in personalized medicine and drug development, transforming the pharmaceutical sector by speeding up the identification of new drug candidates and facilitating more accurate treatment strategies. Target identification, virtual screening, medication design, and patient stratification have all advanced significantly as a result of the integration of AI techniques like machine learning and deep learning. Target validation and identification is a noteworthy use of AI in drug research. Large-scale omics data, such as transcriptomics, proteomics, and genomes, can be analyzed by AI algorithms to find possible treatment targets and clarify illness causes. To forecast the effectiveness of drug-target interactions, for example, Ching et al. [20] used deep learning models, which made it possible to identify new target-drug correlations. This strategy may broaden the focus of drug development initiatives and hasten the identification of new therapeutic targets. AI tools are being applied to virtual screening to find high-potency compounds for drug development by computationally screening massive chemical libraries. Goh, Hodas, and Vishnu [21] showed how deep learning models may be used to forecast the characteristics and activities of tiny compounds, making it easier to find potential therapeutic options. The benefit of this strategy is that it shortens the duration and lowers the expense of experimental screening, making drug discovery procedures more effective. Furthermore, AI helps create new chemical structures with desirable features, which is a critical aspect of medication creation. Deep generative models have been used to create novel compounds with certain pharmacological properties. Examples of these models are variational autoencoders (VAEs) and generative adversarial networks (GANs). For example, a generative model was used in a work [22] to generate compounds with desirable features for drug discovery. This strategy could improve drug design processes' effectiveness and success rate. AI can evaluate vast amounts of patient data, such as genetic information, clinical imaging, and electronic health records, to enable more accurate diagnosis, treatment planning, and prognosis in customized medicine. According to a study by Esteva et al. [9], dermatologists' level of accuracy in classifying skin cancer photos was matched by deep learning models. This demonstrates how AI can support clinical judgment and enhance patient outcomes.

1.2.2.1 Predictive modeling for drug development

To choose compounds with the highest chance of success and make well-informed judgments, predictive modeling is essential to the drug development process. Predictive models can find possible drug candidates, forecast their pharmacokinetic features, evaluate safety profiles, and enhance formulation techniques by utilizing machine learning algorithms and large-scale data analysis. Predictive modeling can bring about significant cost-savings and speed up the search for new treatments when it is used in drug development. Predicting drug-target interactions is one way that predictive modeling is used in drug development. To anticipate possible drug-target interactions, machine learning algorithms can evaluate and combine data from a variety of sources, such as chemical structure, protein sequence, and information on known interactions. For instance, Ozturk, Ozkirimli, and Ozgur's [23] DeepDTA is a deep learning model that predicts drug-target binding affinities with accuracy, making it easier to identify possible therapeutic candidates. Predictive modeling is useful not only for anticipating drug-target interactions but also for evaluating the pharmacokinetic characteristics of potential new drugs. Important factors, including absorption, distribution, metabolism, and excretion (ADME), can be estimated by computational models based on machine learning methods. These models can direct the design of therapeutic candidates with enhanced bioavailability and decreased toxicity as well as assist in identifying molecules with the best pharmacokinetic properties. The selection of drug candidates with favorable pharmacokinetic features is aided by the application of machine learning algorithms to predict human liver microsomal stability, as evidenced by a study by Ekins et al. [24]. Predictive modeling can also be used to evaluate the safety profiles of potential medications. Based on structural traits and compound properties, machine learning algorithms trained on toxicity data can forecast possible side effects, such as hepatotoxicity or cardiotoxicity. For example, a work by Sedykh et al. [25] used machine learning approaches to construct a predictive model for hepatotoxicity, allowing the early identification of substances with possible safety problems. In the process of developing new drugs, predictive modeling also helps with formulation optimization. Models can predict the solubility, stability, and bioavailability of drug candidates by utilizing machine learning algorithms and formulation data. This helps with the selection of suitable formulations and delivery systems. To aid in the formulation optimization process, Wang, Van Eerdenbrugh, Karlsson, Kou, and Bergström [26] employed machine learning models to forecast the stability of amorphous solid dispersions.

1.2.2.2 Precision medicine and patient-specific treatments

Precision medicine, which is often referred to as customized medicine, endeavors to customize medical interventions for individual patients according to their distinct attributes, including genetics, lifestyle, and environment. Precision medicine is made possible by advanced technologies such as AI and machine learning, which analyze enormous volumes of patient data to provide insights for individualized diagnosis, treatment selection, and therapeutic response prediction (see Figure 1.4). By enhancing patient outcomes and cutting costs, the use of AI in precision medicine has the potential to completely transform the health care industry. Genomic research is one area where AI is significantly advancing precision medicine. Genetic variations linked to particular diseases and treatment outcomes can be identified through the use of machine learning algorithms in the analysis of genomic data. For instance, Yang et al. [27] developed the Genomics of Drug Sensitivity in Cancer (GDSC) database, which uses machine learning methods

Figure 1.4 Precision medicine workflow with AI.

to correlate the genetic profiles of cancer cell lines with their susceptibility to different anticancer medicines. Based on a patient's genetic composition, this information can help doctors choose tailored treatments. The creation of treatment response prediction models is another way that AI is being used in precision medicine. To forecast a patient's reaction to a particular medication, machine learning algorithms can evaluate a variety of patient data, such as genetic information, clinical records, and lifestyle factors. For example, a machine learning model was created in a study by Geeleher, Cox, and Huang [28] to predict how breast cancer patients would react to neoadjuvant chemotherapy. This model enables tailored treatment decisions based on anticipated results. AI is essential for medical imaging analysis as well, as it helps to identify illness patterns and create treatment strategies tailored to individual patients. To identify and categorize anomalies in medical image analysis, CNNs are frequently utilized.

For instance, a study by Esteva et al. [9] used pictures of skin lesions to show how well a CNN classified skin cancer. Based on imaging data, this technology may help physicians diagnose illnesses and choose the best courses of action. AI plays a key role in the analysis of electronic health records (EHRs) for precision medicine, in addition to genomics and medical imaging. EHR data can be mined by machine learning algorithms to find trends, correlations, and disease risk factors. To enable preemptive treatments and individualized care, Rajkomar et al. [29] conducted a study in which they applied machine learning techniques to predict patient deterioration using EHR data.

1.3 FINANCE AND BANKING

1.3.1 AI for fraud detection and prevention

AI has shown to be an effective tool for both identifying and stopping fraud in a variety of businesses. AI-based fraud detection systems have the potential to greatly improve fraud prevention efforts by evaluating vast amounts of data, spotting patterns, and spotting abnormalities. Promising outcomes have been observed in terms of increasing accuracy, decreasing false positives, and adjusting to changing fraud patterns when machine learning algorithms and AI technologies are integrated into fraud detection systems. The finance industry is one where AI has significantly improved fraud detection. Large volumes of financial transaction data may be analyzed in real time by sophisticated machine learning techniques, which makes it possible to identify potentially fraudulent activity and suspicious activity. For instance, Phua, Lee, Smith-Miles, and Gayler [30] investigated the application of machine learning algorithms, such as SVMs and neural networks, for the identification of credit card fraud. The outcomes showed how well these algorithms worked at correctly identifying fraudulent transactions. Additionally, to combat medical fraud and abuse, AI-based fraud detection systems are being used in the health care sector. AI systems can detect patterns in health care claims data, such as duplicate claims, abnormal billing, or needless medical treatments, that point to fraudulent activity. Based on patient treatment records, a study by Pires, Garcia, and Santos [31] used machine learning algorithms to identify health care fraud. The study illustrated how AI can lower health care fraud and increase the accuracy of fraud detection. AI fraud detection is also progressing significantly in e-commerce and on internet platforms. Due to the increase in online transactions, AI algorithms are now able to detect fraudulent behaviors, including account takeovers and payment fraud, by analyzing user behavior, transaction history, and other pertinent data. For example, Li et al. [32] suggested an AI-based fraud detection solution for e-commerce platforms that uses machine learning methods to accurately identify fraudulent orders. AI is being used not only to stop transactional fraud but also to identify and stop insurance fraud. AI algorithms have the ability to detect possibly fraudulent insurance claims by examining past claims data and spotting unusual patterns. To accurately identify fraudulent claims, Su, Wen, and Chen [33] built an AI-based fraud detection model for motor insurance by employing machine learning approaches.

1.3.1.1 Machine learning for anomaly detection

The isolation forest approach, put forth by Liu, Ting, and Zhou [34], is a well-liked machine learning algorithm for anomaly identification. This algorithm constructs isolation trees to isolate anomalies from normal instances, making it suitable for detecting outliers and anomalies in datasets. The isolation forest algorithm has been successfully applied in various domains, including network intrusion detection [35] and credit card fraud detection [36]. The one-class SVM is another often used machine learning method for anomaly detection. Schölkopf, Platt, Shawe-Taylor, Smola, and Williamson [37] proposed the one-class SVM technique, which learns a boundary around normal instances. This boundary allows anomalies to be identified as instances that reside outside it. One-class SVM has been applied in diverse fields, such as intrusion detection [38] and fraud detection [39]. Promising outcomes in anomaly identification have also been observed using deep learning techniques, specifically autoencoders. Neural networks that have been taught to reconstruct input data are called autoencoders, and variations in reconstruction error may be a sign of anomalies. This method has been used in a number of fields, such as industrial fault detection [40] and cybersecurity [41]. Ensemble methods, such as random forests, have also been employed for anomaly detection. By combining multiple anomaly detection models, ensemble methods can enhance detection accuracy and robustness. A study by Zimek, Schubert, and Kriegel [42] compared various ensemble methods for anomaly detection and highlighted their effectiveness in improving performance.

1.3.1.2 Behavioral biometrics and authentication

One significant application of behavioral biometrics is in continuous authentication systems. These systems continuously monitor and analyze user behaviors during their interactions with digital devices to verify their identities. Researchers have proposed various techniques, such as keystroke dynamics [43] and mouse dynamics [44], for continuous user authentication. By continuously monitoring and analyzing behavioral patterns, these systems can detect anomalies or unauthorized access attempts. Furthermore, behavioral biometrics can also be used for user identification in mobile devices. Tap patterns, swipe gestures, and touch dynamics have been explored as potential authentication methods for mobile devices [45]. These approaches provide an additional layer of security beyond traditional PINs or passwords, as they rely on the unique behavioral characteristics of the users. Research in behavioral biometrics and authentication is ongoing, with a focus on improving accuracy and reliability. SVMs and neural networks are two examples of machine learning methods that are frequently used to train models and efficiently identify behavioral patterns [46]. Additionally, researchers are exploring the combination of multiple behavioral biometrics modalities to create robust and multimodal authentication systems.

1.3.2 AI-based investment strategies and risk assessment

The financial industry has shown a great deal of interest in AI-based risk assessment and investing techniques because of their potential to improve portfolio management and decision-making. Sentiment analysis in financial markets and predictive analytics for portfolio management are two important areas where AI is having an impact in this field.

1.3.2.1 Predictive analytics for portfolio management

Using machine learning algorithms, predictive analytics examines past data to find trends that can be used to forecast future asset performance and market trends. The application of AI methods for risk assessment and portfolio optimization has been the subject of numerous studies. For instance, Zhou, Li, Wu, and Zhang [47] created a hybrid model incorporating machine learning algorithms and statistical techniques for portfolio selection, while Chen, Song, Bai, and Wei [48] suggested a deep learning–based model for predicting stock prices. These studies show how AI may enhance portfolio management by optimizing investment methods and producing more precise predictions.

1.3.2.2 Sentiment analysis in financial markets

Sentiment analysis is the process of gleaning information about investor behavior and market sentiment from textual data, including news articles, social media posts, and financial reports. Enormous amounts of unstructured data are analyzed and interpreted using AI techniques, including NLP and machine learning. Empirical studies have demonstrated the predictive power of sentiment research in identifying market trends and stock market movements. For example, Bollen, Mao, and Zeng [49] developed a stock market volatility prediction model using Twitter data. In a similar vein, Zhang, Fuehres, and Gloor [50] used sentiment analysis on financial news stories to forecast changes in stock prices. These studies demonstrate how sentiment analysis powered by AI has the ability to offer insightful information for financial decision-making.

1.4 TRANSPORTATION AND LOGISTICS

1.4.1 Autonomous vehicles and intelligent transportation systems

1.4.1.1 Self-driving cars and their impact on transportation

Autonomous vehicles have surfaced as a transformative technology in the transportation domain, possessing the capability to completely alter mobility and urban landscapes. Research by Anderson, Kalra, Stanley, Sorensen, Samaras, and Oluwatola [51] discusses the technical advancements and policy implications of self-driving cars, highlighting their potential benefits in terms of reducing traffic congestion, improving road safety, and enhancing transportation efficiency.

1.4.1.2 AI-powered traffic management systems

AI-driven traffic management systems employ machine learning algorithms to evaluate traffic data in real time, enhance traffic flow, and boost the effectiveness of transportation. For instance, Huang, Xu, Zhang, and Zhao [52] propose an AI-based traffic prediction model that leverages deep learning techniques to forecast traffic congestion patterns, enabling proactive traffic management and route optimization.

1.4.2 Optimization algorithms for supply chain management

1.4.2.1 Route optimization and inventory management

Because they maximize inventory levels, cut down on delivery costs, and optimize transportation routes, optimization algorithms are essential to supply chain management. Ahuja, Magnanti, and Orlin [53] discuss various optimization techniques, including linear programming, network flow models, and heuristic algorithms, for solving transportation and inventory management problems in supply chains.

1.4.2.2 Predictive maintenance and asset tracking

Machine learning algorithms are used by AI-based predictive maintenance and asset monitoring systems to evaluate sensor data and forecast equipment faults, allowing for preemptive maintenance and reducing downtime. Research by Wang, Shen, Zhang, and Zhang [54] presents an AI-driven framework for predicting maintenance needs in supply chain systems, leading to improved asset utilization and cost-savings.

1.5 EDUCATION

1.5.1 Intelligent tutoring systems and personalized learning

1.5.1.1 Adaptive learning platforms

Adaptive learning platforms leverage AI algorithms to deliver personalized educational content and adapt the learning experience to the individual needs of students. Kebritchi, Hirumi, and Bai [55] discuss the benefits of adaptive learning platforms in terms of increased engagement, improved learning outcomes, and enhanced retention rates.

1.5.1.2 Natural language processing for automated grading

NLP techniques enable automated grading of student assignments and provide timely feedback. Research by Attali and Burstein [56] explores the effectiveness of automated essay scoring systems based on NLP algorithms, highlighting their reliability and consistency compared to human grading.

1.5.2 AI-based student support and career guidance

1.5.2.1 Virtual assistants for student services

Virtual assistants powered by AI technologies, such as chatbots, are increasingly used in educational institutions to provide support and guidance to students. For instance, research by Arnold and Pistilli [57] explores the use of virtual assistants for student advising, showcasing their potential in assisting students with course selection, registration, and other administrative tasks.

1.5.5.2 Predictive analytics for student success

By analyzing student data and projecting academic outcomes, predictive analytics makes use of AI and machine learning algorithms to enable focused interventions and support. Baker and Siemens [58] discuss the application of predictive analytics in education and its potential for identifying at-risk students and implementing early intervention strategies. AI has transformative potential in the field of education, enabling intelligent tutoring systems, personalized learning experiences, and AI-based student support services. Adaptive learning platforms and automated grading systems enhance individualized learning, while virtual assistants and predictive analytics contribute to effective student support and guidance. AI in education has the potential to improve student success, efficiency, and educational results.

1.6 ENVIRONMENTAL SUSTAINABILITY

1.6.1 AI for climate change modeling and prediction

AI has become a potent instrument in the study of climate change, providing novel methods for deciphering complex climate data, refining prediction models, and deepening our knowledge of climate dynamics.

1.6.1.1 Machine learning for weather forecasting

Large amounts of climate data have been analyzed, and important insights have been extracted through the use of machine learning algorithms. For example, studies by Chevuturi, Hossain, and Anagnostou [59] show how machine learning algorithms, including random forest and SVMs, can be used to anticipate precipitation patterns and extreme weather occurrences, which can help with the assessment of the effects of climate change. AI techniques, including neural networks and deep learning, have been integrated into climate prediction models to improve their accuracy and predictive capabilities. A study by Rasp and Lerch [60] utilizes CNNs to develop a weather model that outperforms traditional methods, offering more accurate short-term climate predictions. AI enables the integration and fusion of diverse climate data sources, including satellite imagery, climate models, and ground-based observations. Research by Hsieh, Lee, and Zwiers [61] explores the use of AI techniques, such as Bayesian data assimilation, to assimilate data from different sources and improve the accuracy of climate model simulations. AI techniques have also been employed to quantify and propagate uncertainties in climate modeling and prediction. Research by Rasp, Pritchard, Gentine, and Lerch [62] introduces a deep ensemble approach that captures model uncertainty and produces probabilistic climate predictions, enhancing decision-making under uncertain climate conditions. AI is essential for determining climate risk and creating adaptable plans. To enable targeted adaptation strategies, Steinhaeuser, Williams, and Vargo [63] conducted research using machine learning to assess historical climate data and identify places most vulnerable to future climate change.

1.6.1.2 AI in environmental monitoring and conservation

AI methods have been used to evaluate remote sensing data, such as satellite photography and aerial photos, including computer vision and machine learning. For example, research by Cheng, Chen, and Zhou [64] shows how deep learning algorithms may be used for change detection and land cover classification, allowing for efficient monitoring of habitat deterioration and deforestation. AI systems have been created to automatically identify species using photos or audio recordings, facilitating population monitoring and biodiversity assessments. CNNs are used in a study by Norouzzadeh et al. [65] to identify animals in camera trap photos, allowing for accurate and efficient species monitoring. AI algorithms can assist in detecting and predicting environmental threats, such as wildfires, illegal logging, and poaching. Research by Cruz-Mota et al. [66] applies machine learning techniques to identify potential areas of illegal deforestation, supporting targeted conservation efforts. AI enables the development of predictive models for species distribution, habitat suitability, and conservation planning. For example, research by Elith et al. [67] demonstrates the use of machine learning algorithms, including random forests, for species distribution modeling, aiding in prioritizing conservation actions. AI techniques facilitate the integration and analysis of diverse environmental data sources, such as climate data, biodiversity databases, and socioeconomic data. The application of AI-based decision support systems for sustainable land use planning and conservation management is investigated in the research by Singh et al. [68].

1.6.2 Smart energy management and resource optimization

Achieving sustainability and efficiency in our energy systems requires effective resource optimization and energy management. In this field, AI has become a potent instrument that facilitates optimization, automation, and intelligent decision-making.

1.6.2.1 AI-driven energy grid optimization

A deep reinforcement learning–based strategy is put out by Shi et al. [69] for the best scheduling of power generation and storage in microgrids that include renewable energy sources. The study demonstrates how AI algorithms can effectively manage the operation of distributed energy resources, leading to improved energy utilization and cost-efficiency. Li et al. [70] present an AI-driven approach for optimizing energy distribution networks through intelligent control and predictive analytics. The study demonstrates how AI methods, such as machine learning and optimization algorithms, can improve the stability and efficiency of energy distribution networks. For short-term load forecasting in smart grids, Chen et al. [71] offer a hybrid model that combines deep learning and long short-term memory (LSTM) networks. The study shows how AI systems can estimate energy consumption with high accuracy, allowing for resource efficiency and proactive load balancing. AI techniques are applied to grid operation optimization to overcome the issues posed by the integration of renewable energy sources and energy storage systems. Zeng et al. [72] present an AI-based approach for optimal operation and scheduling of integrated energy systems, enabling efficient utilization of renewables and storage resources. AI algorithms are applied to optimize energy trading and demand response mechanisms. Liu et al. [73] propose an AI-based framework for optimal bidding strategies in electricity markets, considering factors such as price volatility and market uncertainties. The study emphasizes how AI may boost energy trading's financial viability.

1.6.2.2 Demand response systems and energy efficiency

Li et al. [74] provide a study that suggests a demand response framework supported by AI that makes use of machine learning algorithms to forecast and optimize patterns of energy usage in smart grids. The study demonstrates how well AI works to balance loads, lower peak demand, and increase energy efficiency. An AI-based strategy for intelligent energy management in buildings is presented by Fumo et al. [75]. Predictive analytics and sophisticated control algorithms are the main tools used in this study to optimize energy use. The outcomes show considerable operational efficiency gains and energy savings. A reinforcement learning–based strategy for energy-efficient building control is put forth by Li et al. [76]. The study shows how AI systems may adaptively learn the best control strategies to reduce energy use while preserving consumer comfort. An AI-based predictive analytics approach for energy efficiency in industrial settings is presented by Jin et al. [77] in their work. The study demonstrates how machine learning algorithms can evaluate past data to offer suggestions for maximizing energy use, cutting waste, and raising overall productivity. Chang et al. [78] present an AI-driven energy management system for smart homes that integrates energy storage, renewable energy sources, and demand response mechanisms. The research showcases the capabilities of AI in dynamically optimizing energy consumption and reducing electricity costs.

1.7 ETHICAL AND SOCIAL IMPLICATIONS

1.7.1 Considerations of bias and fairness in AI applications

Dwork et al. [79] introduce the concept of "fairness through awareness" and emphasize the need to address potential biases in training data and algorithms. The study emphasizes the importance of fairness across different groups and provides guidelines for achieving fairness in machine learning. Racial and gender biases in facial recognition systems are discussed by Buolamwini and Gebru [80]. The study shows that darker-skinned and female faces had higher mistake rates, emphasizing the need to correct algorithmic biases and enhance fairness in AI systems. Bolukbasi et al. [81] study how gender bias affects word embeddings, which are an essential part of models used in NLP. The work indicates how social biases can be amplified and encoded by AI systems and also suggests ways to counteract these biases. The fairness of AI algorithms utilized in the criminal justice system is examined by Larson et al. [82]. In this context, the research delves into the differences in risk assessment tools based on race and highlights the significance of fairness metrics and transparency in the development and application of AI systems. The "Ethically Aligned Design" document, released in 2019 by the Institute of Electrical and Electronics Engineers (IEEE), offers a thorough framework for tackling ethical issues in AI research [83]. It includes a specific focus on bias and fairness, stressing the need for transparency, accountability, and inclusive design practices.

1.7.2 Privacy and security concerns in multidisciplinary AI systems

It is critical to address the privacy and security concerns related to these systems as interdisciplinary applications of AI continue to develop. Large volumes of data from numerous sources are frequently accessed by AI technology, which raises questions around data privacy and potential security flaws. The notion of "differential privacy," as presented by Abadi et al. [84], attempts to safeguard private data while facilitating efficient dataset analysis. The idea of "adversarial examples" is introduced by Szegedy et al. [85], who also show how AI systems are susceptible to malevolent attacks. The study highlights the need for robust defenses to protect AI models from adversarial manipulations that can compromise security and privacy. Rajkomar et al. [86] discuss privacy challenges in AI applications for health care. The research emphasizes the importance of data anonymization, secure data sharing, and compliance with privacy regulations to ensure patient privacy and confidentiality. Buczak and Guven [87] analyze the security risks associated with AI systems integrated with the internet of things (IoT).

1.7.3 Transparency and accountability in AI decision-making

Transparency in AI decision-making enables stakeholders to learn more about the underlying mechanisms and contributing aspects of the results. It helps build trust and enables users to assess the reliability and fairness of AI systems. Accountability ensures that individuals or organizations are held responsible for the consequences of AI decisions, promoting ethical behavior and preventing potential harm. To achieve transparency and accountability in AI decision-making, approaches such as explainable AI, fairness assessment, and the development of regulatory frameworks and auditing processes have been proposed. These measures aim to enhance the understanding, fairness, and accountability of AI systems, fostering their responsible use and promoting public trust.

1.8 CONCLUSION

1.8.1 Recap of multidisciplinary applications of AI

First, we discussed the importance of a multidisciplinary approach in AI applications. Drawing from various disciplines, such as computer science, mathematics, psychology, and social sciences, this approach brings together different perspectives to tackle complex challenges and leverage the full potential of AI technologies. Next, we delved into specific applications of AI in different sectors. In the medical field, AI has demonstrated remarkable capabilities in medical diagnosis and treatment, aiding health care professionals in accurate and timely decision-making. We examined how machine learning algorithms are used for disease detection, enabling early diagnosis and intervention. Furthermore, we explored AI-assisted surgical procedures, where AI technologies enhance surgical precision, reduce risks, and improve patient outcomes. We talked about how AI is accelerating the medication development process in the fields of personalized medicine and drug discovery. The time and expense involved in bringing new drugs to market are decreased by the effective identification of possible drug candidates, made possible by predictive modeling approaches. Additionally, we highlighted the concept of precision medicine and patient-specific treatments, where AI-driven approaches enable tailored and effective health care interventions based on individual characteristics and genetic profiles. Moving on to the financial sector, we explored the use of AI-based investment strategies and risk assessment. By examining past data and spotting trends, predictive analytics helps portfolio managers forecast market trends and make wise investment choices. We also talked about sentiment analysis in the financial markets, where AI tools are used to examine news and social media sentiment to determine market mood and make smart investment decisions. We looked at the effects of driverless cars and intelligent transportation systems in the fields of logistics and transportation. AI-powered traffic management systems optimize traffic flow and lessen congestion, while self-driving cars offer the promise of safer and more effective mobility. We also talked about supply chain management optimization algorithms, which improve inventory control, reduce logistics processes, and optimize routes. Techniques for asset tracking and predictive maintenance guarantee effective resource management and reduce downtime. We investigated individualized learning and intelligent tutoring systems in the realm of education. AI is used by adaptive learning platforms to customize instructional materials and delivery strategies for each student, improving the learning process. NLP for automated grading enables efficient and objective assessment of student work. Additionally, we discussed AI-based student support and career guidance, where virtual assistants provide personalized assistance and predictive analytics aid in identifying factors influencing student success. Finally, we touched upon AI's role in addressing environmental sustainability. AI for climate change modeling and prediction enhances our understanding of climate patterns, enabling better mitigation and adaptation strategies. Smart energy management and resource optimization leverage AI to optimize energy grids, enable demand response systems, and enhance energy efficiency.

1.8.2 Future directions and potential impact of AI in various fields

AI has enormous potential to change several industries and influence our society in the future. We may anticipate an increase in AI's influence in a variety of fields as it develops. Here, we look at some potential future paths and applications of AI across a range of industries:

1. Health care: AI has the power to completely change how health care is provided. More advanced AI algorithms that can evaluate intricate medical data, forecast disease outcomes, and facilitate individualized treatment regimens are likely to emerge in the future. AI-powered virtual assistants that give patients real-time medical advice and monitoring might become ubiquitous. Furthermore, by speeding up the identification of novel compounds and simplifying the clinical trial procedure, AI may play a significant role in drug discovery.

2. Transportation: Self-driving vehicles are likely to become more prevalent in the future, leading to safer and more efficient transportation systems. AI algorithms will continue to improve autonomous vehicle capabilities, enabling them to navigate complex environments and interact with other vehicles and pedestrians seamlessly. Furthermore, traffic flow will be optimized by AI-driven traffic control systems, which will lessen congestion and boost overall transportation effectiveness.
3. Education: By enabling tailored learning experiences, AI will have a big impact on education. With the help of AI algorithms, adaptive learning platforms will continue to develop, customizing course materials to each student's unique requirements and preferences. AI-driven virtual tutors and assistants will give individualized instruction, real-time support, and question-answering. Furthermore, AI will play a role in automating administrative tasks, allowing educators to focus more on teaching and student engagement.
4. Finance: AI will continue to reshape the financial industry, improving risk assessment, fraud detection, and customer service. Advanced AI algorithms will enable more accurate predictions in financial markets, aiding investment decisions and portfolio management. Furthermore, chatbots and virtual assistants driven by AI will improve client experiences by offering tailored financial advice and assistance.
5. Environmental Sustainability: AI has a significant role to play in solving environmental issues. More precise climate change modeling made possible by AI algorithms would empower decision-makers to create efficient mitigation plans. AI-powered systems will also improve resource management, lower waste, and maximize energy use, all of which will lead to a more sustainable future.
6. Manufacturing and Industry: AI-driven automation will continue to transform manufacturing and industry. Intelligent robots and machines will collaborate with human workers, enhancing productivity and efficiency. AI algorithms will optimize production processes, enabling predictive maintenance and minimizing downtime. Furthermore, AI-powered quality control systems will ensure higher product standards and reduce defects.
7. Social Sciences: AI can help the social sciences by finding patterns and insights in massive amounts of data through analysis. It can help with societal trend prediction, public policy analysis, and social research. Algorithms powered by AI have the potential to enhance social services including crime prevention, welfare distribution, and tailored health care.

REFERENCES

[1] Y. LeCun, Y. Bengio, and G. Hinton, "Deep learning," Nature, vol. 521, no. 7553, 2015, pp. 436–444.

[2] D. Jurafsky and J. H. Martin, "Speech and language processing: An introduction to natural language processing, computational linguistics, and speech recognition" (3rd ed.). Pearson, 2020.
[3] L. Fei-Fei, J. Deng, and K. Li, "Computer vision," in S. Russell & P. Norvig (Eds.), Artificial intelligence: A modern approach (4th ed.), Pearson, 2021, pp. 1019–1072.
[4] O. Khatib, "Robotics: A perspective," Annual Review of Computer Science, vol. 1, no. 1, 2018, pp. 167–194.
[5] N. Bostrom, "Superintelligence: Paths, dangers, strategies." Oxford University Press, 2014.
[6] P. Domingos, "A few useful things to know about machine learning," Communications of the ACM, vol. 55, no. 10, 2012, pp. 78–87.
[7] E. Brynjolfsson and A. McAfee, "The second machine age: Work, progress, and prosperity in a time of brilliant technologies." W. W. Norton & Company, 2014.
[8] AI100 Standing Committee and Study Panel, "Artificial intelligence and life in 2030: One hundred year study on artificial intelligence." Stanford University, 2016.
[9] A. Esteva et al., "Dermatologist-level classification of skin cancer with deep neural networks," Nature, vol. 542, no. 7639, 2017, pp. 115–118.
[10] E. Choi et al., "Retrospective prediction of clinical events using time-series electronic health record data," Nature Medicine, vol. 25, no. 3, 2019, pp. 452–459.
[11] X. Liu et al., "A comparison of deep learning performance against healthcare professionals in detecting diseases from medical imaging: A systematic review and meta-analysis," The Lancet Digital Health, vol. 1, no. 6, 2018, pp. e271–e297.
[12] H. Suresh, N. Hunt, and A. Johnson, "Clinical applications of machine learning in healthcare," Proceedings of the IEEE, vol. 106, no. 8, 2018, pp. 120–134.
[13] V. Gulshan et al., "Development and validation of a deep learning algorithm for detection of diabetic retinopathy in retinal fundus photographs," JAMA, vol. 316, no. 22, 2016, pp. 2402–2410.
[14] R. Fakoor, F. Ladhak, A. Nazi, and M. Huber, "Using deep learning to enhance cancer diagnosis and classification," in Proceedings of the 30th International Conference on Machine Learning (ICML-13), 2013, pp. 1097–1105.
[15] Z. I. Attia, S. Kapa, F. Lopez-Jimenez, P. M. McKie, D. J. Ladewig, G. Satam, and P. A. Friedman, "Screening for cardiac contractile dysfunction using an artificial intelligence–enabled electrocardiogram," Nature Medicine, vol. 25, no. 1, 2019, pp. 70–74.
[16] T. Kooi et al., "Large scale deep learning for computer aided detection of mammographic lesions," Medical Image Analysis, vol. 35, 2017, pp. 303–312.
[17] I. Ahmad, S. R. Kulkarni, Y. Wang, J. W. Catto, and O. M. Aboumarzouk, "Artificial intelligence and robotics in urology: Current clinical applications," Current Urology Reports, vol. 19, no. 12, 2018, p. 107.
[18] O. Charron, F. Lalys, and P. Jannin, "Medical image analysis using artificial intelligence and machine learning for computer-assisted interventions," Medical Image Analysis, vol. 56, 2019, pp. 207–208.
[19] F. Jiang et al., "Artificial intelligence in healthcare: Past, present and future," Stroke and Vascular Neurology, vol. 5, no. 4, 2020, pp. 230–243.
[20] T. Ching et al., "Opportunities and obstacles for deep learning in biology and medicine," Journal of the Royal Society Interface, vol. 15, no. 141, 2018, p. 20170387.
[21] G. B. Goh, N. O. Hodas, and A. Vishnu, "Deep learning for computational chemistry," Journal of Computational Chemistry, vol. 38, no. 16, 2017, pp. 1291–1307.

[22] M. H. Segler et al., "Generating focused molecule libraries for drug discovery with recurrent neural networks," ACS Central Science, vol. 4, no. 1, 2018, pp. 120–131.
[23] H. Ozturk, E. Ozkirimli, and A. Ozgur, "DeepDTA: Deep drug-target binding affinity prediction," Bioinformatics, vol. 34, no. 17, 2018, pp. i821–i829.
[24] S. Ekins et al., "Exploiting machine learning for predicting human liver microsomal stability," Journal of Cheminformatics, vol. 11, no. 1, 2019, pp. 1–16.
[25] A. Sedykh et al., "Human hepatotoxicity prediction by GUSAR, case study for industrial chemicals in the European inventory of existing commercial chemical substances," Chemical Research in Toxicology, vol. 32, no. 6, 2019, pp. 1244–1258.
[26] W. Wang, B. van Eerdenbrugh, G. Karlsson, J. Kou, and C. A. Bergström, "Machine learning in the formulation development of amorphous solid dispersions: A tutorial," AAPS PharmSciTech, vol. 21, no. 3, 2020, pp. 1–19.
[27] W. Yang et al., "Genomics of Drug Sensitivity in Cancer (GDSC): A resource for therapeutic biomarker discovery in cancer cells," Nucleic Acids Research, vol. 41, no. D1, 2013, pp. D955–D961.
[28] P. Geeleher, N. J. Cox, and R. S. Huang, "Clinical drug response can be predicted using baseline gene expression levels and in vitro drug sensitivity in cell lines," Genome Biology, vol. 17, no. 1, 2016, pp. 1–12.
[29] A. Rajkomar et al., "Scalable and accurate deep learning with electronic health records," npj Digital Medicine, vol. 1, no. 1, 2018, pp. 18.
[30] C. Phua, V. Lee, K. Smith-Miles, and R. Gayler, "A comprehensive survey of data mining-based fraud detection research," arXiv preprint arXiv: 1009.6119, 2010.
[31] I. M. Pires, N. M. Garcia, and M. F. Santos, "Healthcare fraud: A systematic review," BMC Health Services Research, vol. 16, no. 1, 2016, pp. 1–12.
[32] G. Li et al., "An artificial intelligence-driven approach to optimal operation of energy distribution networks," IEEE Transactions on Smart Grid, vol. 11, no. 2, 2019, pp. 1534–1544.
[33] M. Su, C. Wen, and Y. Chen, "Automobile insurance fraud detection using machine learning algorithms," Computers, Materials & Continua, vol. 63, no. 1, 2020, pp. 193–210.
[34] F. T. Liu, K. M. Ting, and Z. H. Zhou, "Isolation forest," in Proceedings of the 2008 Eighth IEEE International Conference on Data Mining, 2008, pp. 413–422.
[35] Z. Gao, C. Wang, and C. Li, "Network intrusion detection using isolation forest and improved random subspace ensemble," Sensors, vol. 19, no. 4, 2019, p. 818.
[36] Z. M. Yaseen and W. M. Kadir, "A hybrid framework for credit card fraud detection based on machine learning techniques," IEEE Access, vol. 9, 2021, pp. 1697–1712.
[37] B. Schölkopf, J. C. Platt, J. Shawe-Taylor, A. J. Smola, and R. C. Williamson, "Estimating the support of a high-dimensional distribution," Neural Computation, vol. 13, no. 7, 2001, pp. 1443–1471.
[38] S. Ghosh and K. Veeraruna, "Anomaly detection in computer networks using one-class support vector machines," International Journal of Computer Applications, vol. 47, no. 5, 2012, pp. 14–19.
[39] A. L. P. Rau, J. Mendes-Moreira, and A. M. Jorge, "Fraud detection in telecommunications using one-class support vector machines," Expert Systems with Applications, vol. 42, no. 22, 2015, pp. 8772–8782.
[40] V. Venkataramanan, A. Mukherjee, S. Balakrishnan, and S. Mishra, "Anomaly detection using deep autoencoders for industrial fault diagnosis," IET Cyber-Physical Systems: Theory & Applications, vol. 2, no. 3, 2020, pp. 144–152.

[41] Y. Zhou, H. Xu, Y. Du, and X. Jin, "Autoencoder-based anomaly detection for industrial big data in cyber-physical systems," IEEE Access, vol. 7, 2019, pp. 66577–66586.
[42] A. Zimek, E. Schubert, and H. P. Kriegel, "A survey on unsupervised outlier detection in high-dimensional numerical data," Statistical Analysis and Data Mining, vol. 6, no. 5, 2013, pp. 363–387.
[43] F. Monrose and A. D. Rubin, "Keystroke dynamics as a biometric for authentication," Future Generation Computer Systems, vol. 16, no. 4, 2000, pp. 351–359.
[44] S. Kumar and D. Zhang, "Continuous user authentication using mouse dynamics," Information Sciences, vol. 272, 2014, pp. 173–190.
[45] J. R. Kwapisz, G. M. Weiss, and S. A. Moore, "Activity recognition using cell phone accelerometers," ACM SigKDD Explorations Newsletter, vol. 12, no. 2, 2011, pp. 74–82.
[46] A. Roy, I. Ahmad, and M. S. Roy, "A comprehensive review of behavioral biometric authentication systems using machine learning techniques," Journal of Network and Computer Applications, vol. 154, 2020, p. 102562.
[47] Y. Zhou, G. Li, S. Wu, and Y. Zhang, "Portfolio selection with hybrid machine learning and statistical methods," European Journal of Operational Research, vol. 276, no. 1, 2019, pp. 319–330.
[48] Q. Chen, X. Song, W. Bai, and Y. Wei, "A deep learning approach to stock market prediction," Information Sciences, vol. 415, 2017, pp. 280–291.
[49] J. Bollen, H. Mao, and X. Zeng, "Twitter mood predicts the stock market," Journal of Computational Science, vol. 2, no. 1, 2011, pp. 1–8.
[50] Y. Zhang, H. Fuehres, and P. Gloor, "Predicting stock market indicators through Twitter 'I hope it is not as bad as I fear'," Procedia Computer Science, vol. 138, 2018, pp. 491–498.
[51] J. M. Anderson, N. Kalra, K. D. Stanley, P. Sorensen, C. Samaras, and O. A. Oluwatola, "Autonomous vehicle technology: A guide for policymakers." Rand Corporation, 2014.
[52] H. J. Huang, Y. Xu, X. Zhang, and D. Zhao, "Deep learning-based traffic prediction for intelligent transportation systems," Transportation Research Part C: Emerging Technologies, vol. 90, 2018, pp. 166–180.
[53] R. K. Ahuja, T. L. Magnanti, and J. B. Orlin, "Network flows: Theory, algorithms, and applications." Prentice Hall, 2008.
[54] H. Wang, Z. Shen, H. Zhang, and Z. Zhang, "Predictive maintenance for supply chain systems using AI-driven dynamic factor analysis," Computers & Industrial Engineering, vol. 129, 2019, pp. 422–435.
[55] M. Kebritchi, A. Hirumi, and H. Bai, "The effects of modern mathematics computer games on mathematics achievement and class motivation," Computers & Education, vol. 55, no. 2, 2010, pp. 427–443.
[56] Y. Attali and J. Burstein, "Automated essay scoring with e-rater® v.2," Journal of Technology, Learning, and Assessment, vol. 4, no. 3, 2006, pp. 1–30.
[57] K. E. Arnold and M. D. Pistilli, "Course signals at Purdue: Using learning analytics to increase student success," in Proceedings of the 2nd International Conference on Learning Analytics and Knowledge, 2012, pp. 267–270.
[58] R. S. Baker and G. Siemens, "Educational data mining and learning analytics," in Handbook of research on educational communications and technology, Wiley Online Library, 2014, pp. 439–456.

[59] A. Chevuturi, F. Hossain, and E. N. Anagnostou, "Machine learning approaches for rainfall and flood predictions: An overview," Water, vol. 10, no. 4, 2018, p. 436.
[60] S. Rasp and S. Lerch, "Neural networks for postprocessing ensemble weather forecasts," Monthly Weather Review, vol. 146, no. 11, 2018, pp. 3885–3900.
[61] W. W. Hsieh, C. K. Lee, and F. W. Zwiers, "Bayesian parameter estimation and uncertainty quantification for climate model predictions," Journal of Climate, vol. 31, no. 5, 2018, pp. 1881–1903.
[62] S. Rasp, M. S. Pritchard, P. Gentine, and S. Lerch, "Deep learning of representations for convection," Journal of Advances in Modeling Earth Systems, vol. 11, no. 2, 2019, pp. 453–472.
[63] K. Steinhaeuser, A. P. Williams, and T. Vargo, "High-resolution probabilistic projections of temperature changes over the United States," Nature Communications, vol. 9, no. 1, 2018, pp. 1–8.
[64] C. Cheng, Q. Chen, and X. Zhou, "Remote sensing image scene classification: Benchmark and state of the art," Proceedings of the IEEE, vol. 108, no. 6, 2020, pp. 869–885.
[65] M. S. Norouzzadeh et al., "Automatically identifying, counting, and describing wild animals in camera-trap images with deep learning," Proceedings of the National Academy of Sciences, vol. 115, no. 25, 2018, pp. E5716–E5725.
[66] J. J. Cruz-Mota et al., "Neural network classification of deforestation successional stages in the Brazilian Amazon rainforest using multitemporal spectral and spatial information," Remote Sensing, vol. 8, no. 2, 2016, p. 144.
[67] J. Elith et al., "A statistical explanation of MaxEnt for ecologists," Diversity and Distributions, vol. 13, no. 2, 2008, pp. 123–135.
[68] P. Singh et al., "An AI-based decision support system for sustainable land use planning," Environmental Science & Policy, vol. 104, 2020, pp. 165–175.
[69] D. Shi et al., "Deep reinforcement learning for optimal scheduling of power generation and storage in renewable energy-integrated microgrids," IEEE Transactions on Sustainable Energy, vol. 11, no. 4, 2020, pp. 2365–2376.
[70] G. Li et al., "An artificial intelligence-driven approach to optimal operation of energy distribution networks," IEEE Transactions on Smart Grid, vol. 11, no. 2, 2019, pp. 1534–1544.
[71] L. Chen et al., "Short-term load forecasting using a hybrid model of deep learning and LSTM network in smart grids," Energies, vol. 14, no. 7, 2021, p. 1911.
[72] Y. Zeng et al., "Artificial intelligence-based optimal operation and scheduling of integrated energy systems," Applied Energy, vol. 263, 2020, p. 114665.
[73] Y. Liu et al., "Optimal bidding strategy for power retailers in an electricity market: An artificial intelligence approach," Applied Energy, vol. 238, 2019, pp. 1173–1184.
[74] Y. Li et al., "An AI-enabled demand response framework for smart grids considering user comfort and system reliability," Applied Energy, vol. 261, 2020, p. 114408.
[75] J. Fumo et al., "AI for intelligent energy management in buildings," Applied Energy, vol. 255, 2019, p. 113881.
[76] Y. Li et al., "Reinforcement learning-based optimal control for energy-efficient buildings," Applied Energy, vol. 284, 2021, p. 116543.
[77] H. Jin et al., "Artificial intelligence for industrial energy efficiency: A survey," Applied Energy, vol. 213, 2018, pp. 219–234.

[78] W. Chang et al., "AI-driven energy management system for smart homes considering renewable energy sources, storage systems, and demand response," Applied Energy, vol. 260, 2020, p. 114302.
[79] C. Dwork et al., "Fairness through awareness," in Proceedings of the 3rd Innovations in Theoretical Computer Science Conference, 2012, pp. 214–226.
[80] J. Buolamwini and T. Gebru, "Gender shades: Intersectional accuracy disparities in commercial gender classification," in Proceedings of the 1st Conference on Fairness, Accountability, and Transparency, 2018, pp. 77–91.
[81] T. Bolukbasi et al., "Man is to computer programmer as woman is to homemaker? Debiasing word embeddings," in Advances in Neural Information Processing Systems, 2016, pp. 4349–4357.
[82] J. Larson et al., "How fairness considerations matter when developing algorithms for criminal justice," in the Proceedings of the Conference on Fairness, Accountability, and Transparency, 2016, pp. 11–20.
[83] IEEE Global Initiative on Ethics of Autonomous and Intelligent Systems, "Ethically Aligned Design: A Vision for Prioritizing Human Well-being with Artificial Intelligence and Autonomous Systems," 2019. [Online]. Available: https://standards.ieee.org/content/dam/ieee-standards/standards/web/documents/other/ead_v2.pdf.
[84] M. Abadi et al., "Deep learning with differential privacy," in Proceedings of the 2016 ACM SIGSAC Conference on Computer and Communications Security, 2016, pp. 308–318.
[85] C. Szegedy et al., "Intriguing properties of neural networks," in Proceedings of the 2013 IEEE Conference on Computer Vision and Pattern Recognition, 2013, pp. 1912–1919.
[86] A. Rajkomar et al., "Scalable and accurate deep learning with electronic health records," npj Digital Medicine, vol. 1, no. 1, 2018, p. 18.
[87] A. L. Buczak and E. Guven, "A survey of data mining and machine learning methods for cyber security intrusion detection," IEEE Communications Surveys & Tutorials, vol. 18, no. 2, 2016, pp. 1153–1176.

Chapter 2

Advancements in deep learning

Unveiling future trends and applications across AI for intelligent computing

K. Gurucharan, S. S. Kiran, Dr. Mallavarapu Rajanbabu, and P. Srujana

2.1 INTRODUCTION

Artificial intelligence (AI) refers to a method for teaching a computer, a computer-controlled machine, or software the facility to think in the same manner as intelligent people do. AI is achieved by first researching in what way the human intelligence works, as well as by what method people learn, choose, and function while trying to resolve a problem, besides applying the outcomes toward the growth of intelligent software as well as structures. AI is a science and technology that combines disciplines such as computer science, biochemistry, neuroscience, linguistics, mathematics, and engineering to produce intelligent machines. The advances in computer capabilities related to human cognition, such as thinking, learning, and problem-solving, is a primary focus of AI.

One or more of the key areas may contribute to the development of an integrated system. The applications of AI include entertainment, natural language processing (NLP), decision support systems, imaging systems, voice recognition, handwriting detection, and autonomous robots as shown in Figure 2.1. The inspiration for artificial neural networks came from the human intelligence's neural networks. The human brain is very intricate. Scientists and engineers studied the brain carefully and devised an architecture that would work in our digital world of binary computers. Figure 2.2 depicts one such typical building.

An input layer with many sensors is used to gather data from the outside environment. We have an output layer on the right-hand side that provides us the network's anticipated outcome. Several layers are concealed between these two. Each new layer increases the difficulty of training the network, but it also improves the outcomes in the majority of cases.

DOI: 10.1201/9781003496410-3

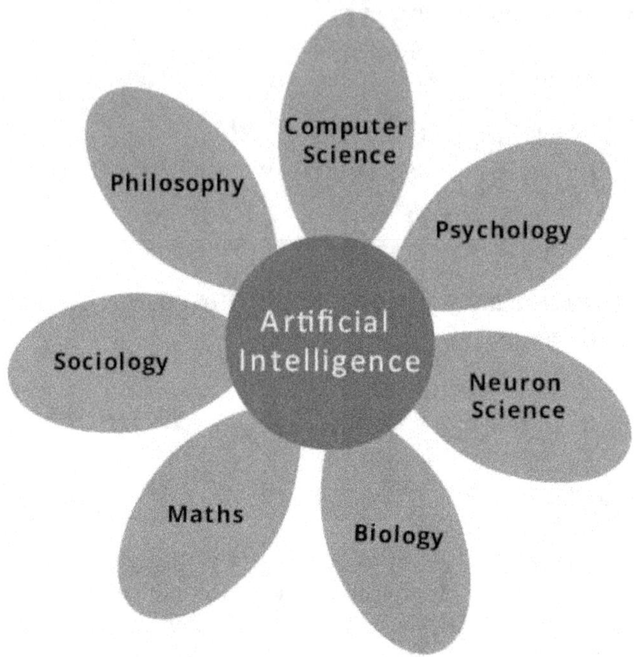

Figure 2.1 Artificial intelligence.

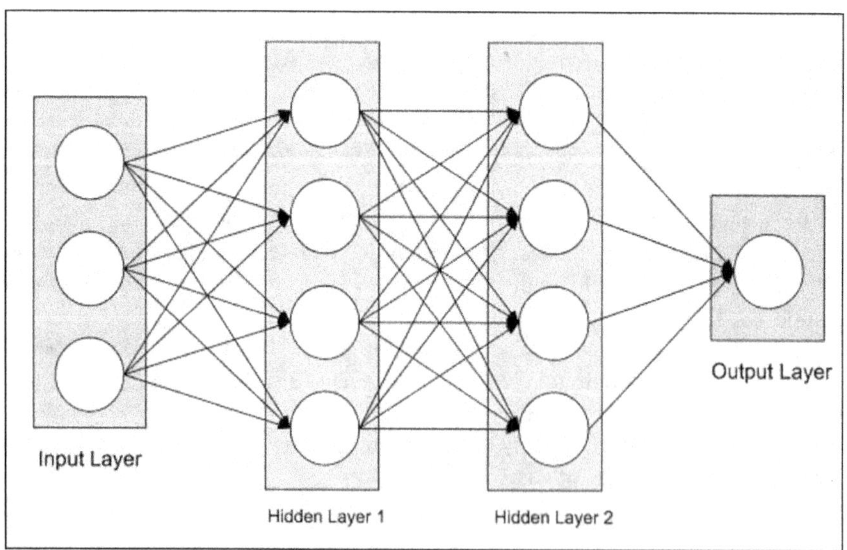

Figure 2.2 Architecture of neural network.

2.1.1 Machine learning

Machine learning (ML) makes use of AI to let computers learn from their own mistakes instead of needing to be specifically programmed for each task. (In other words, machines learn without human intervention.) The kind of data available determines the algorithms used and attempted to automate the job. ML is a subdivision of AI that lets computers secure ideas and information without being explicitly programmed. It begins with actual observations to prepare for data characteristics and trends, resulting in improved outcomes and results in the future. DL uses ML techniques to represent high-level abstractions in data via multiple nonlinear regressions.

2.1.2 Categories of machine learning

The following are the several types of deep learning approaches: unsupervised, supervised, semi-supervised, or partly supervised. Furthermore, there seems to be a category of learning method known as deep reinforcement learning (DRL), which is often addressed as part of unsupervised or semi-supervised learning approaches, as illustrated in Figure 2.3.

2.1.2.1 Supervised learning

Supervised learning is equivalent to training a child to walk. One should hold the child's hand, demonstrate how to put his or her foot forward, walk for a demonstration, and so on, until the youngster is able to walk independently. Supervised learning is a kind of learning that makes use of labeled data that has a set of inputs and outputs (x_t, y_t). For instance, if for input x_t, the intellectual model might predict $\hat{y}_t = f(x_t)$, the model shall obtain a loss value $L\left(y_t, \hat{y}_t\right)$. The model shall at that point repetitively fine-tune the system constraints to get a nearer match to intended output. The model shall be capable of getting accurate responses towards queries from the surroundings after effective training. In addition to convolutional neural networks (CNN), supervised deep learning approaches include the recurrent

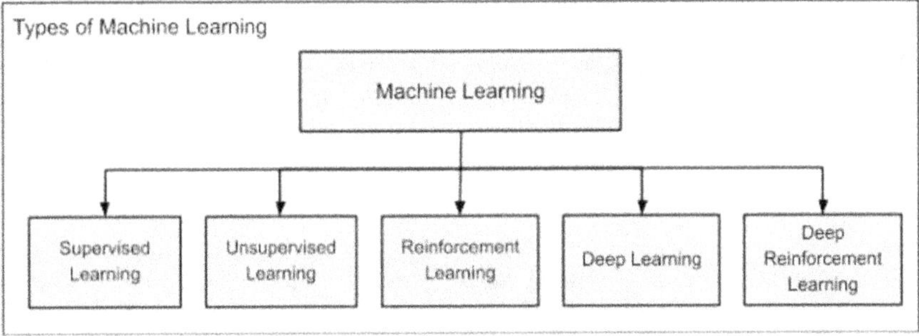

Figure 2.3 Categories of machine learning.

neural network (RNN), deep neural network (DNN), gated recurrent unit (GRU), and long short-term memory (LSTM).

2.1.2.2 Unsupervised learning

Unsupervised learning networks remain capable of learning devoid of the need of data labels. In this instance, the model discovers undiscovered connections or structure inside the input data by learning the internal representation or key characteristics.

2.1.2.3 Deep reinforcement learning

Reinforcement learning (RL) is more difficult to learn than traditional supervised approaches since we don't have a simple loss function. RL and supervised learning differ in the following ways: when you are trying to optimize a function, there is no total access to it, and you must interact with it in a state-based context. Prior actions determine the value of input *xt*. Depending on the problem's size and breadth, it's possible to choose the appropriate RL for the job at hand. If the problem includes a large number of variables that need to be optimized, DRL is the ideal approach. When there are fewer parameters to optimize, a derivation-free RL technique is advantageous.

2.1.2.4 Deep learning

Using just images, text, or speech as input, a model learns to perform classification problems via deep learning. Neural network architecture is often used to accomplish deep learning. Network depth is proportional to its layer count, with an increase in layers leading to increased network complexity. Traditional neural networks have two or three layers, while deep neural networks may contain hundreds. Inspired by organic nerve systems, a DNN comprises many nonlinear processing layers, utilizing basic components working in parallel.

A deep CNN's significant improvement in computing capacity, however, requires architectural innovations. Making use of spatial and channel data, architectural complexity in addition to breadth, and information processing using several paths, in particular, have all received much attention. Biological image classification, computer vision, adaptive testing, tumor identification, face recognition, NLP, object identification, voice recognition, and handwriting recognition are the most frequently utilized deep learning domains.

2.2 INTRODUCTION TO NEURAL NETWORKS

What we've learned: as a general rule, "training" is used in this way. A neural network may be trained using a variety of approaches, including gradient descent and backpropagation. The mathematical model of the neuron in the human brain is shown in Figure 2.4.

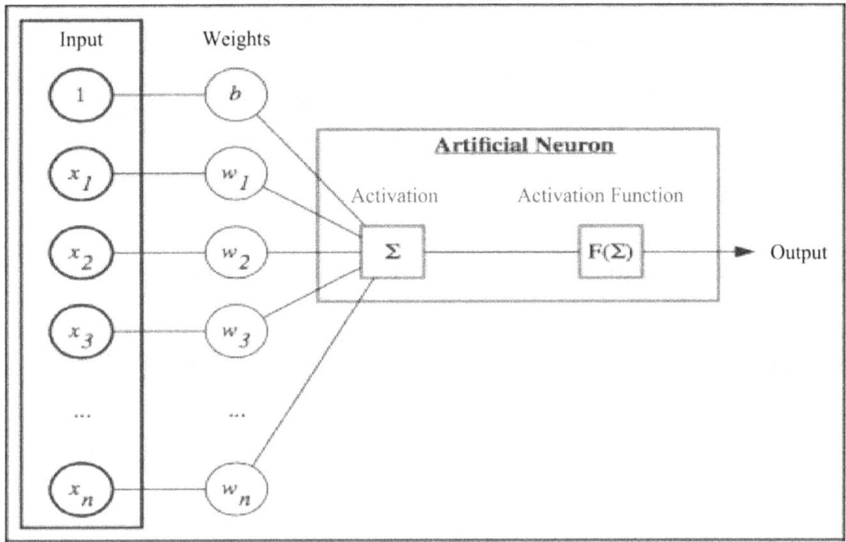

Figure 2.4 Mathematical model of a simple neuron in the human brain.

A neural net is a model or a system that processes data or inputs, rather than a preprogrammed program. There are two major features of a neural network: a neuronal network architecture may be summarized as follows. This determines the types of connections that exist between neurons (such as feed-forward or recurrent), the number of layers (such as multilayered or single-layered as shown in Figure 2.1), and so on.

The neuron is defined as

$$y = f\left(\sum_i a_j w_j + b\right) \tag{2.1}$$

Earlier, we calculate the weighted sum $\sum a_j w_j$ of the inputs a_j and the weights w_j (also known as an activation value); a_j represents either the input data numerical values or the outputs of other neurons (i.e., if the neuron is part of a neural network), w_j indicates the strength of the inputs or, alternatively, the strength of connections among the neurons in the network. When it comes to the bias weight, it's a particular value with a constant input of 1. A transfer function is created by taking the weighted sum as an input and applying it to *f*, which is an activation function.

2.2.1 Vanishing and exploding gradient problem

Over time, the partial derivative of the error function is updated proportionally to each weight in a neural network. In other cases, this derivative term is so small that updates are impossible. The update is achieved in the deep layers of the neural network by multiplying different partial derivatives. The total update gets extremely tiny and approaches nil if these partial derivatives are very small. Weights will not be able to update in this scenario, resulting in delayed or no convergence. The vanishing gradient problem is the name given to this situation. If the derivative term is big, then the updates will be significant as well. The algorithm will exceed the minimum in this instance and will be unable to converge. The exploding gradient problem is the name given to this situation. One way to prevent these issues is to choose the proper activation function. The following are the most frequent activation functions.

2.2.2 Different types of activation functions

Activation functions determine the output of a neural network, such as yes or no. It goes from 0 to 1, from −1 to 1, and so on. The activation functions can be classified into linear and nonlinear activation functions as shown in Figure 2.5.

2.2.2.1 Linear activation function

The function is either a line or linear function as shown in Figure 2.6. As a result, the function's output will not be limited by any ranges.

The equation for the linear activation function is

$$f(x) = x \tag{2.2}$$

It makes no difference to the complexity or different characteristics of the data that is supplied to the neural networks.

2.2.2.2 Nonlinear activation functions

It allows the model to generalize or adapt to a wide range of inputs while still distinguishing between outcomes. The range or curves of nonlinear activation functions are used to categorize them.

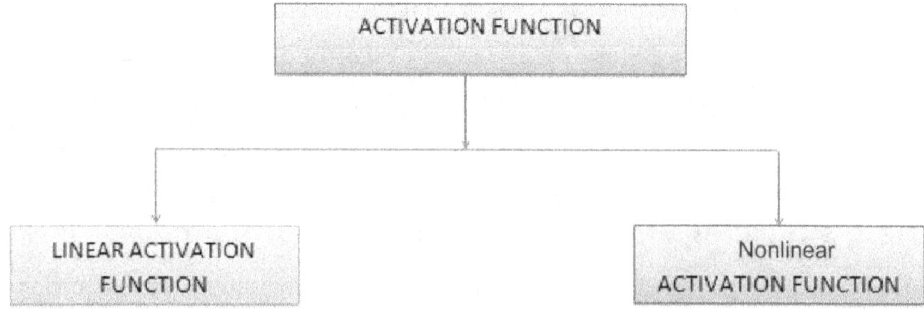

Figure 2.5 Types of activation functions.

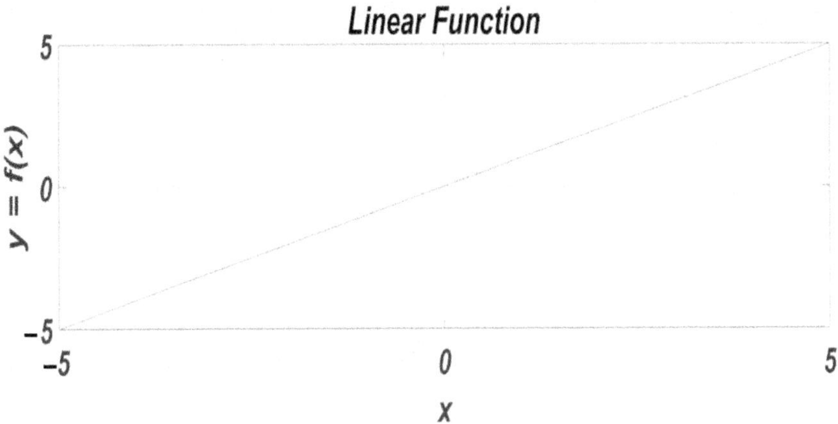

Figure 2.6 Output function of linear activation function.

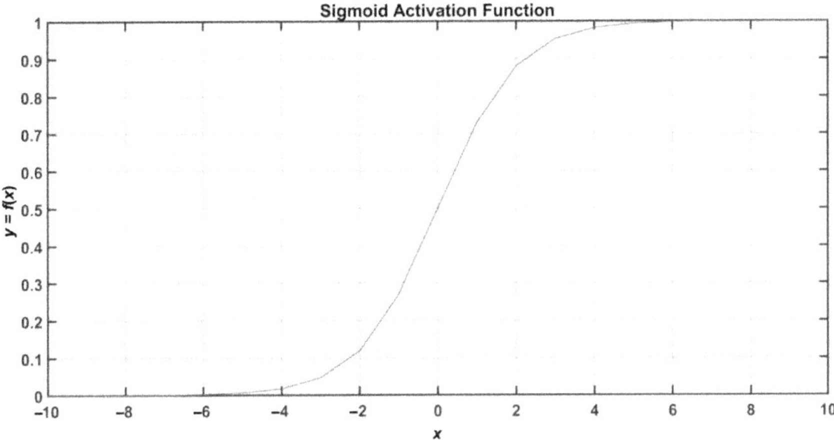

Figure 2.7 Sigmoid activation function.

2.2.2.3 Sigmoid function

A sigmoid is a mathematical function with an "S" shape (as shown in Figure 2.7) whose formula is as follows:

$$y(x) = \frac{1}{1+e^{-x}} = \frac{e^{x}}{1+e^{x}} \tag{2.3}$$

Advantages:

The sigmoid function is a continuous and differentiable function.
It will restrict output to a range of 0–1 and provide very specific predictions for binary classification.

Disadvantages:

It has the potential to create a vanishing gradient issue.
It isn't quite centered on zero.
Expensive in terms of computation.

2.2.2.4 *Softmax function*

The softmax function is a multiclass sigmoid extension. I use it in the final layer of multiclass categorization. As illustrated in Figure 2.8, it converts a list of k-valued real numbers into a probability distribution with k probabilities matching the input numbers. Prior to applying softmax, certain vector components may be negative or greater than 1 and therefore not sum to 1, but after applying softmax, each component will be in the range of 0–1 and so sum to one. The softmax function is defined as follows:

$$y(x_i) = \frac{e^{x_i}}{\sum_{K=1}^{K} e^{x_k}} \; for\, i = 1,2,3,\ldots K \tag{2.4}$$

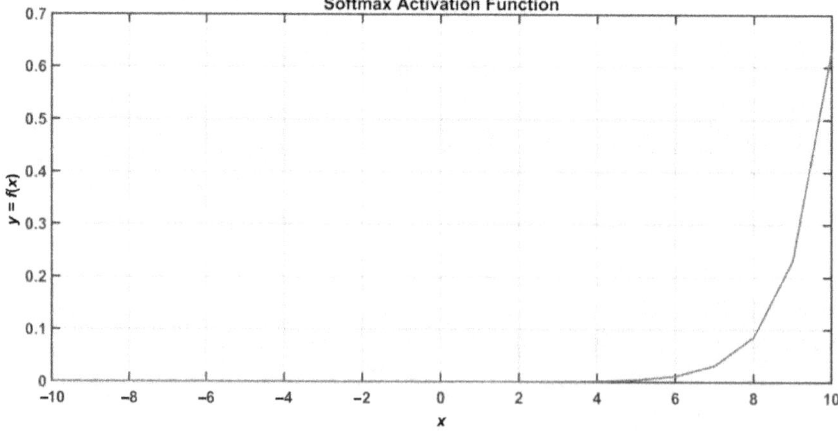

Figure 2.8 Softmax activation function.

Advantage:

It may be utilized in the output layer of neural networks to do multiclass classification.

Disadvantage:

It takes a long time to compute since there are so many exponent terms to calculate.

2.2.2.5 *Hyperbolic tangent (tanh)*

Hyperbolic tangent or in short "tanh" is represented as follows:

$$tanh(x) = \frac{e^{x} - e^{-x}}{e^{x} + e^{-x}} \tag{2.5}$$

It's a lot like the sigmoid function. It has a range of −1 to +1 and is centered at zero as shown in Figure 2.9.

Advantages:

Everywhere, it is continuous and distinct.
It revolves around zero.
It will set a restriction on output between −1 and +1.

Disadvantages:

It may result in a vanishing gradient issue.
It is computationally costly.

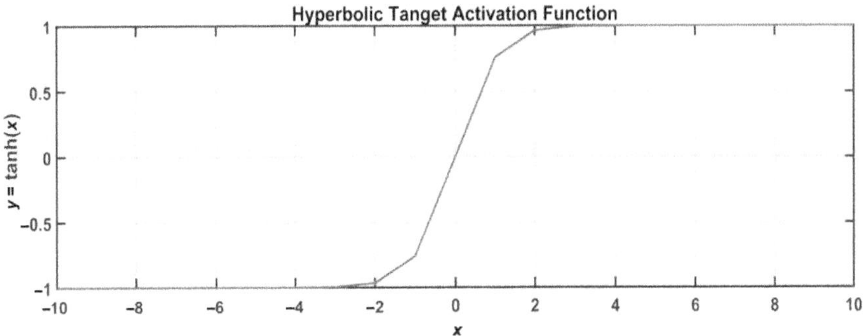

Figure 2.9 Hyperbolic tangent activation function.

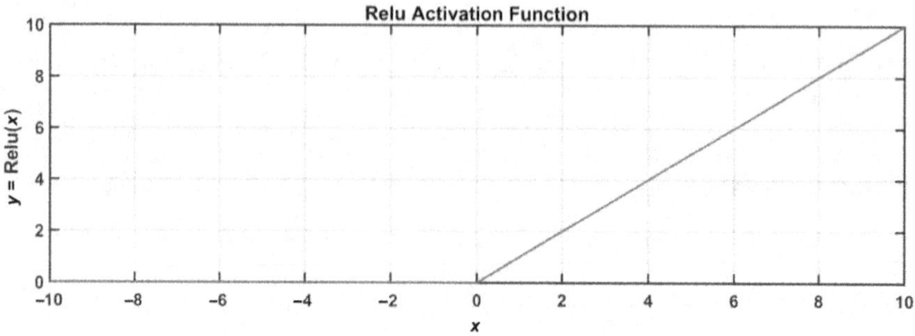

Figure 2.10 ReLU activation function.

2.2.2.6 ReLU

The rectified linear function often called as just a rectifier or ReLU is given as follows:

$$f(x) = max(0, x) \tag{2.6}$$

Figure 2.10 shows the simulation result of the ReLU equation (equation 2.6).

Advantages:

Simple to calculate; does not result in disappearing gradients.

Due to the fact that not all neurons are engaged, the network is sparse, making it quick and efficient.

Disadvantages:

Causes the problem of an exploding gradient.

Not centered on zero.

Because it always returns 0 for negative numbers, it may destroy certain neurons indefinitely.

We may adjust the saturation threshold value, or the maximum value that the function will return, to avoid the expanding gradient issue in ReLU activation.

2.2.2.7 *Leaky ReLU*

The enhancement of ReLU function is the leaky ReLU. The dying ReLU situation occurs when the ReLU function kills some neurons in each cycle. Instead of

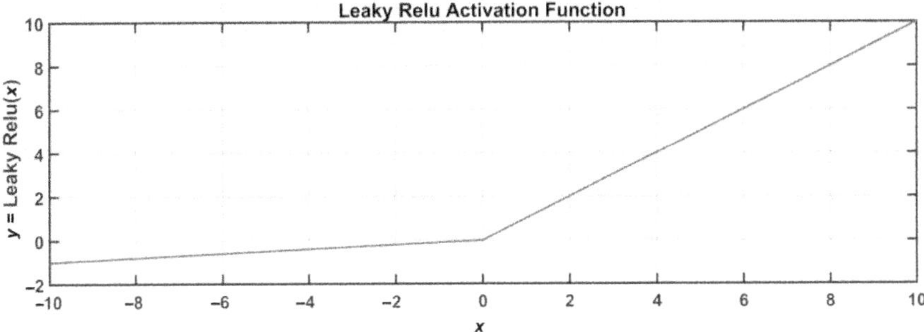

Figure 2.11 Leaky ReLU activation function.

returning 0 for negative values, a leaky ReLU will calculate output using a very tiny component of input, ensuring that no neuron is killed.

$$f(x) = \begin{cases} x \text{ } if \text{ } x > 0 \\ 0.01 \text{ } x \text{ } otherwise \end{cases} \tag{2.7}$$

Figure 2.11 shows the Leaky ReLU Activation Function, in which the output is same as the input when the value is greater than zero, and it is equal to 0.01 times the input value otherwise.

2.2.2.8 Parameterized ReLU

Instead of setting a rate for the negative axis, the parameterized ReLU passes it as a new trainable parameter that the network learns on its own to gain quicker convergence as shown in Figure 2.12.

$$f(x) = \begin{cases} x \text{ } if \text{ } x > 0 \\ ax \text{ } otherwise \end{cases} \tag{2.8}$$

The result of equation 2.8 is shown in Figure 2.9 with$\alpha = 0.01$.

Advantages:

The network will figure out the best alpha value on its own.
Doesn't result in a vanishing gradient issue.

Disadvantages:

Calculation is difficult.
The issue determines how well you perform.

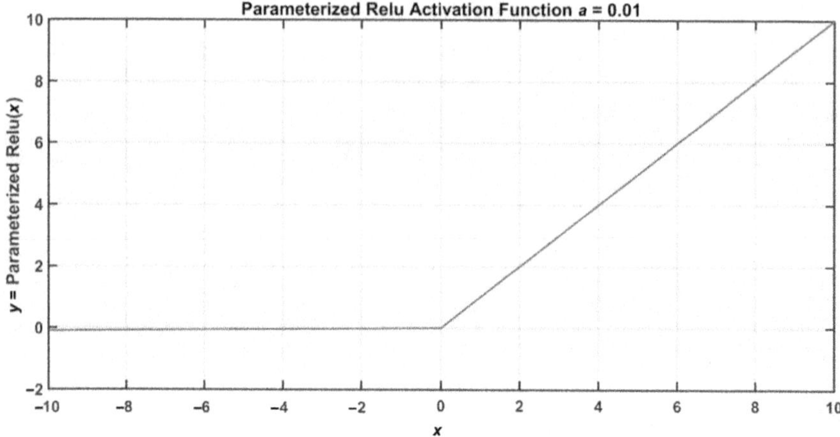

Figure 2.12 Parameterized ReLU activation function.

2.2.2.9 *Swish*

Using the sigmoid function, we can get the swish function by multiplying x by it.

$$f(x) = \frac{x}{1 + e^{-\beta x}} \tag{2.9}$$

The result of equation 2.9 is shown in Figure 2.13 with $\beta = 0.01$.

Google's brain team proposed the swish feature. Across many difficult datasets, their tests demonstrate that swish works quicker than ReLU of deep models.

Advantages:

Does not create the issue of disappearing gradients.
It has been shown to be somewhat superior than ReLU.

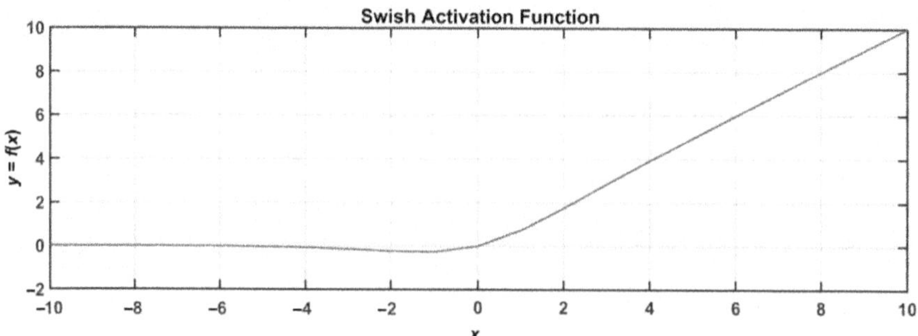

Figure 2.13 Swish activation function.

Disadvantage:

Expensive in terms of computation.

2.2.2.10 ELU

Another variant of the ReLU is the exponential linear unit (ELU), which attempts to get activations closer to zero to speed up learning. It outperforms the ReLU in terms of classification accuracy. ELUs have negative values, bringing the activation mean closer to zero.

$$f(x) = \begin{cases} x & if\ x > 0 \\ \alpha\left(e^{x} - 1\right) & otherwise \end{cases} \tag{2.10}$$

2.3 TRAINING A NEURAL NETWORK

Due to the fact that each neural network is the approximation of a function, they will not be equal to the intended function but will differ by a factor called error. The purpose of training is to minimize this error. We attempt to minimize the error associated with the weights, since the error is a function of the weights in the network. The error function is dependent on a large number of weights and, thus, on a large number of variables. In mathematics, the collection of points where this function is zero is referred to as a hyper-surface, and we aim to identify a minimum on this surface by selecting a point and then tracing a curve in the direction of the minimum.

2.3.1 Backpropagation

Backpropagation is a common training technique for artificial neural networks, particularly DNNs.

Backpropagation, as shown in Figure 2.14, is required to compute the gradient, and the weights of the weight matrices must be adjusted. The gradient of the loss function is used to modify the weights of the neural network's neurons (i.e. nodes). A gradient descent optimization technique is employed for this. Backward propagation of mistakes is another name for it.
Backpropagation networks may be divided into two categories:

Backpropagation in a static state.
Recurrent backpropagation is a kind of backpropagation that occurs repeatedly.

In backpropagation in a static state, static input to static output mapping is generated via a backpropagation network. For example, optical character recognition issues that need static classification may be solved using this technique.

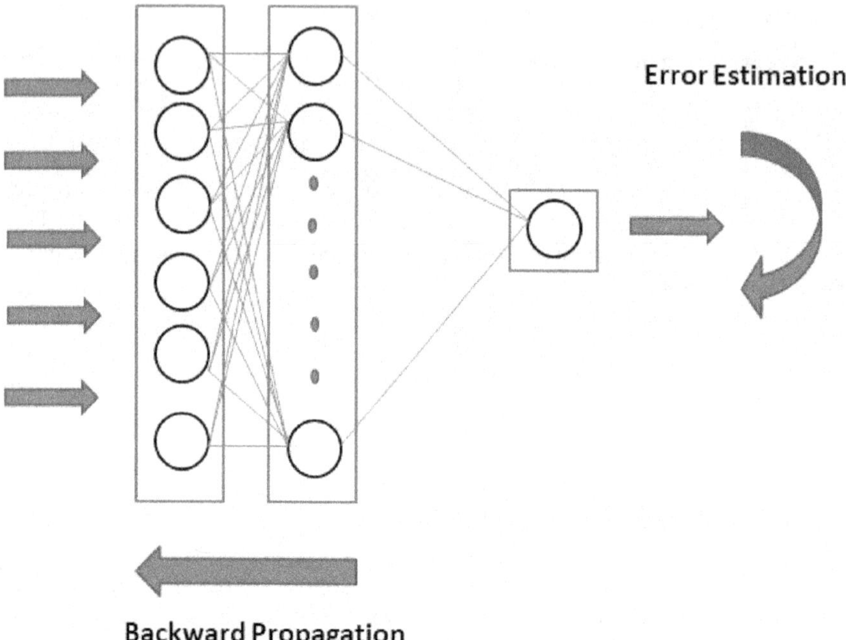

Figure 2.14 Illustration of backpropagation.

In a kind of backpropagation known as recurrent backpropagation (RBP), the technique is employed until a certain value is found in data mining. The blunder is then computed and sent in the reverse direction.

Static back-mapping propagation is quick, while recurrent backpropagation is nonstatic; this is a significant distinction.

2.4 ARCHITECTURAL EVOLUTION OF DEEP CNNS

Among biologically inspired AI methods, CNNs are now the most frequently utilized algorithms. Hubel and Wiesel's neurobiological experiments [1] are the foundation of CNN's history. Their work paved the way for a slew of cognitive models, with CNN eventually displacing almost all of them. Various attempts have been made throughout the years to enhance the performance of CNNs.

2.4.1 Beginning of CNN

LeCun et al. demonstrated the world's first multilayer CNN, ConvNet, in 1989, with its origins in Fukushima's Neocognitron [2, 3]. In contrast to its predecessor, Neocognitron's unsupervised reinforcement learning method, LeCun suggested supervised training of ConvNet using the backpropagation technique [4, 5]. Thus, LeCun's work laid the groundwork for current 2D CNNs. Handwritten digit and zip code recognition issues were solved successfully with this ConvNet [6, 7]. LeCun introduced an enhanced version of ConvNet, dubbed LeNet-5, in 1998, and pioneered the use of CNNs for character classification in document recognition applications [6]. CNN's commercial usage in ATMs and banks began in 1993 and 1996, respectively, because of its excellent performance in optical character and fingerprint identification. During this time, LeNet-5 set many records for optical character recognition, but it struggled with other image recognition issues.

CNN training was difficult because of the complexity of the architecture and the amount of processing required. In the early 2000s, it was widely believed that the backpropagation technique used for CNN training failed to converge to the error surface's global minima. As a consequence, CNN was shown to be less successful in extracting features than handcrafted features [8]. Furthermore, there was no comprehensive dataset of various picture classifications accessible at the time. Due to their comparatively high performance, other statistical techniques, particularly support vector machines (SVMs), were more popular than CNNs at the time [9, 10, 11]. Meanwhile, a few research organizations continued to work with CNNs, attempting to improve their performance. Researchers, Simard, Steinkraus, and Platt [12] enhanced CNN architecture in 2003 and demonstrated excellent performance on a hand-digit benchmark dataset when compared to SVMs and MNIST [12, 13]. Along with optical character recognition, this performance boost has propelled CNN research by allowing it to extend its applications to include character recognition for other scripts, the image sensor placement for facial recognition in video conferencing, and street crime control [14, 15, and 16]. Similarly, CNN-based systems for consumer tracking have been industrialized in marketplaces [17, 18, and 19]. Furthermore, the potential of CNNs in additional applications including medical picture segmentation, anomaly detection, and robot vision was investigated [20, 21].

2.4.2 Revival of CNN

Deep convolutional networks feature a complex structure and a lengthy training period. In the early 2000s, training DNNs required just a few parallel processing methods and minimal hardware resources. A gradient explosion and exponential decay are possible outcomes of training deep CNNs with a conventional activation function like the sigmoid. The CNN optimization issue has been the subject of considerable research since 2006. Several intriguing initialization and training methods have been published in this respect to overcome the challenges faced in deep CNN training and learning invariant features. In 2006, the idea of greedy layer-wise pretraining was proposed [22], which rekindled interest in deep learning research. Both supervised and unsupervised pretraining were shown to be more effective than random initialization in experimental investigations. The sigmoid activation function, according to Glorot and Bengio [23] and other researchers, is unsuitable for training deep networks with random weight initialization. This discovery led to the adoption of non-sigmoid activation functions such as ReLU, tanh, and others [23]. One of the reasons that pushed deep CNNs into the spotlight was the resurgence of deep learning [24]. Instead of subsampling, Ranzato, Huang, Boureau, and LeCun [25] utilized max-pooling, which produced excellent results by learning invariant features. Researchers began utilizing graphic processing units (GPUs) to speed DNN and CNN architecture training in late 2006 [26, 27]. In 2007, NVIDIA released the CUDA programming framework, which enables more efficient use of the GPU's parallel processing capabilities [28]. The main cause for the resurgence of CNN research was the use of GPUs for artificial neural networks and their training, as well as other advances in technology [29, 30]. ImageNet, a huge database of annotated pictures comprising millions of photos belonging to a wide range of classifications, was created by Fei-Fei Li's group at Stanford in 2010. This database [31] was used in conjunction with the annual ImageNet Large-Scale Visual Recognition Challenge (ILSVRC), which assessed and rated the performance of different models. Similarly, Stanford published the PASCAL 2010 VOC dataset for object identification the same year.

The availability of large amounts of training data as well as technology improvements have aided in the progress of CNN research. However, parameter optimization techniques and novel architectural concepts are the primary driving factors that have expedited research and given birth to the usage of CNNs in image classification and identification applications [32, 33, 34]. AlexNet, which outperformed conventional computer vision (CV) techniques in the 2012 ILSVRC [35] by lowering the error rate from 25.8 to 16.4 points, was the biggest advancement in CNN performance. As a consequence of these efforts, the computational cost of CNNs was significantly reduced. CNN performance was also improved via the investigation of depth of layers and parameter optimization techniques. The same was true of architectural concepts, with a slew of new structural reformulations presented in an effort to solve the faults of prior architecture proposals. Deep CNNs' popularity makes it very difficult to specify filter size, stride, padding, and other hyper-parameters for each layer individually. Convolutional layers with a fixed design are used to overcome this issue. Thus, unique layer design gave way to a modular, standard layer design. CNNs' modular design facilitates customizing them for a variety of tasks [36]. In this regard, the Google group developed a new concept of branching and blocking inside a layer [37–52]. It's worth noting that throughout this time period, two distinct kinds of architecture were in use: deep and narrow and deep and broad.

2.5 EXISTING ARCHITECTURES FOR IMAGE CLASSIFICATION

2.5.1 LeNet

LeCun proposed LeNet [53]. It is well known for its historical significance as the first CNN to demonstrate cutting-edge performance on hand-digit recognition tasks. It can identify digits without being influenced by minor distortions, rotation, or changes in position and scale. LeNet is a feed-forward neural network that consists of five convolutional and pooling layers alternated with two fully connected layers. GPUs were not widely utilized to speed up training in the early 2000s, and even CPUs were sluggish [54]. The fundamental foundation of the picture, in which adjacent pixels are linked to one another and feature patterns are spread throughout the whole image, was explored by LeNet. Consequently, a useful technique for extracting similar features from many locations is convolution with learnable parameters. Sharable parameter learning has altered the traditional training perspective, which treats each individual image pixel as an independent feature with no regard to how it connects to its surrounding pixels. By eliminating many of the previously required parameters, LeNet was the first CNN architecture to learn features from raw pixels (Figure 2.15).

2.5.2 AlexNet

LeNet, the first deep CNN, could only recognize hand digits and scored poorly in other picture classes. AlexNet is the first deep CNN architecture to achieve breakthrough accuracy in picture classification and recognition. It improved the CNN's learning ability by deepening it and optimizing its parameters [35]. Figure 2.16 shows AlexNet's basic architectural layout. In the early 2000s, hardware

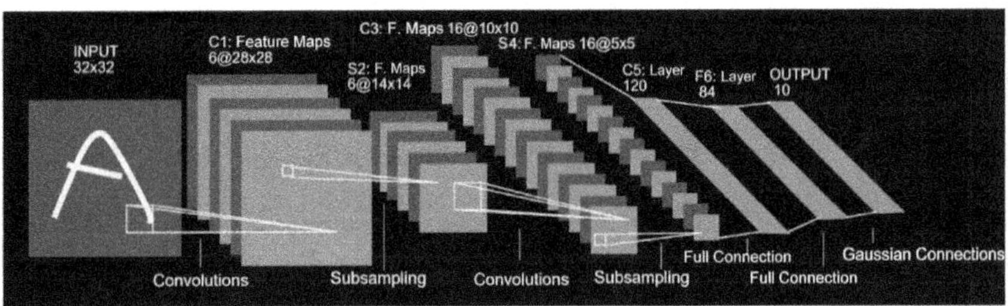

Figure 2.15 Lenet architecture.

Source: Courtesy: Google images.

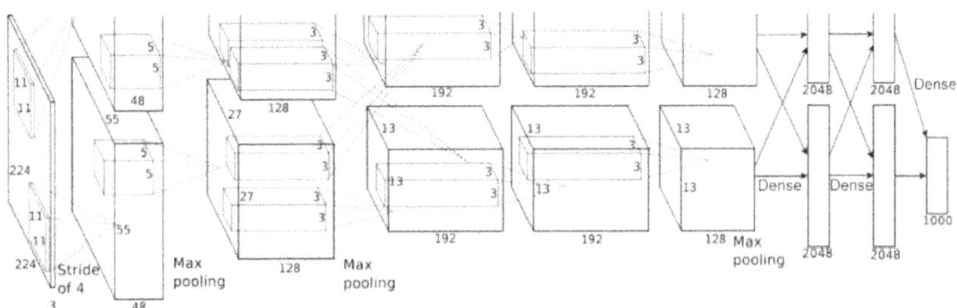

Figure 2.16 Architecture of alexNet.

Source: Courtesy: Google images.

restrictions limited the scale of deep CNN systems. To overcome hardware restrictions, AlexNet was trained on two NVIDIA GTX 580 GPUs in parallel. CNN's depth was expanded from five to eight layers to accommodate a wider variety of image types.

Increased depth causes overfitting, even when it increases generality across picture resolutions. Hinton's notion [55] was employed to tackle this problem by skipping random transformational units during training to force the model to acquire more robust features. ReLU was also employed as a non-saturating activation function to improve convergence rate by reducing vanishing gradient. Using overlapping subsampling and local response normalization reduced overfitting and increased generalization. AlexNet's fast learning approach has ushered in a new era of research in CNN architectural breakthroughs.

2.5.3 VGG

CNNs have helped image recognition tasks. As a result, visual geometry group (VGG) was built 19 layers deep to imitate depth and network representational capabilities [56]. Small-size filters, according to ZfNet, a frontline network in the 2013 ILSVRC competition, were found to be crucial for capturing intricate details within images while efficiently managing computational resources. For the 11×11 and 5×5 filters, VGG substituted a stack of 3×3 filters, demonstrating how stacking tiny (3×3) filters may create the illusion of a big filter size (5×5 and 7×7). Because they contain fewer parameters, tiny filters need less processing resources. This led CNN to start utilizing smaller-size filters in their research. VGG reduces the network's overall complexity by adding 1×1 convolutions between each convolutional layer.

Network tuning uses max-pooling after the convolutional layer and padding to maintain spatial resolution [25]. Using VGG, we were able to classify and locate images. This year's ILSVRC winner earned attention for its simplicity, homogenous topology, and enhanced depth.

2.5.4 GoogLeNet

GoogLeNet, also known as Inception-V1, won the 2014 ILSVRC. The GoogLeNet design sought high accuracy at cheap expense. It proposed an inception block in CNN, utilizing the divide, transform, and merge approach to integrate multiscale convolutional alterations. The inception block is built similarly to how GoogLeNet's network-in-network (NIN) approach substitutes convolutional layers with micro neural networks. This block collects spatial data at various sizes (1×1 to 5×5). By splitting, modifying, and combining photos, GoogLeNet was able to handle a problem that required learning multiple variants within the same category. GoogLeNet adds a 1×1 convolutional filter bottleneck layer before utilizing large kernels. It also used sparse connections to reduce duplication and save money by removing unnecessary feature maps. To reduce connection density, global average pooling was utilized in the final layer instead of a completely connected layer. These parameter tunings decreased the number of parameters from 138 million to 4 million. Regulatory concerns included batch normalization and RmsProp as an optimizer. Auxiliary learners were created by GoogLeNet to aid with convergence.

2.5.5 Applications of CNNs

These and other ML applications have been implemented using CNNs [1, 57, 58]. However, CNN needs much data to learn. It has shown significant results in medical picture segmentation; natural image identification of people, text, and pedestrians; and traffic sign recognition. These regions have plenty of labeled data.

2.5.5.1 Use of convolutional neural networks in computer vision

This artificial system can analyze visual input such as photographs and movies and extract valuable information from them. CV cares about it. Face, posture, activity, and other applications are all covered by CV. Face recognition is one of the most difficult CV tasks. Face recognition systems must cope with lighting, posture, and emotion variations. In addition to recognizing blocked faces, Deep CNN was recommended [59]. Another research uses a multitasking cascaded CNN to recognize faces [59]. Compared to modern approaches, Farfade, Saberian, and Li's approach [59] performed well. It's tough to evaluate human posture since the body's position varies so much. Using heterogeneous deep CNNs, we estimated posture [60].

2.5.5.2 Application of CNN in natural language processing

NLP allows computers to understand language. For example, language modeling and analysis employ CNNs, despite the fact that RNNs are more suited for this kind of application in general. With CNN as an alternate representation learning approach, language modeling or phrase molding has taken a new route. Sentence modeling is used to comprehend phrase semantics and offer creative and appealing solutions that suit consumer demands. Traditional data retrieval ignores the sentence's substance in favor of terms or attributes. A CNN-based architecture [61] can handle multilayer perceptron (MLP)-related tasks like name-entity recognition and semantic role modeling.

2.5.5.3 Application of CNN in segmentation and object detection

R-CNN is currently commonly utilized to detect objects. An object's borders may be recognized and scored using a fully connected CNN. One method uses deep CNNs to identify objects [62]. The method works with PASCAL VOC 2007 and 2012 [62]. Other CNN designs for semantic and instance-based segmentation include FCN, SegNet, Mask R-CNN, and U-Net.

2.5.5.4 *Application of CNN in image classification*

Medical pictures [63, 64] are one of CNN's main uses, particularly for the detection of cancer [65] utilizing histological images. CNN is used for breast cancer picture detection and the results were compared to a network trained on a dataset using handmade descriptors.

2.6 CONCLUSION

Rapid use of deep learning algorithms shows their success and adaptability. Deep learning's achievements and higher accuracy rates indicate the technology's evolution and future research possibilities. This chapter introduces DNNs and their structures, methodologies, properties, and limits. Also discussed are the significant distinctions between DNNs and standard ML, as well as the big obstacles ahead. In addition to a historical review of significant deep learning architectures, this book examines the methodologies and frameworks utilized, as well as the impact of each application in the real world. Deep learning is a unique technique to hierarchical layer processing that improves existing ML applications. Deep learning has found use in image processing and speech recognition.

Future aspects of deep learning include deep networks functioning in a complicated nonstatic strident environment, optimizing deep network characteristics to improve performance, and unsupervised academic compatibility of DNNs. Deep reinforcement learning is the future. Deep networks will need to be able to make inferences, be efficient, and be accurate. Deep generative models with superior temporal modeling abilities for parametric speech recognition will need to be built.

REFERENCES

[1] G. Kapila, B. Vandana, A. Khaitan, A. Francis Avinash, and C. H. Ajay Kumar, "Apple fruit classification and damage detection using pre-trained deep neural network as feature extractor," in Saini, H. S., Singh, R. K., Tariq Beg, M., Mulaveesala, R., and Mahmood, M.R. (eds.), Innovations in electronics and communication engineering. Lecture notes in networks and systems, vol. 355, Springer, Singapore, 2022. doi: 10.1007/978-981-16-8512-5_26.

[2] K. Fukushima, "Neocognitron: A hierarchical neural network capable of visual pattern recognition," Neural Networks, vol. 1, pp. 119–130, 1988.

[3] K. Fukushima, and S. Miyake, "Neocognitron: A self-organizing neural network model for a mechanism of visual pattern recognition," in Competition and cooperation in neural nets, Springer, 1982, pp. 267–285.

[4] S. Linnainmaa, "The representation of the cumulative rounding error of an algorithm as a Taylor expansion of the local rounding errors," Master's Thesis (in Finnish), Univ Helsinki, 1970, pp. 6–7.

[5] Y. LeCun et al., “Backpropagation applied to handwritten zip code recognition,” Neural Computation, vol. 1, pp. 541–551, 1989.
[6] Y. LeCun, Y. Bengio, and G. Hinton, “Deep learning,” Nature, vol. 521, pp. 436–444, 2015, doi: 10.1038/nature14539.
[7] X. Zhang, and Y. LeCun, “Text understanding from scratch,” arXiv preprint arXiv:1502.01710, 2015.
[8] J. Schmidhuber, “New millennium AI and the convergence of history,” in Challenges for computational intelligence, Springer, 2007, pp. 15–35.
[9] T. Joachims, “Text categorization with support vector machines: Learning with many relevant features,” in European conference on machine learning, Springer, 1998, pp. 137–142.
[10] D. Decoste, and B. Schölkopf, “Training invariant support vector machines,” Machine Learning, vol. 46, pp. 161–190, 2002.
[11] C.-L. Liu et al., “Handwritten digit recognition: Benchmarking of state-of-the-art techniques,” Pattern Recognition, vol. 36, pp. 2271–2285, 2003.
[12] P. Y. Simard, D. Steinkraus, and J. C. Platt, “Best practices for convolutional neural networks applied to visual document analysis,” in null, p. 958, 2003.
[13] K. Chellapilla, S. Puri, and P. Simard, “High performance convolutional neural networks for document processing,” in Tenth International Workshop on Frontiers in Handwriting Recognition, 2006.
[14] A. Abdulkader, “Two-tier approach for Arabic offline handwriting recognition,” in Tenth International Workshop on Frontiers in Handwriting Recognition, 2006.
[15] L. Deng, “The MNIST database of handwritten digit images for machine learning research [best of the web],” IEEE Signal Processing Magazine, vol. 29, pp. 141–142, 2012.
[16] D. C. Ciresan, U. Meier, L. M. Gambardella, and J. Schmidhuber, “Deep, big, simple neural nets for handwritten,” Neural Computation, vol. 22, pp. 3207–3220, 2010.
[17] C. Garcia, and M. Delakis, “Convolutional face finder: A neural architecture for fast and robust face detection,” IEEE Transactions on Pattern Analysis and Machine Intelligence, doi: 10.1109/TPAMI.2004.97.
[18] A. Frome et al., “Large-scale privacy protection in Google Street View,” in Proceedings of the IEEE International Conference on Computer Vision, 2009.
[19] Y. LeCun et al., “Convolutional networks and applications in vision,” in ISCAS, IEEE, 2010, pp. 253–256.
[20] B. Fasel, “Facial expression analysis using shape and motion information extracted by convolutional neural networks,” in Proceedings of the 12th IEEE Workshop on Neural Networks for Signal Processing, IEEE, 2002, pp. 607–616.
[21] M. Matsugu et al., “Convolutional spiking neural network model for robust face detection,” in Neural Information Processing, 2002, ICONIP’02. Proceedings of the 9th International Conference on, 2002, pp. 660–664.
[22] G. E. Hinton, S. Osindero, and Y.-W. Teh, “A fast learning algorithm for deep belief nets,” Neural Computation, vol. 18, pp. 1527–1554, 2006.
[23] X. Glorot, and Y. Bengio, “Understanding the difficulty of training deep feedforward neural networks,” in Proceedings of the Thirteenth International Conference on Artificial Intelligence and Statistics, 2010, pp. 249–256.
[24] Y. Bengio, P. Lamblin, D. Popovici, and H. Larochelle, “Greedy layer-wise training of deep networks,” in Advances in neural information processing systems, The MIT Press, 2007, pp. 153–160.

[25] M. Ranzato, F. J. Huang, Y. L. Boureau, and Y. LeCun, "Unsupervised learning of invariant feature hierarchies with applications to object recognition," in Proceedings of the IEEE Computer Society Conference on Computer Vision and Pattern Recognition, IEEE, 2007, pp. 1–8.
[26] D. C. Cireşan et al., "High-performance neural networks for visual object classification," arXiv preprint arXiv:1102.0183, 2011.
[27] G. Nguyen et al., "Machine learning and deep learning frameworks and libraries for large-scale data mining: A survey," Artificial Intelligence Review, vol. 52, pp. 77–124, 2019, doi: 10.1007/s10462-018-09679-z.
[28] J. Nickolls, I. Buck, M. Garland, and K. Skadron, "Scalable parallel programming with CUDA," in ACM SIGGRAPH 2008 Classes on—SIGGRAPH '08, ACM Press, New York, 2008, p. 1.
[29] K.-S. Oh, and K. Jung, "GPU implementation of neural networks," Pattern Recognition, vol. 37, pp. 1311–1314, 2004.
[30] D. C. Ciresan et al., "Multi-column deep neural networks for image classification," in IEEE Computer Society Conference on Computer Vision and Pattern Recognition, 2012.
[31] O. Russakovsky et al., "ImageNet large scale visual recognition challenge," International Journal of Computer Vision, 2015, doi: 10.1007/s11263-015-0816-y.
[32] J. Gu et al., "Recent advances in convolutional neural networks," Pattern Recognition, vol. 77, pp. 354–377, 2018, doi: 10.1016/j.patcog.2017.10.013.
[33] T. Sinha, B. Verma, and A. Haidar, "Optimization of convolutional neural network parameters for image classification," in 2017 IEEE Symposium Series on Computational Intelligence (SSCI), 2018, pp. 1–7, doi: 10.1109/SSCI.2017.8285338.
[34] Q. Zhang et al., "Recent advances in convolutional neural network acceleration," Neurocomputing, vol. 323, pp. 37–51, 2019, doi: 10.1016/j.neucom.2018.09.038.
[35] A. Krizhevsky, I. Sutskever, and G. E. Hinton, "ImageNet classification with deep convolutional neural networks," Advances in neural information processing systems, pp. 1–9, 2012, doi: 10.1061/(ASCE)GT.1943-5606.0001284.
[36] K. Simonyan, and A. Zisserman, "Very deep convolutional networks for large-scale image recognition," ICLR, pp. 398–406, 2015.
[37] C. Szegedy et al., "Going deeper with convolutions," in 2015 IEEE Conference on Computer Vision and Pattern Recognition (CVPR), IEEE, 2015, pp. 1–9.
[48] G. Huang, Z. Liu, L. Van Der Maaten, and K. Q. Weinberger, "Densely connected convolutional networks," in Proceedings—30th IEEE Conference on Computer Vision and Pattern Recognition, CVPR 2017, 2017, pp. 2261–2269, doi: 10.1109/CVPR.2017.243.
[39] T. Y. Lin et al., "Feature pyramid networks for object detection," in Proceedings—30th IEEE Conference on Computer Vision and Pattern Recognition, CVPR, 2017.
[40] R. Girshick, "Fast R-CNN," in Proceedings of the IEEE International Conference on Computer Vision, 2015, pp. 1440–1448.
[41] J. Long, E. Shelhamer, and T. Darrell, "Fully convolutional networks for semantic segmentation," in: 2015 IEEE Conference on Computer Vision and Pattern Recognition (CVPR), IEEE, 2015, pp. 3431–3440.
[42] O. Vinyals, A. Toshev, S. Bengio, and D. Erhan, "Show and tell: Lessons learned from the MSCOCO image captioning challenge," IEEE Transactions on Pattern Analysis and Machine Intelligence, doi: 10.1109/TPAMI.2016.2587640.

[43] G. Huang et al., "Deep networks with stochastic depth," in European conference on computer vision, Springer, 2016, pp. 646–661.
[44] C. Szegedy, S. Ioffe, and V. Vanhoucke, "Inception-v4, inception-ResNet and the impact of residual connections on learning," arXiv preprint arXiv:1602.07261, pp. 262–263, 2016, doi: 10.1007/s10236-015-0809-y.
[45] J. Hu, L. Shen, and G. Sun, "Squeeze-and-excitation networks," in 2018 IEEE/CVF Conference on Computer Vision and Pattern Recognition, IEEE, 2018, pp. 7132–7141.
[46] A. G. Roy, N. Navab, and C. Wachinger, "Concurrent spatial and channel 'squeeze & excitation' in fully convolutional networks," in Lecture Notes in Computer Science (including subseries Lecture Notes in Artificial Intelligence and Lecture Notes in Bioinformatics), vol. 11070 LNCS, 2018, pp. 421–429, doi: 10.1007/978-3-030-00928.
[47] S. Woo, J. Park, J. Lee, and I. S. Kweon, "CBAM: Convolutional block attention module," in Lecture Notes in Computer Science (including subseries Lecture Notes in Artificial Intelligence and Lecture Notes in Bioinformatics), vol. 11211 LNCS, 2018, pp. 3–19, doi: 10.1007/978-3-030-01234-2_1.
[48] A. Khan, A. Sohail, and A. Ali, "A New Channel Boosted Convolutional Neural Network Using Transfer Learning," arXiv preprint arXiv:1804.08528, 2018.
[49] M. F. Shakeel et al., "Detecting driver drowsiness in real time through deep learning based object detection," in Lecture Notes in Computer Science (including subseries Lecture Notes in Artificial Intelligence and Lecture Notes in Bioinformatics), 2019.
[50] N. Frosst, and G. Hinton, "Distilling a neural network into a soft decision tree," in CEUR Workshop Proceedings, 2018.
[51] A. G. Howard et al., "MobileNets: Efficient convolutional neural networks for mobile vision applications," arXiv preprint arXiv:1704.04861, 2017.
[52] Y. Xiong, H. J. Kim, and V. Hedau, "ANTNets: Mobile convolutional neural networks for resource efficient image classification," arXiv preprint arXiv:1904.03775, 2019.
[53] Y. LeCun et al., "Learning algorithms for classification: A comparison on handwritten digit recognition," Neural Networks Statistical Mechanics Perspective, vol. 261, p. 276, 1995.
[54] S. Potluri et al., "CNN based high performance computing for real-time image processing on GPU," in Proceedings of the Joint INDS'11 & ISTET'11, pp. 1–7, 2011.
[55] G. E. Dahl, T. N. Sainath, and G. E. Hinton, "Improving deep neural networks for LVCSR using rectified linear units and dropout," in Acoustics, Speech and Signal Processing (ICASSP), 2013 IEEE International Conference on, pp. 8609–8613, 2013.
[56] M. D. Zeiler, and R. Fergus, "Visualizing and understanding convolutional networks," arXiv preprint arXiv:1311.2901, 2013, pp. 30:225–231, doi: 10.1111/j.1475-4932.1954.tb03086.x.
[57] A. Khan et al., "A recent survey on the applications of genetic programming in image processing," arXiv preprint arXiv:1901.07387, 2019.
[58] Z. Batmaz, A. Yurekli, A. Bilge, and C. Kaleli, "A review on deep learning for recommender systems: Challenges and remedies," Artificial Intelligence Review, vol. 52, pp. 1–37, 2019, doi: 10.1007/s10462-018-9654-y.

[59] S. S. Farfade, M. J. Saberian, and L.-J. Li, "Multi-view face detection using deep convolutional neural networks," in Proceedings of the 5th ACM on International Conference on Multimedia Retrieval—ICMR '15, ACM Press, New York, New York, 2015, pp. 643–650.

[60] S. Li, Z.-Q. Liu, and A. B. Chan, "Heterogeneous multi-task learning for human pose estimation with deep convolutional neural network," in 2014 IEEE Conference on Computer Vision and Pattern Recognition Workshops, IEEE, 2014, pp. 488–495.

[61] R. Collobert, and J. Weston, "A unified architecture for natural language processing: Deep neural networks with multitask learning," in Proceedings of the 25th International Conference on Machine Learning, ACM, 2008, pp. 160–167.

[62] S. Gidaris, and N. Komodakis, "Object detection via a multi-region and semantic segmentation-aware U model," in IEEE International Conference on Computer Vision, 2015, pp. 1134–1142, doi: 10.1109/ICCV.2015.135.

[63] G. Levi, and T. Hassner, "Sicherheit und Medien," Sicherheit und Medien, 2009, doi: 10.1109/CVPRW.2015.7301352.

[64] Z. M. Long, S. Q. Guo, G. J. Chen, and B. L. Yin, "Modeling and simulation for the articulated robotic arm test system of the combination drive," in 2011 International Conference on Mechatronics Materials Engineering (ICMME) 2011, vol. 151, pp. 480–483, 2012, doi: 10.4028/www.scientific.net/AMM.151.480.

[65] D. C. Cireşan, A. Giusti, L. M. Gambardella, and J. Schmidhuber, "Mitosis detection in breast cancer histology images with deep neural networks BT—Medical image computing and computer-assisted intervention—MICCAI 2013," in Proceedings MICCAI, pp. 411–418, 2013.

Chapter 3

Comprehensive comparative analysis of artificial intelligence, machine learning, and deep learning

Hanane Lamaazi and Elezabeth Mathew

3.1 INTRODUCTION

Artificial intelligence (AI), machine learning (ML), and deep learning (DL) have evolved into buzzwords in the world of contemporary technology and now predominate talks about the future [1, 2]. These sectors are at the cutting edge of innovation, altering industries and significantly affecting our daily lives. This introduction aims to give readers a basic grasp of AI, ML, and DL, their interactions, and their relevance in the modern world [3]. The goal of the vast and all-encompassing area of AI in computer science is to build intelligent agents that can simulate cognitive functions in humans [4]. AI systems strive to display traits including perception, problem-solving, learning, and decision-making [5]. The search for AI has roots in ancient mythology about mechanical entities with humanlike abilities. Still, it is frequently said that the Dartmouth Workshop in 1956 was the first to take it seriously [6]. AI covers many methods, including rule-based systems and neural networks, and it is used in many industries, including banking and health care. Developing methods and models that allow computers to learn from data and make predictions or judgments without being explicitly programmed is the focus of the AI subfield of machine learning [7]. Large datasets are analyzed, trends are found, and ML systems gradually improve in performance as they "learn" to generalize from past experiences to new ones. Unforeseen data is a fundamental idea in machine learning [5, 7]. There are three primary subcategories of machine learning: supervised learning (label-assisted data), unsupervised learning (label-free data), and reinforcement learning (learning by environment interaction) [7, 8]. It is used in translation, picture recognition, recommendation systems, and more. To achieve its objectives, AI frequently uses ML methods. For instance, ML algorithms are used by chatbots powered by AI to comprehend and produce human language [9]. Based on medical data, AI systems in the health care industry may also use ML for disease diagnosis and prediction [10]. A part of machine learning called deep learning has become quite popular in recent years. It is defined using deep neural networks, artificial neural networks

DOI: 10.1201/9781003496410-4

with many layers inspired by the human brain's structure and operation [8]. These deep networks are ideally suited for picture and speech recognition tasks because they can autonomously learn hierarchical input representations [10]. AI has been revolutionized because of deep learning, which has made breakthroughs in fields like computer vision, natural language processing (NLP), and autonomous cars possible. The link between DL and ML is extremely strong. Deep neural networks are ML models with more layers to learn hierarchical representations from data automatically. DL uses ML concepts to create more complicated models for challenging tasks.

This chapter explores the differences and common points between AI, ML, and DL concepts. It explores the key components of an AI-based system, from data processing to decision-making. Choosing an optimal model depends on the application objective. This chapter provides examples of real-life use cases and AI, ML, and DL integration. Moreover, the chapter offers a variety of existing implementation tools and frameworks used to develop smart and autonomous solutions. Choosing the optimal metrics is a mandatory and important step in evaluating the efficiency of any developed model. This chapter covers the most used evaluation metrics and provides a comprehensive comparative study that helps researchers select the most relevant metrics to evaluate their approaches.

3.2 FUNDAMENTALS OF ARTIFICIAL INTELLIGENCE

The origins of AI can be found in tales and fables from antiquity, which included mechanical creatures with humanlike characteristics. However, the Dartmouth Workshop in 1956, where John McCarthy, Marvin Minsky, Nathaniel Rochester, and Claude Shannon first used the phrase "artificial intelligence," is frequently cited as the official birthplace of AI as a scientific field [11]. The foundation for AI research and development was laid during this session. Fundamentally, AI aims to imitate human intelligence and problem-solving skills in robots [12]. The three fundamental tenets of AI are as follows:

- **Learning:** AI programs study data to learn. Machine learning, a component of AI, entails the creation of algorithms that allow computers to learn from their experiences and advance. This learning can be supervised, unsupervised, or reinforced.
- **Reasoning:** AI systems make decisions logically and apply logic to address issues. This covers methods like expert systems, rule-based reasoning, and symbolic reasoning.
- **Perception:** AI systems use sensors, cameras, microphones, and other sensory inputs to collect data from their environment. Applications of perception-based AI include computer vision, speech recognition, and NLP.

AI includes a diverse range of methods and strategies [13]. Among the essential methods are the following:

1. **Machine Learning**: AI systems can extract patterns and forecasts from data thanks to ML techniques. Algorithms for regression, classification, and clustering are included.
2. **Neural Networks**: The key element of deep learning is the modeling of the human brain. Numerous-layered deep neural networks (also known as "deep learning") have revolutionized AI and made significant progress in speech and image recognition.
3. **Natural Language Processing**: The goal of NLP is to make it possible for computers to comprehend, analyze, and create human language. This has resulted in applications such as sentiment analysis, language translation, and chatbots.
4. **Computer Vision**: AI systems can now analyze and interpret visual data thanks to computer vision, paving the way for features like object identification, facial recognition, and autonomous driving.

AI has enormous potential, but it also has problems and moral conundrums. Concerns include the influence of AI on the workforce, privacy concerns, and prejudice in AI systems. Finding a balance between technological progress and ethical considerations for the AI community continues to be quite difficult [14].

3.2.1 Interrelationships

AI, ML, and DL are closely related to one another. The aim of building intelligent machines is the overarching concept of AI, while ML and DL are subsets of AI that offer the methods and tools to do it. While AI seeks to imitate human intellect, ML and DL give it a chance through data-driven learning and sophisticated neural networks [15]. The basis for DL, in turn, is ML. DL is a subset of ML that uses neural networks with numerous hidden layers to simulate complicated relationships in data. Tasks that were previously thought to be prohibitively difficult, including speech and picture recognition, have significantly advanced thanks to DL. It's important to remember that ML comprises a wider range of methods than DL alone [16]. Here's a simplified representation of their interrelationship illustrated in Figure 3.1:

Where:

- AI encompasses both ML and DL.
- ML is a subset of AI, focusing on data-driven learning.
- DL is a subset of ML, specifically using deep neural networks for complex tasks.

Within the larger field of AI, each layer represents a more specialized and narrowly focused domain. ML and DL offer the procedures and strategies to reach various levels of machine intelligence based on data and neural network topologies, whereas AI aspires toward intelligent machines [16]. Continued cooperation and innovation are promised for the future of AI, ML, and DL. Researchers and practitioners are investigating the best approaches to integrate the strengths of these professions. For example:

- ***Transfer Learning***: It is becoming possible to learn more quickly and perform better thanks to the development of techniques for transferring information from one activity or topic to another.
- ***Explainable AI***: DL models and other AI systems are being worked on to increase their interpretability and transparency, which will increase their credibility [17].
- ***Federated Learning***: Decentralized learning techniques are becoming more common to protect data privacy while promoting group learning and model advancements [17].

3.2.2 AI instances

There are numerous uses of AI in various industries, such as speech recognition that is used to identify and categorize images and audio [18], personalized recommendations for products and content that users are likely to be interested in, e-commerce websites, and streaming services [19]. It is also used for predictive maintenance systems to examine data from sensors and other sources to anticipate when equipment may fail, hence, minimizing downtime and maintenance expenses [20]. In medical diagnosis, AI systems are used to examine patient data and medical imaging to give clinicians more precise diagnoses and treatment options [21]. Virtual personal assistants (VPA) like Siri or Alexa were developed to comprehend user demands, such as playing music, setting reminders, and responding to inquiries, using NLP [19]. In image recognition, AI systems recognize objects, people, and scenes in photographs and applications, including photo organization, security systems, and autonomous robotics [18].

3.3 MACHINE LEARNING: THE FOUNDATION

One of the areas of AI with the biggest potential for change is the discipline of machine learning. Its roots and tenets have enabled a fundamental transformation in how computers see the environment and communicate with one another [4]. The idea of ML is not new; it dates back to the middle of the 20th century. Early innovators like Alan Turing and Arthur Samuel built the foundation for ML, by creating algorithms that could change and get better over time. Machine learning did not, however, really take off until the digital era, when huge datasets and computing power became widely available [4, 5]. The Dartmouth Workshop launched AI and ML research formally in 1956. John McCarthy and his associates presented the concept of computers that might learn from experience during this workshop. Rule-based systems, expert systems, and statistical techniques have all been phases of ML's development since then. ML has reached new heights due to the development of neural networks in the 1980s and the following renaissance of DL in the 21st century [6]. Building algorithms and models that allow computers to recognize patterns in data and make predictions or judgments based on those patterns is at the heart of ML.

ML frameworks rely on a systematic workflow that contains the main steps required to develop a solution for a domain-specific problem. Regardless of the domain of the application, those steps are mandatory. However, what makes a key different from another one is the framework design, choice of the algorithm, type of collected data, and desired output. The main steps are illustrated in Figure 3.2 and described as follows:

- Input: Data or input from various sources, including sensors, databases, or external systems, is used to start the process. This information may be presented as text, pictures, numbers, or any other appropriate way.
- Preprocessing: Preprocessing is frequently needed before AI algorithms use the data. To prepare the data for analysis, this stage entails cleaning, converting, and organizing.
- Feature Extraction: Sometimes extracting pertinent traits or attributes from the data is necessary. Feature extraction streamlines the data while preserving crucial details.
- Algorithms: The data is sent into AI algorithms following preprocessing. These algorithms can vary depending on the AI objective, including ML models, rule-based systems, or DL networks.
- Training: A training step is involved if the AI system is based on DL or ML. The algorithm is trained using previous data to generate predictions or choices, and its parameters are changed.
- Inference: The AI model may forecast, decide, or classify fresh, unexplored data after training. This is the stage where AI shows its intelligence by analyzing or producing new ideas utilizing the learned patterns.
- Output: The final product can take many shapes depending on the AI application. It might be a suggestion, forecast, choice, visualization, or any other kind of usable result [14].

While ML has come a long way, it still has issues, including data privacy, algorithmic bias, and moral dilemmas. With improvements in fields like explainable AI and federated learning, the industry is continuously expanding and tackling some problems. In summary, ML is the cornerstone of contemporary AI [18–21].

3.4 DEEP LEARNING: THE NEURAL NETWORK REVOLUTION

Deep Learning has become a revolutionary force in the constantly changing field of AI, sparking a neural network revolution that is transforming industries, advancing science, and expanding the capabilities of AI [1]. Although DL originated in the middle of the 20th century, the phrase "deep learning" only became popular in the 21st century. Individuals like Frank Rosenblatt and Marvin Minsky pioneered artificial neural networks (ANNs) and perceptrons. However, enthusiasm waned due to their initial restrictions and computational limits [1, 3]. Several significant advances have contributed to the rebirth of DL, including:

- **Theoretical Foundations:** The mathematical underpinnings of neural networks were uncovered by researchers like Geoffrey Hinton, Yann LeCun, and Yoshua Bengio, who made fundamental contributions to the DL theory.
- **Big Data:** Deep neural networks could be trained successfully thanks to the abundance of labeled data made possible by the digital age.
- **Computational Power:** Deep network training was sped up by improvements in hardware, especially graphic processing units (GPUs), which enabled previously impossible jobs to be completed.

ANNs, based on the human brain's structure and operation, are at the heart of deep learning. These networks comprise layers of connected artificial neurons, each taking increasingly abstract representations from the input data. The recurrent neural network (RNN) for sequential data and convolutional neural network (CNN) for image analysis are the two neural network architectures most commonly associated with deep learning [18]. One distinguishing feature of DL is the depth of these networks. Deep networks, as opposed to shallow networks, have several hidden layers, enabling them to simulate complicated and nuanced patterns in data. The success of DL can be primarily ascribed to the ability of these deep neural networks to capture hierarchical information representations, enabling tasks like speech synthesis, image recognition, and NLP [22].

Although DL has had great success, there are still certain difficulties. Deep neural networks' "black-box" nature raises questions about bias and interpretability. Large labeled-dataset requirements and limited computational resources are other drawbacks [18–22].

3.5 TOOLS AND FRAMEWORKS

To assess any developed system's efficiency, it must be implemented using an adequate tool and platform. Choosing optimal implementation tools or platforms relies on a set of factors, including the application domain, the availability and wide use of the tools, the community, cost, and user expertise. In this section, a set of existing tools/platforms related to the application of AI, ML, and DL solutions are presented:

3.5.1 AI implementation tools

Various industries have developed AI tools that companies and researchers can use to develop AI solutions (see Table 3.1). ***IBM Watson*** provides AI tools that deploy NLP for human speech analysis. It is used to understand the syntax and the meaning of words and answer human questions quickly. Another portfolio of AI services is the ***Microsoft Azure AI*** [23]. It is a flexible service that benefits from research and AI practices, allowing data scientists and developers to implement their AI solutions. It has customized APIs that provide high-quality language, speech, vision, and decision-making AI models. It also provides an interoperable AI supercomputing infrastructure with common development environment platforms such as Visual Studio Code and Jupyter Notebooks and open source frameworks such as PyTorch and TensorFlow. Similarly, Google has developed ***Google Cloud AI*** to offer AI services and tools, such as Dialogflow and TensorFlow, which researchers and developers can customize to meet their solution needs. Examples of Google AI solutions are AI call center operators, DocAi for document processing, Translation Hub for speech and text translation using large-scale languages, and Vertex Vision AI [24], which allows the implementation and monitoring of computer vision applications using a simplified user interface. It also analyzes images and videos by integrating various components such as vision warehouse, data streams, and live video analytics. *Amazon Web Services (AWS)* and *OpenAI GPT-3* are other alternatives to these AI services. AWS offers various services for building conversational interfaces, such as Amazon Lex and ML, like Amazon SageMaker. Conversely, GPT-3 relies on language modeling through an API with various AI applications for text generation and chatbots.

3.5.2 Machine learning implementation tools

Scikit-Learn: Scikit-Learn is a popular ML library that supports supervised and unsupervised learning algorithms, including linear and logistic regression, decision trees, clustering, and k-means [25]. It is characterized by the simplicity and efficiency of the provided tools used for predictive data analysis. Also, it is accessible and easy to use and deploy in various contexts. It is built on NumPy, SciPy, and matplotlib. The library is open source and commercially usable under a BSD license [26]. Scikit-Learn offers a set of algorithms for the following:

- ***Classification*** includes k-nearest neighbor and random forest to identify an object's category (e.g., image recognition).
- ***Regression***, such as gradient boosting and logistic regression, to predict a continuous-valued attribute associated with an object (e.g., stock prices).

- ***Clustering***, such as k-means and HDBSCAN, automatically groups similar objects into sets.
- ***Dimensionality reduction***, such as feature selection and principal component analysis (PCA), to reduce the number of random variables to consider.
- ***Model selection***, such as cross-validation and grid search, is used to select parameters and compare and validate models.

Gradient Boosting: It is a powerful and popular ML technique in various applications such as classification, ranking, regression, and recommendation systems [27]. It is a category of ensemble learning that combines the predictions from multiple inaccurate learners to create a strong model with high-accuracy prediction. It is characterized by handling categorical and numerical features and can work with unbalanced datasets. However, choosing random hyper-parameter tuning during the implementation can lead to overfitting; high resources, as it is computationally expensive; and high training time, especially for large data [28].

3.5.3 Deep learning implementation tools

DL has some common frameworks with ML, such as TensorFlow, Keras, and PyTorch. However, other frameworks, such as CNTK, Caffe, MxNet, and Theano, are dedicated to developing DL solutions.

- **TensorFlow:** TensorFlow is an open source DL library. It is used to write a program in Python and run it on your CPU or GPU without having to write at the C++ or CUDA level to utilize GPUs.
- **PyTorch:** PyTorch is a framework for ML that relies on the Torch library. This software is useful for various applications, including NLP and computer vision. Initially created by Meta AI, PyTorch is now a part of the Linux Foundation umbrella. It is open source, with a modified BSD license.
- **Keras:** Keras is an open source library for working with ANNs in Python. It bridges the TensorFlow library and supports various back ends until version 2.3, including Theano, Microsoft Cognitive Toolkit, PlaidML, and TensorFlow. Keras follows best practices to reduce cognitive load, providing clear and simple APIs, minimizing the actions you must take for common use cases, and giving actionable error messages [29].
- **Microsoft Cognitive Toolkit (CNTK):** It is a framework that enables users to design new AI solutions using a single model or combination of various popular model types like feed-forward DNNs, CNNs, and RNNs/LSTMs. It incorporates stochastic gradient descent learning with automatic parallelization and differentiation across multiple servers and GPUs. CNTK is under an open source license, and it is available for public use.
- **Caffe:** The Caffe framework was developed by the Berkeley Vision and Learning Center in collaboration with community contributors. It is an open

source framework under the BSD license that provides a faster and more modular DL framework. It is used for computer vision tasks, image classification, and framing. It is a C++ library with a Python interface. The Caffe framework is the foundation for Google's DeepDream project [30].

- **MXNet:** It is an open source framework with a "forgetful backdrop" for saving memory, and it is particularly helpful for RNNs that process long sequences. It's characterized by its scalability, with multi-GPU and multi-machine training support. It allows writing customized layers in high-level languages.
- **Theano:** It is a powerful library for numerical computation that can be used in parallel with other libraries, such as Keras. Its primary purpose is to simplify the implementation station of DL models for research and development purposes. The speed of Theano makes it an ideal choice for complex tasks such as DL and other computationally demanding projects.

The choice of tools and frameworks depends on the specific project requirements, the team's expertise, and the implementation's goals. Many practitioners use a combination of these tools to achieve their objectives efficiently and effectively.

3.6 EVALUATION PARAMETERS USED FOR AI, ML, DL

Evaluating the effectiveness of solutions-based AI, ML, and DL models is important to ensure they align with the desired objective and satisfy all the requirements while resolving the studied problem. Many evaluation parameters are deployed to validate the efficiency and assess the performance of the models. Choosing the optimal parameters depends on the objectives of the studied problem; AI, ML, DL specifications; tasks, and the solution design (see Table 3.2). This section presents the most common evaluation metrics used to evaluate the AI, ML, and DL models, which are described as follows:

- Accuracy: They are used to measure the correctness of classifier prediction. It is calculated as the ratio of the correct predictions to the total number of predictions.
- Precision: Also known as positive predictive value, which computes the proportion of instances that are truly classified under a specific class out of all the instances classified under that class (True Positives (TP) + False Positives (FP)). Precision is a valuable measure for determining the level of confidence we can have in considering a prediction [31].
- Recall (Sensitivity or True Positive Rate): Recall, also known as sensitivity, reveals the number of correctly classified instances belonging to a class (True Positives (TP) + False Negatives (FN)). It measures a model's ability to classify specific class instances accurately.

- F1-Score: The F1-score measures the model accuracy and is defined as a harmonic mean combining precision and recall, and it always falls within the range of 0 to 1.
- It is crucial to understand that the closer the value is to 1, the better our model performs. It is also essential to note that the F1-score is heavily influenced by both recall and precision [32, 33].
- Specificity (True Negative Rate): Specificity, also known as true negative rate (TNR), is used to measure the model's capability to predict the true negatives of each available class. It refers to the percentage of true negative cases correctly identified, which means that a certain percentage of actual negative cases can be predicted as positive, known as false positives. Note that the sum of specificity and false positive rate will always be equal to 1. A high specificity indicates that the model accurately identifies most negative cases. In contrast, a low specificity suggests that the model incorrectly labels many negative cases as positive [34].
- Area under the Receiver Operating Characteristic Curve (AUC-ROC): It is mostly used for binary classification problems. It is used to estimate the accuracy of a model where a model with a good measure of separability has an AUC close to or equal to 1, while a bad one has a value close to 0. AUC-ROC relationship helps to measure the model's ability to differentiate between classes [35].
- Mean Absolute Error (MAE): It measures the average of the remainder of the dataset. It is calculated as the average absolute difference between the predicted and actual values in the dataset.
- Mean Squared Error (MSE): It is used to measure the variance of the remainder of a dataset. It represents the average squared differences between predicted and original values in regression tasks [36].
- Root Mean Squared Error (RMSE): It evaluates the target variable's consistency by measuring the remainder's standard deviation. It is calculated as the squared error of the mean squared error. RMSE is used in regression models to estimate the loss function. It has the same units as the dependent variable, making it the widely used metric compared to MSE [36].
- R-squared (Coefficient of Determination): It is considered a scale-free score where the value of R-square is less than one, regardless of the large or small values. It is calculated as the proportion of the variance in the dependent variable [37].
- Mean Average Precision (mAP): It measures the performances of object detection models such as Fast Mask R-CNN, YOLO, and R-CNN. It considers both precision and recall, as well as false positives and negatives. This makes it a suitable metric for a wide range of detection applications [38, 39].
- IoU (Intersection over Union): It measures the overlap between the predicted and the ground truth bounding boxes [40]. It is mostly used in object detection models to measure localization accuracy and errors. The higher the overlap region, the greater the IoU [41].

- BLEU (Bilingual Evaluation Understudy): It measures the quality of predicted text based on the similarity of the human-generated reference text and the machine-generated text. It varies from 0 to 1, where more than 0.3 is considered a good score [42].
- Perplexity (in NLP): It measures how well a language model predicts a text sample. It is deployed to compare different language models, fine-tune a single model parameter, or identify problems in a dataset. The lower the perplexity is, the better the model prediction. However, perplexity can be used only as a preliminary measure, not for final decision-making, because it does not present accuracy. A model can have low perplexity while providing a high error rate, negatively affecting the accuracy of the model predictions [43, 44].

3.7 CASE STUDIES AND APPLICATIONS

Nowadays, AI, ML, and DL have become indispensable tools to develop smart solutions. It is integrated into a set of domains, such as health care, NLP, finance, retail, autonomous vehicles, entertainment, and agriculture. Choosing the optimal AI, ML, and DL model depends on domain requirements and objectives. IBM Watson for oncology, disease outbreak prediction, and medical image analysis are examples of solutions-based AI, ML, and DL used in health care to improve patient well-being and ensure the early detection of chronic diseases. In the financial domain, algorithmic trading, AI-powered chatbots, and fraud detection have helped many banks and financial institutions answer queries and support customers, make decisions for high-frequency trading, and identify fraudulent transactions. However, virtual assistants, language translation, and sentiment analysis ensure a better understanding and response to natural language queries; provide accurate translation between different languages; and highly analyze public sentiments regarding brands, products, and political issues using social media data. Other examples of successful integration of AI, ML, and DL in our lives are self-driving and driver behavior analysis, object detection, personalized and content recommendations, inventory management, visual search, music composition and video game design, precision agriculture, pest and disease detection, and crop yield prediction.

3.8 FUTURE DIRECTIONS AND CHALLENGES

AI, ML, and DL are promising techniques for developing intelligent systems. Integrating these techniques into our daily lives makes understanding and problem-solving much easier. Health care, advancements in medical diagnosis, drug discovery, and personalized treatment plans rely heavily on AI. Autonomous systems, including self-driving cars and drones, will continue to progress, with advancements in safety features and increased autonomy playing significant roles. Also, AI will play a crucial role in climate change by modeling and preventing natural disasters. Similarly, in education, AI helps to develop personalized learning platforms such as virtual classrooms, online assignments, and adaptive assessments. Such automated and intelligent systems can be targeted by cyberattacks. In this regard, AI detects and identifies cyber threats, finds anomalies, and recognizes unknown patterns.

Considering all the previously mentioned benefits, AI, ML, and DL techniques have several challenges, including data quality, model bias, interoperability, interpretation, security, and scalability. In the training steps, if the data is biased or has low quality, it can lead to biased AI models. A good preprocessing of the data and a good algorithm implementation are mandatory to achieve better results. Also, developing AI systems is domain specific, and deploying them all together is very challenging. Developing complex AI and DL models and training large models makes the decisions hard to interpret and can be computationally exhaustive. Opting for scalable and energy-efficient models is crucial. Finally, ensuring the security of AI systems, especially autonomous systems, against cyberattacks is still a big concern. Addressing these challenges is important to ensure AI technologies' sustainable and efficient progress.

3.9 CONCLUSION

The world of AI, ML, and DL has grown exponentially and has become the main tool for developing smart solutions that do not require human intervention. Integrated into all domains of daily life, these emerging techniques provide various tools, resources, and frameworks used to solve complex problems and drive innovation. Choosing the optimal tools or frameworks depends on the user's needs and solution objectives, in addition to the user's expertise.

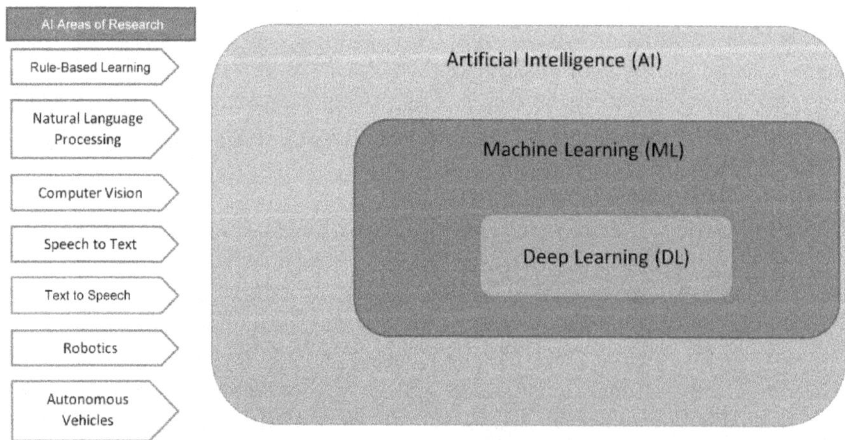

Figure 3.1 AI, ML, and DL simplified in one graph.

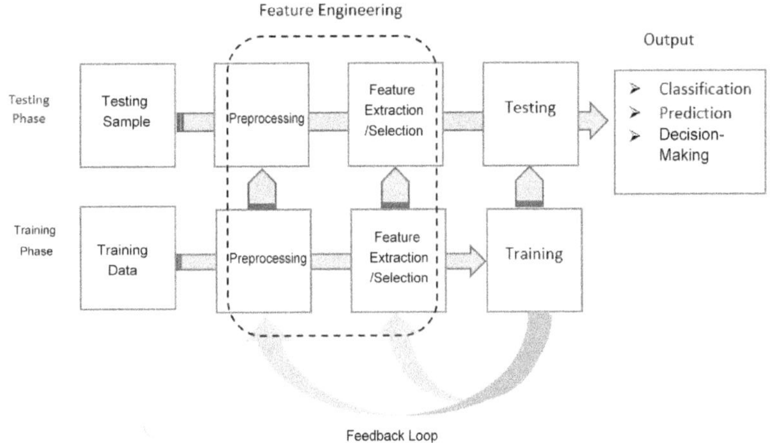

Figure 3.2 Typical machine learning system.

Table 3.1 Comparison of the most common tools and frameworks for developing solutions-based on AI, ML, and DL

			Use case			*Programming*			
Tools/frameworks		*Purpose*	*Difficult*	*Moderate*	*Easy*	*Python*	*C++*	*Scala*	*Various*
AI	***IBM Watson***	AI services& Tools		✓					✓
	Microsoft Azure AI	AI services and cloud-based tools		✓					✓
	Google Cloud AI	AI services and ML tools		✓					✓
	Amazon AI			✓					✓
	OpenAI-GPT3	Language model via API			✓				NA / (API)
ML	***Scikit-Learn***	ML libraries			✓	✓			
	XGBoost				✓				✓
DL	***TensorFlow***	DL framework		✓		✓	✓		
	PyTorch			✓		✓			
	Keras			✓		✓			
	Caffee			✓		✓	✓		
	MXNet			✓		✓		✓	
	Theano			✓					✓
	CNTK			✓		✓	✓		

Note: AI, artificial intelligence; API, application programming interface; DL, deep learning; ML, machine learning.

Table 3.2 Comparison of the most common evaluation parameters used for AI, ML, and DL

Evaluation parameter	*Formula*		*Use case*				*Objective*
			Classification	*Regression*	*Object Detection*	*NLP*	
Accuracy	$Accuracy = \frac{TP + TN}{TP + TN + FP + FN}$	(3.1)	✓				Predict one of the multiple predefined classes.
F1-score	$\mathrm{F1}_{\mathrm{Score}} = 2 * \frac{Precision * Recall}{Precision + Recall}$	(3.2)	✓				Find the balance between precision and recall.
Recall	$Recall = \frac{TP}{TP + FN}$	(3.3)	✓				Evaluate a model's ability to identify all positive instances correctly.
Precision	$Precision = \frac{TP}{TP + FP}$	(3.4)	✓				Assess the quality of positive predictions.
Specificity	$Specificity = \frac{TN}{TN + FN}$	(3.5)	✓				Correctly identify negative instances.
AUC-ROC	$AUC = \frac{\sum_{i=1}^{N} Rank(TP) - \lvert TP \rvert * (\lvert TP \rvert + 1)/2}{TP + FN}$	(3.6)	✓				Assess model discrimination.
MAE	$MAE = \frac{1}{N} \sum_{i=1}^{N} \lvert Y_i - \hat{Y} \rvert$	(3.7)		✓			Evaluate the accuracy of the model.
MSE	$MSE = \frac{1}{N} \sum_{i=1}^{N} \left(Y_i - \hat{Y}\right)^2$	(3.8)		✓			Assess the goodness of fit in the model.
RMSE	$RMSE = \sqrt{MSE} = \sqrt{\frac{1}{N} \sum_{i=1}^{N} \left(Y_i - \hat{Y}\right)^2}$	(3.9)		✓			Interpretability of the error metric.

R-square	$R^2 = 1 - \frac{\sum_{i=1}^{N}\left(Y_i - \hat{Y}\right)^2}{\sum_{i=1}^{N}\left(Y_i - \bar{Y}\right)^2}$	(3.10)	✓			Evaluate the goodness of fit in the models.
mAP	$mAP = \frac{1}{n}\sum_{i=1}^{k=n} AP_k$ AP_k = The average precision of class k n = The number of classes	(3.11)		✓		Assess the performance of the model.
IoU	$IoU = \frac{\text{Area of Overalp}}{\text{Area of Union}} = \frac{TP}{TP + FP + FN}$	(3.12)		✓		Evaluate the accuracy.
BLEU	$Brevity\ Penalty * e^{\sum_{i=1}^{n} \omega_n log(Precision)}$	(3.13)			✓	Assess the performance.
Perplexity	$2^{\frac{1}{N}} \sum_{i=1}^{N} \log p(W_i)$ where: W = Test set *B* = total number of words *P* (*Wi*) = Probability of each word *Wi*	(3.14)			✓	Evaluate language models in tasks.

Note: AI, artificial intelligence; DL, deep learning; ML, machine learning; NLP, natural language processing.

REFERENCES

[1] R. Fallah Madvari, "Artificial intelligence (AI), machine learning (ml) and deep learning (dl) on health, safety, and environment (HSE)," *Archives of Occupational Health*, vol. 6, no. 4, pp. 1321–1322, 2022.

[2] M. Woschank, E. Rauch, and H. Zsifkovits, "A review of further directions for artificial intelligence, machine learning, and deep learning in smart logistics," *Sustainability*, vol. 12, no. 9, p. 3760, 2020.

[3] M. Soori, B. Arezoo, and R. Dastres, "Artificial intelligence, machine learning and deep learning in advanced robotics, a review," *Cognitive Robotics*, vol. 3, pp. 54–70, 2023.

[4] R. V. Yampolskiy, *Artificial intelligence safety and security*. CRC Press, 2018.

[5] M. H. Jarrahi, "Artificial intelligence and the future of work: Human-ai symbiosis in organizational decision making," *Business Horizons*, vol. 61, no. 4, pp. 577–586, 2018.

[6] J. Howard, "Artificial intelligence: Implications for the future of work," *American Journal of Industrial Medicine*, vol. 62, no. 11, pp. 917–926, 2019.

[7] T. K. Hua, "A short review on machine learning," *Author a Preprints*, 2022.

[8] T. O. Ayodele, "Types of machine learning algorithms," *New Advances in Machine Learning*, vol. 3, pp. 19–48, 2010.

[9] S. Singh, and H. K. Thakur, "Survey of various AI chatbots based on technology used," in *2020 8th international conference on reliability, Infocom technologies and optimization (trends and future directions) (ICRITO)*. IEEE, 2020, pp. 1074–1079.

[10] P. Singh, N. Singh, K. K. Singh, and A. Singh, "Diagnosing of disease using machine learning," in *Machine learning and the internet of medical things in healthcare*. Elsevier, 2021, pp. 89–111.

[11] M. Kuipers, and R. Prasad, "Journey of artificial intelligence," *Wireless Personal Communications*, pp. 1–16, 2022.

[12] Z.-H. Zhou, "Abductive learning: Towards bridging machine learning and logical reasoning," *Science China Information Sciences*, vol. 62, pp. 1–3, 2019.

[13] I. Antonopoulos, V. Robu, B. Couraud, D. Kirli, S. Norbu, A. Kiprakis, D. Flynn, S. Elizondo-Gonzalez, and S. Wattam, "Artificial intelligence and machine learning approaches to energy demand-side response: A systematic review," *Renewable and Sustainable Energy Reviews*, vol. 130, p. 109899, 2020.

[14] P. Zhang, W. Dou, and H. Liu, "Hierarchical data structures for flowchart," *Scientific Reports*, vol. 13, no. 1, p. 5800, 2023.

[15] P. P. Angelov, E. A. Soares, R. Jiang, N. I. Arnold, and P. M. Atkinson, "Explainable artificial intelligence: An analytical review," *Wiley Interdisciplinary Reviews: Data Mining and Knowledge Discovery*, vol. 11, no. 5, p. e1424, 2021.

[16] E. Mathew, and S. Abdulla, "Integrating AI in e-procurement of hospitality industry in the UAE," in *Artificial intelligence and machine learning for EDGE computing*. Elsevier, 2022, pp. 145–167.

[17] A. Raza, K. P. Tran, L. Koehl, and S. Li, "Designing ECG monitoring healthcare system with federated transfer learning and explainable AI," *Knowledge-Based Systems*, vol. 236, p. 107763, 2022.

[18] A. Amberkar, P. Awasarmol, G. Deshmukh, and P. Dave, "Speech recognition using recurrent neural networks," in *2018 international conference on current trends towards converging technologies (ICCTCT)*. IEEE, 2018, pp. 1–4.

[19] Y. Huang, "Design of personalised English distance teaching platform based on artificial intelligence," *Journal of Information & Knowledge Management*, vol. 21, no. Supp. 2, p. 2240017, 2022.
[20] Y. Ran, X. Zhou, P. Lin, Y. Wen, and R. Deng, "A survey of predictive maintenance: Systems, purposes and approaches," arXiv preprint arXiv:1912.07383, 2019.
[21] P. Szolovits, R. S. Patil, and W. B. Schwartz, "Artificial intelligence in medical diagnosis," *Annals of Internal Medicine*, vol. 108, no. 1, pp. 80–87, 1988.
[22] W. Konen, "General board game playing for education and research in generic AI game learning," in *2019 IEEE conference on games (CoG)*. IEEE, 2019, pp. 1–8.
[23] M. A. AI, "Azure AI: Drive business results and improve customer experiences with AI solutions," https://azure.microsoft.com/en-us/solutions/ai.
[24] G. C. AI, "Vertex AI vision," https://cloud.google.com/vertex-ai-vision.
[25] B. Topuz, and N. Ç. Alp, "Machine learning in architecture," *Automation in Construction*, vol. 154, p. 105012, 2023.
[26] scikit learn, "scikit-learn: Machine learning in Python," https://scikit-learn.org/stable/.
[27] A. Natekin, and A. Knoll, "Gradient boosting machines, a tutorial," *Frontiers in Neurorobotics*, vol. 7, p. 21, 2013.
[28] C. Bentéjac, A. Csörgő, and G. Martínez-Muñoz, "A comparative analysis of gradient boosting algorithms," *Artificial Intelligence Review*, vol. 54, pp. 1937–1967, 2021.
[29] Keras, "Keras: Simple, flexible, powerful," 2015, https://keras.io/.
[30] Yangqing, E. Jia, and Shelhamer, "Cafee: Deep learning framework by BAIR," 2017, https://caffe.berkeleyvision.org/.
[31] W. Choukri, H. Lamaazi, and N. Benamar, "RPL rank attack detection using Deep Learning," in *2020 international conference on innovation and intelligence for informatics, computing and technologies (3ICT)*. IEEE, 2020, pp. 1–6.
[32] Priyanka, "Beyond accuracy: Recall, precision, F1-score, ROC-AUC," 2022, https://medium.com/@priyankads/beyond-accuracy-recall-precision-f1-score-roc-auc-6ef2ce097966.
[33] W. Choukri, H. Lamaazi, and N. Benamar, "A novel deep learning-based framework for blackhole attack detection in unsecured RPL networks," in *2022 international conference on innovation and intelligence for informatics, computing, and technologies (3ICT)*. IEEE, 2022, pp. 457–462.
[34] A. Kumar, "Machine learning—Sensitivity vs specificity difference," 2023, https://vitalflux.com/ml-metrics-sensitivity-vs-specificity-difference/.
[35] TURING, "Understanding AUC-ROC curves and their usage for classification in Python," www.turing.com/kb/auc-roc-curves-and-their-usage-for-classification-in-python.
[36] W. Choukri, H. Lamaazi, and N. Benamar, "Abnormal network traffic detection using deep learning models in IoT environment," in *2021 3rd IEEE middle east and north Africa communications conference (MENACOMM)*. IEEE, 2021, pp. 98–103.
[37] A. Chugh, "MAE, MSE, RMSE, coefficient of determination, adjusted R squared—which metric is better?," 2020, https://medium.com/analytics-vidhya/mae-mse-rmse-coefficient-of-determination-adjusted-r-squared-which-metric-is-better-cd0326a5697e.

[38] R. J. Tan, “Breaking down mean average precision (mAP),” 2019, https://towardsdatascience.com/breaking-down-mean-average-precision-map-ae462f623a52.
[39] D. Shah, “Mean average precision (mAP) explained: Everything you need to know,” 2022, www.v7labs.com/blog/mean-average-precision.
[40] V. S. Subramanyam, “IOU (Intersection over Union),” 2021, https://medium.com/analytics-vidhya/iou-intersection-over-union-705a39e7acef.
[41] D. Shah, “Intersection over Union (IoU): Definition, calculation, code,” 2023, www.v7labs.com/blog/intersection-over-union-guide: text=Intersection.
[42] Priyanka, “Evaluation metrics in natural language processing—BLEU,” https://medium.com/@priyankads/evaluation-metrics-in-natural-language-processing-bleu-dc3cfa8faaa5.
[43] F. Chiusano, “Two minutes NLP—Perplexity explained with simple probabilities,” 2022, https://medium.com/nlplanet/two-minutes-nlp-perplexity-explained-with-simple-probabilities-6cdc46884584.
[44] TECHSLANG, “Perplexity in NLP: Definition, pros, and cons,” 2022, www.techslang.com/perplexity-in-nlp-definition-pros-and-cons/::text=Perplexity. Watson. Introducing watsonx, url = www.ibm.com/watso.

Chapter 4

Applications of artificial intelligence in smart distributed processing and big data mining

Fazal Wahab, Anwar Shah, Inam Ullah, Deepak Adhikari, and Ijaz Khan

4.1 INTRODUCTION

An enormous increase in data collection is currently being observed around the world in a wide range of business and academic domains. This flood of data, also known as big data, has forced the development of more robust methods for analyzing and processing it. Big data mining and distributed processing play an essential part in the process of identifying relevant patterns and insights hidden inside enormous datasets, which in turn enables businesses to make decisions based on the collected data. The purpose of this chapter is to shed light on the importance of artificial intelligence (AI) in large data mining and distributed processing in today's computer-driven world and investigate the various difficulties and opportunities related to these two topics. Following are the uses of AI in big data mining and distributed processing.

4.1.1 Applications of artificial intelligence in big data and distributed processing

The integration of AI and big data has dramatically altered how businesses analyze large datasets. Businesses can gain useful insights and spur innovation with the help of AI approaches for processing, analyzing, and visualizing big data. AI is essential for extracting big data's total value since it improves data processing efficiency, paves the way for real-time analytics, and smoothens the way for distributed processing. Big data mining and distributed processing are already being revolutionized by AI, and this trend will only accelerate as more and more businesses adopt AI-driven strategies.

4.1.2 Improving the effectiveness of data processing

Today's data presents significant problems for conventional data processing techniques due to its sheer amount, variation, and rapidity. However, AI algorithms

DOI: 10.1201/9781003496410-5

can automate and speed up several steps in the data processing pipeline [1]. Preprocessing and cleansing data using AI-powered techniques like machine learning (ML) and natural language processing (NLP) can reduce noise and improve data quality. In addition, AI algorithms allow for effective data indexing and querying, which leads to quicker and more precise information retrieval.

4.1.3 Making possible cutting-edge analytics

The goal of "big data mining" is to discover actionable insights hidden in massive datasets. AI algorithms excel at managing intricate data processing jobs, which opens new avenues of discovery for enterprises [2]. For instance, ML algorithms can quickly and accurately identify trends and outliers, forecast future results, and recommend next steps. Analytics powered by AI allows businesses to make evidence-based decisions, enhance existing operations, and identify untapped opportunities.

4.1.4 Enabling on-demand information processing

Real-time data processing is essential for many applications in the age of quickly changing data streams, such as banking, health care, and internet of things (IoT) systems. The processing of massive amounts of data streams relies heavily on AI methods like stream mining and deep learning (DL). With the help of these algorithms, businesses will be able to detect and respond to unusual or unexpected patterns in data in real time. Proactive decision-making, minimizing hazards, and productivity are all aided by AI-powered real-time data processing [3].

4.1.5 Processing in a distributed manner

Robust mechanisms for processing and interpreting data in parallel are necessary since big data is commonly stored across multiple systems. Distributed computing frameworks and parallel processing algorithms are examples of AI technologies that improve the scalability and performance of data-intensive applications. With AI, businesses may use distributed computing to efficiently process big datasets by splitting up data processing duties among several nodes or clusters [4].

4.1.6 Data extraction and visualization automation

Extraction and representation of data manually can be laborious and prone to mistakes. Automation made possible by AI methods would significantly improve the efficiency with which both structured and unstructured data may be extracted from a wide variety of sources [5]. Algorithms in the field of computer vision, for instance, can analyze and interpret visual data. In contrast, those in the field of NLP can extract information from text materials. The gathered information can subsequently be presented in interactive and user-friendly formats using data visualization tools powered by AI, improving data comprehension and subsequent decision-making.

The proliferation of big data techniques that have found extensive use in a variety of contexts has emerged as one of the most influential driving forces behind the development of modern civilization, particularly in the context of the augmented and virtual environment [1]. The production of data has not ceased since the age of pictorial symbols and proto-writing, and the quantity of data stored in digital formats has been expanding exponentially since our new and contemporary age of digital data. This trend can be traced up to the present day. In this chapter, a detailed assessment of recent breakthroughs in big data technologies is provided.

Knowledge discovery in databases (KDD) presents an efficient ability to discover relevant and usable information originating from a collection of data. The rapid development of information technology and the resulting collection of data allowed for the development of this ability. KDD has many applications in real life and has led to the development of several different data mining tasks [2], including clustering, association rule mining (ARM), classification, and outline detection, to name a few. The newly discovered information can be categorized in a general sense as frequent sets of items and relationships, sequential patterns, linear guidelines, graphs, and various intriguing trends [7]. This is possible since diverse requirements in various domains and applications lead to the discovery of this new information. A relational database neatly organized and constrained in its storage schema is characteristic of small data. The conventional approaches to analyzing tiny quantities of data are not overly complicated and are based on the model of relational databases between the various subjects [4]. The relatively insignificant data kept in the data warehouse can typically be comprehended in its entirety. The term "big data" came into existence as a result of the large, non-relational, complex, and unstructured amount of data that exist in our digital world and come from a wide range of fields and resources, including commercial, social networks, industrial, and many more. The production of understandable small amounts of data is the primary goal of big data. Large-scale data availability opens the door for significant developments to be made in several industries. Data, in general, and significant data approaches complement traditional data tools. Big data is still the fundamental concept and one of the most recent topics of debate in the entire data universe [5].

In the scientific and research disciplines in which we work, the enormous quantity of data has changed the way in which we conducted research in the past, as well as the way in which we conduct research and analysis to control tremendous amounts of intricate and often unavailable data. Big data mining is a new scientific breakthrough that has been improved upon to serve us better [6]. It is an information-intensive technique. Due to the increasing development in resource sharing, distributed systems are being developed, and these distributed systems can be utilized to accelerate calculations [7]. The practice of obtaining useful information from vast amounts of data using complex techniques is known as data mining or DM. It has a wide variety of practical applications in the real world. However, conventional DM algorithms assume that the data is gathered in one place, that it is memory-resident, and that it is unchanging [8]. When you just have a few resources available, it might be challenging to manage enormous amounts of data and process them. For instance, vast amounts of data can be produced in a short length of time and stored at several different sites [9]. The availability of large amounts of data paves the way for significant advancements to be made in a variety of sectors. In general, data techniques and massive data techniques are complementary to standard information analyzers. Despite this, big data is the main topic of discussion in the data world today [10].

The decision-making process in e-business has gotten more complicated and difficult, mainly because of the presence of extensive records and the resulting information overload [11–13]. To implement more effective new technologies across a variety of applications, there is an essential requirement to have a solid understanding of the evolution of big data technology. In recent years, there has been much emphasis on various uses of virtual reality. The metaverse is one of the most innovative concepts in the area of virtual reality. Digital twins and other related technologies are used to generate a reflection of the real world within the distributed, three-dimensional metaverse [14].

4.2 CHALLENGES IN BIG DATA MINING

DM presents numerous difficulties because of the data's heterogeneity, velocity, volume, and veracity. The sheer magnitude of big data necessitates the development of innovative methods and algorithms that are capable of efficiently processing vast amounts of information. Traditional DM techniques often struggle to keep up with big data's massive size and intricate nature. Furthermore, some challenges must be solved to guarantee fast processing and analysis due to the real-time nature of data generation. Multiple data types may comprise structured, semi-structured, and unstructured data, further complicating the mining process. Verifying the data's integrity, which is tied to its veracity, dependability, and quality, remains a significant challenge [15].

Researchers and experts in the business world have been building DM algorithms that can scale to manage massive amounts of data to address these issues. Large amounts of data may be processed quickly and efficiently using techniques like distributed processing frameworks, parallel computing, and cutting-edge ML algorithms [16–19].

4.3 FRAMEWORKS OF DISTRIBUTED PROCESSING

The use of distributed processing frameworks in large-scale data mining and analytics has become increasingly important in recent years. Processing and analyzing big datasets across clusters or distributed systems is made possible by these

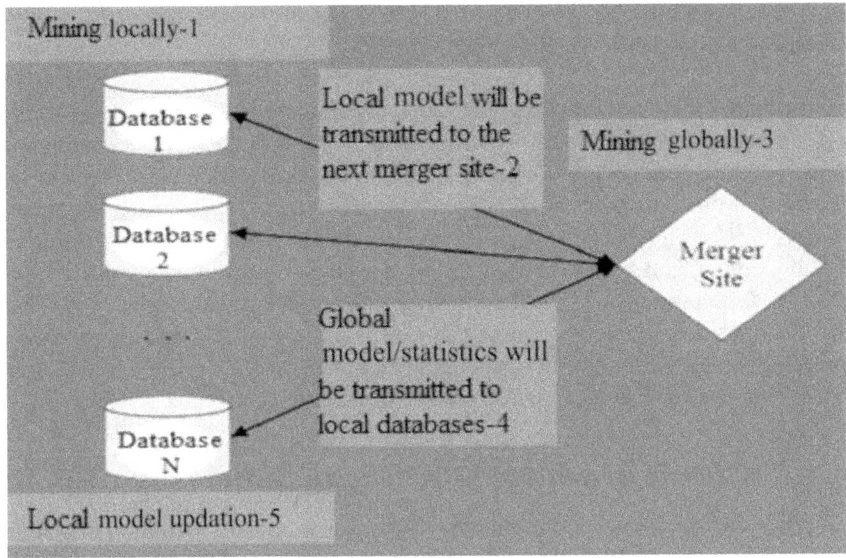

Figure 4.1 Distributed processing.

frameworks, which provide the architecture and tools necessary for doing so. The MapReduce programming style is used for distributed computing, and Apache Hadoop, one of the pioneering frameworks, was the one that introduced it [20]. Users of MapReduce are given the ability to distribute data processing jobs among a group of machines in parallel, which paves the way for the effective processing of big datasets. Another popular framework, Apache Spark, expands the capabilities of Hadoop by providing in-memory computation, which significantly increases the speed at which data is processed [21]. Apache Flink is a framework for processing data streams that are particularly effective at doing so in real time while maintaining a low level of latency and a high level of throughput [22]. Figure 4.1 illustrates the distributed processing environment.

4.4 DISTRIBUTED SYSTEM AND DATA MINING

The study of distributed systems falls under the purview of the subject of computer science known as distributed computing [23]. A distributed system is one in which the individual components of the system are housed on separate computers that are connected via a network. These computers can communicate with one another and coordinate their activities by sending messages to one another from any system [24]. A shared objective can be accomplished through the components' interactions with one another. Distributed systems are distinguished by the presence of three important characteristics: the concurrent operation of their components, the lack of a centralized timepiece, and the breakdown of separate components [26]. There are many other kinds of distributed systems, such as those based on service-oriented architecture (SOA), peer-to-peer technology, and enormously interactive online games [26]. A distributed program is a computer program created specifically to run within a distributed system, and the process of creating such programs is known as distributed programming [27]. The message transmission mechanism has several different types of implementations, some of which are remote procedure call (RPC)-like connectors, message queues, and HTTP [28]. The utilization of different computer systems to work together to solve a computational issue is another definition of distributed computing. The problem is divided into several smaller jobs in distributed computing, and each of these tasks is assigned to one or more machines. These machines stay in communication with one another as they work. The extended architecture of a distributed system can be seen in Figure 4.2.

4.4.1 Scalable distributed processing techniques

To efficiently mine huge data, scalable distributed processing techniques are necessary. These methods make it possible to handle data effectively across multiple dispersed systems, providing both high performance and tolerance for error. Parallel processing, which involves distributing data and processing tasks among numerous nodes, enables concurrent execution, which in turn dramatically reduces

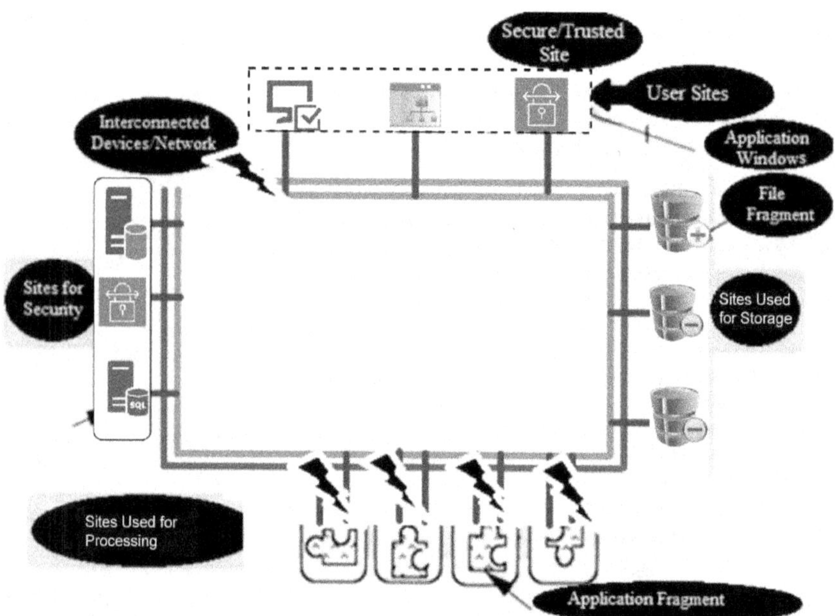

Figure 4.2 Architecture of the distributed system.

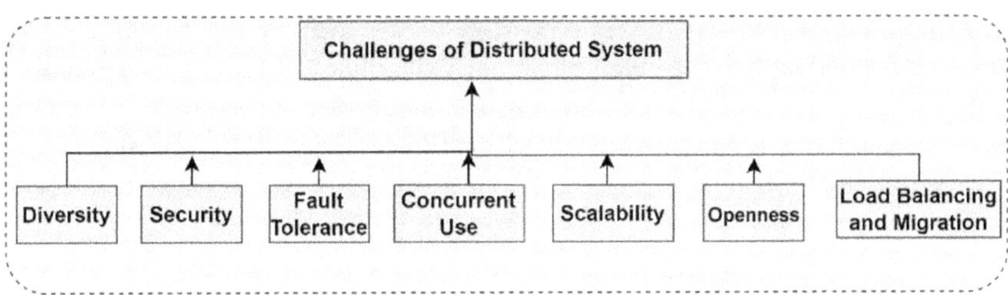

Figure 4.3 Distributed system challenges.

the amount of time required for the processing to be completed. Techniques for data partitioning spread subsets across nodes, enabling retrieval and analysis to occur more quickly. To avoid resource bottlenecks, load balancing algorithms ensure that the workload is distributed evenly across all nodes [19, 20, 29–32]. Enhancing a system's robustness and reliability in the face of failures is accomplished using fault tolerance measures like data replication and task redundancy.

In the quest to improve the scalability and effectiveness of distributed processing for big data mining, researchers are always creating novel strategies. This includes developments in computing parallelism, optimization techniques, and methods for allocating resources.

4.4.2 Challenges in distributed system

In contrast to the more common centralized systems, the idea behind a distributed system refers to a large collection of resources that are dispersed among multiple computers that are all linked together through a network. Sharing can take many forms, including software sharing, hardware sharing, service sharing, and data sharing, to name just a few. The proliferation of collaborative computing, parallel computing, and distributed computing led to the development of distributed system architectures. It is one whereby a networked computer-based component merely exchanges messages with other components to coordinate their actions [33–35]. Figure 4.3 illustrates the challenges of the distributed system.

4.5 OPPORTUNITIES IN BIG DATA MINING

The vast amount of information that is included in big data creates considerable prospects for businesses and organizations operating in a variety of fields. The mining of large amounts of data makes it easier to find valuable insights, trends, and correlations, all of which can help drive strategic decision-making. Discovering hidden patterns, carrying out predictive analysis, and improving business processes are all things that may be accomplished when an organization makes use of sophisticated analytics methods, including DL and ML [20, 29–31]. Mining large amounts of data also makes it possible to provide individualized service to customers, conduct more targeted marketing campaigns, and improve operational efficacy. In today's data-driven economy, businesses that successfully implement mining techniques for large amounts of data have a significant advantage over their competitors [21, 32, 36].

4.5.1 Big data mining applications

The applications of big data mining are numerous. Distributed processing and big data mining both provide a tremendous amount of potential for gaining meaningful insights from large and complicated datasets. Furthermore, big data mining is critical in numerous industries for improving logistics, streamlining supply chains, and increasing operational efficiency.

Business and Financial Sector: Big data mining in finance provides fraud detection, risk assessment, and algorithmic trading. The widespread implementation of statistical analysis in the commercial and financial sectors has led to an increase in the effectiveness of the procedures involved in the making of financial decisions. Nevertheless, technological advances have raised the potential for financial loss from internet attacks as well as online deception, as sensitive information has rapidly become the primary form of currency in the world of digital commerce. Big data technologies are used in financial and business applications to perform a variety of economic and business analyses, resulting in more accurate and complete decision-making models [23]. One of the key factors influencing the financial markets' quest for large profits is technological innovation. This is particularly true of digital technologies, which are widely used in the financial industry and have a substantial impact on a wide range of financial applications.

Marketing Sector: Big data mining in marketing allows companies to study customer behavior, segment their target population, and create customized marketing efforts.

Social Media Platforms: Big data mining is used by social media platforms to monitor user interactions, for sentiment analysis, and for trend identification, allowing businesses to modify their marketing campaigns accordingly

Health Sector: By analyzing massive amounts of health care data, such as genomic and electronic health records, researchers can find trends that aid in early diagnosis, treatment optimization, and population health management [21]. Big data mining in health care aids in patient monitoring, disease prediction, and individualized therapy. According to statistics, the present volume of data from all around the world is expected to triple every year. According to projections, the global market for the storage of health care data is expected to grow from $3.08 billion in 2020 to $6.12 billion in 2027 [37]. Problems arise for medical information systems as a result of the rapid development in data volume caused by an increase in diagnostic techniques such as magnetic resonance imaging (MRI) and computed tomography (CT) scans, as well as an expansion in the overall population of sufferers. The creation of an effective system for processing massive amounts of data quickly has turned into a prerequisite that is necessary for e-health systems. These past few years have seen the development of big-data-driven platforms for personalized health care, with the goals of lowering readmission rates and increasing the speed of real-time response [38]. Discovering novel clinical results has become significantly more accessible because of the many apps that make use of the Clinical Data Warehouse (CDW) database. These applications have combined online analytical processing and advanced network analysis. Jin et al. [39] describe a big-data-driven system that was built with the goal of revealing and analyzing the user behaviors of hospital information systems, as well as the possible features that may be based on behavior analysis. When seen from the angle of velocity, the numerous applications of wearable and sensor-based health tracking gadgets call for a speedy response to satisfy the prerequisites of real-time medical data processing. The speed at which large amounts of medical data are collected and analyzed is critical to the operation of many health care systems that save lives. The article by Ahmed, Geraldes, Ahmed, DeLuca, and Palace [40] describes the development of a wearable medical emergency response system that makes use of a significant number of medical sensors. This system was used to collect and analyze clinical data from patients for the purposes of real-time clinical monitoring, diagnosis, and therapy regimens. The system was built around a slew of wearable sensors [41]. Figure 4.4 represents data mining and distributed processing in e-health (based on Apache 27).

Transportation Sector: Data such as transportation trajectory data, traffic management data, GPS data, and transport system data are among the

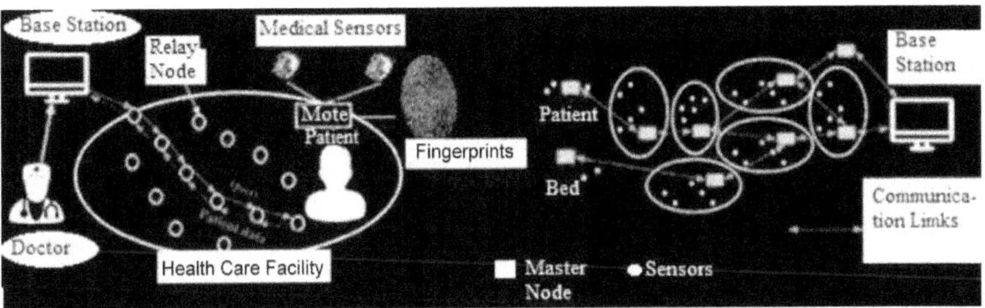

Figure 4.4 Data mining and distributed processing in E-health.

Source: Based on Apache 27.

types of data that exist primarily as an outcome of the quick development of transport networks in contemporary civilization. A range of intelligent transportation systems (ITS) has been developed to fulfill the demands of analyzing and processing the ever-increasing volume of transport data in the big data era due to the growing requirement for efficient transport systems [42, 43]. In the same way the medical industry does, the transportation systems industry is confronting the difficulties of volume, diversity, and validity in dynamic big data. Applications for ITS are concentrated on solving problems caused by large amounts of data by utilizing more advanced forms of big data technology. Big data analytics has been shown in investigations to increase an ITS's capacity for processing data, as well as its operational effectiveness and safety levels [44]. Figure 4.5 illustrates the analytics framework of big data in ITS (based on Shashkina 33).

Telecommunication: In today's world, businesses in the telecommunications industry are focusing on big data mining techniques [45–47]. These businesses are involved in the examination of a great number of records and calls, as well as prices, performance evaluation, enticing consumers, forecasting money, and determining the loyalty of clients. One of the most advanced techniques employed in the telecom sector nowadays is mining mobile user data. It is also rapidly evolving into the primary communication mode in our professional and personal lives [48, 49]. Cellular client data mining is one benefit of keeping mobile data based on actual user behavior [37]. The objective of cell phone consumer data mining is to examine and foresee mobile users' behavior based on the collected data.

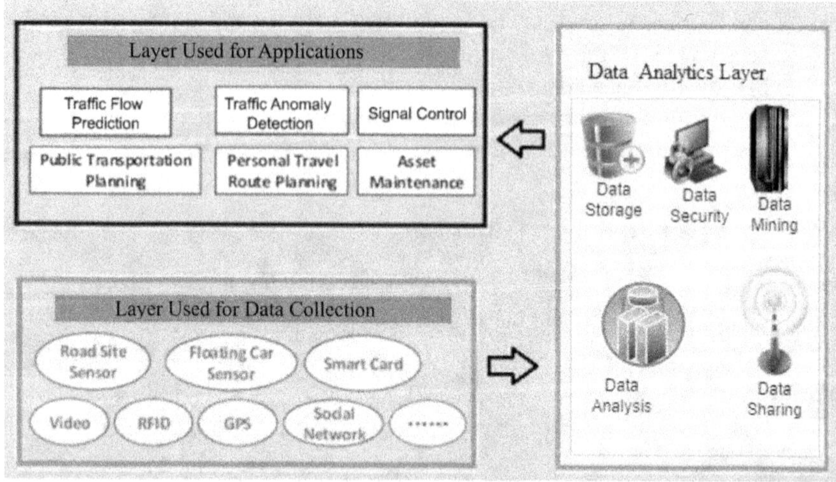

Figure 4.5 Analytics framework of big data in ITS.

Source: Based on Shashkina 33.

4.5.2 Trends in big data technologies

Large-scale data-based technologies are now essential to the success of enterprises, and big data technology has advanced significantly over the years and is now present in almost every aspect of our daily lives [50]. Companies that specialize in online retailing, such as Alibaba, Amazon, and Walmart, among others, leverage cutting-edge data mining techniques to be able to provide clients with individualized shopping experiences and purchasing recommendations. Companies that are involved in shipping and receiving, including FedEx, UPS, and CSX Transport, among others, have adopted sensor-based big data technologies and ML methodology to improve the efficiency of "last-mile delivery."

4.5.3 Tools used in big data mining

Storing a massive amount of data is not particularly difficult; rather, the most challenging aspect of working with big data is figuring out how to turn that massive amount of data into something valuable that makes more sense. The term "big data" describes the gathering and processing of numerous types of data, including organized, semi-organized, unstructured, and even nontraditional databases [51]. These kinds of data might fill a lot of storage space, measured in modern times in petabytes, exabytes, or zettabytes, and require specialized tools to manage such a massive storage capacity.

Many huge businesses today, including Google and Facebook, work with big data. Rather than employing conventional methods of data processing, Google uses specialized big data tools to edit, keep, and examine the stream of data it collects. In addition, there are several distinct big data tools that may be employed to gather, evaluate, and interpret the various complicated and unique sorts of data [52]. Some of the most powerful open source big data tools—Hadoop, MapReduce, and NoSQL—will be briefly introduced in this section.

- (i) **NoSQL:** This program makes use of new and alternative strategies in addition to conventional SQL methods to query and retrieve a sizable, intricate, incomplete, and agnostic dataset [53], which may be maintained distantly on numerous virtual machines at a cloud-based dataset. The term "Not Only SQL" refers to the fact that this tool combines these two parts.
- (ii) **MapReduce Tool:** MapReduce is among the most well-liked open source data mining tools. Using parallel and distributed techniques, developers may manage and produce enormous datasets on a cluster. Numerous computer languages and MapReduce frameworks, such as Java, C++, and C, can accomplish this [38]. MapReduce is one of the most widely used open source data mining techniques.
- (iii) **Hadoop Tool:** Massive volumes of data and programs are processed, developed, and executed in a distributed fashion using the Hadoop application framework, which is based on Java [54, 55]. Users have open access

to a collection of tools that are part of Hadoop. BSD (UNIX), Linux, Windows, and OS X for Apple Macintosh are just a few of the platforms on which Hadoop can be used [53]. Large non-relational dataset processing is one of its critical applications. Hadoop is used by many well-known search engines, including Yahoo and Google.

4.5.4 Technologies used in big data mining

To optimize the accessible and acceptable data required from a dataset, parallel data mining methods have been developed. These tools can be used for the solution of many issues by employing several different techniques, such as decision trees, genetic algorithms, artificial neural networks (ANNs), rule induction, and many more. Several of these technologies are briefly described in this section.

Artificial Neural Network: The ANN is an intriguing plan that can compete with established simulation systems, which are infamous for their costly and labor-intensive processes. Instead of using challenging mathematical equations, an ANN model takes data and learns directly from it to tackle nonlinear and complex issues. The deployment of ANN for automotive applications, however, frequently involves repetition. Continually utilized are the same network sort, method of learning, and training parameters [56].

Decision Trees: The decision tree algorithm is a kind of DM model used to develop induction learning algorithms based on instances [51, 57, 58]. The core nodes of a decision tree are represented by rectangles, and the leaf nodes are represented by ovals. Decision trees have a tree structure that resembles a flowchart. Compared to other classification algorithms, it is the most often used algorithm due to its ease of use and comprehension [59–62]. The decision tree's results indicate how many students would likely pass, fail, or be advanced to the following year. When compared to different classifying methods, decision tree development is rather speedy. Trees are easily converted into SQL queries that efficiently access databases.

Genetic Algorithms: The genetic algorithm is one of the appealing AI paradigms for enhancing the functionality of a system for retrieving information that employs natural evolution strategies for optimizing and searching problems [60]. Numerous fields, including revenue management, bioinformatics, clustering, electrical circuit design, airline artificial creativity, software engineering, and many others, use genetic algorithms.

4.6 CONCLUSION

The integration of AI and big data has dramatically altered how businesses analyze large datasets. Businesses can gain useful insights and spur innovation with the help of AI approaches for processing, analyzing, and visualizing big data. AI is essential for extracting big data's total value since it improves data processing efficiency, paves the way for real-time analytics, and smoothens the way for distributed processing. Big data mining and distributed processing are already being revolutionized by AI, and this trend will only accelerate as more and more businesses adopt AI-driven strategies.

Big data mining and distributed processing both provide a tremendous amount of potential for gaining meaningful insights from large and complicated datasets. However, because of the diversity, amount, pace, and validity of big data, they also present several significant issues. Distributed processing frameworks and scalable algorithms play an essential part in the process of overcoming these obstacles and enabling efficient mining and analysis of big data. Mining large amounts of data has a wide range of applications across a variety of business sectors; these applications improve consumer experiences, revolutionize decision-making procedures, and propel business expansion.

In this survey overview of the current state of big data, AI, and data mining, we quickly review big data and its characteristics, classifications, and approaches. We highlighted the typical small data value and the pressing demand for large data value in a variety of scientific study types. In addition, we presented a brief review of data mining types and techniques for investigating, gathering, and analyzing diverse data types.

Last, this chapter presented some of the most significant information prospects that are available to a wide range of notable sectors, including (industrial, financial, company, telecommunication, science, and engineering). Finally, we concluded that the massive amounts of data traveling around the world comprise the main key to any new digital development, and a parallel algorithm is a suitable option for prominent DM approaches. In addition, we covered the terminology, general designs, and some significant aspects of a distributed system in this study before highlighting the limitations of DM operations in distributed systems. As our main contribution, we investigated the latest advances in distributed data mining and give state-of-the-art information.

REFERENCES

[1] Hasan D. A., Zeebaree S. R., Sadeeq M. A., Shukur H. M., Zebari R.R., Alkhayyat A. H. "Machine Learning-based Diabetic Retinopathy Early Detection and Classification Systems-A Survey," in 2021 1st Babylon International Conference on Information Technology and Science (BICITS), 16–21, 2021.

[2] Ibrahim B. R., Khalifa F. M., Zeebaree S. R., Othman N. A., Alkhayyat A., Zebari R. R., et al. "Embedded System for Eye Blink Detection Using Machine Learning Technique," in 2021 1st Babylon International Conference on Information Technology and Science (BICITS), 58–62, 2021.

[3] Schroeder R., Cowls J. "Big Data, Ethics, and the Social Implications of Knowledge Production," in Data Ethics Workshop, KDD@ Bloomberg, vol. 24, pp. 1–4, 2019.

[4] Jijo B. T., Zeebaree S. R., Zebari R. R., Sadeeq M. A., Sallow A. B., Mohsin S., et al. "A comprehensive survey of 5G mm-wave technology design challenges." Asian Journal of Research in Computer Science, 2021:1–20.
[5] The Four V's of Big Data—IBM. www.ibmbigdatahub.com/infographic/four-vs-big-data (last seen 5 April 2015).
[6] Fazal W., Haizhao Y., Javeed D., Hmoud M., Ahmad S., Khan W., et al. "An AI-driven hybrid-framework for intrusion detection in IoT-enabled E-health." Computational Intelligence and Neuroscience, 2022;2022:11.
[7] Kudyba S. Big Data, Mining, and Analytics: Components of Strategic Decision Making. CRC Press, 2018.
[8] Sadeeq M. A., Zeebaree S. "Energy management for Internet of things via distributed systems." Journal of Applied Science and Technology Trends, 2021;2:59–71.
[9] Fazal W., Ullah I., Shah A., Khan R. A., Choi A., Anwar M. S. "Design and implementation of real-time object detection system based on single shot detector and OpenCV." Frontiers in Psychology, 2022;10:1–17.
[10] Abdullah S. M. S. A., Ameen S. Y. A., Sadeeq S., Zeebaree M. A. "Multimodal emotion recognition using deep learning." Journal of Applied Science and Technology Trends, 2021;2:52–58.
[11] Omer M. A., Zeebaree S. R., Sadeeq M. A., Salim B. W., Mohsin S. X., Rashid Z. N., et al. "Efficiency of malware detection in the android system: A survey." Asian Journal of Research in Computer Science, 2021:59–69.
[12] Gutierrez D. "Big data business impact: Achieving business results through innovation and disruption." 5th Annual Big Data Executive Survey for 2017. Available online: www.businesswire.com/news/home/20170109005058/en/NewVantage-Partners-Releases-5th-Annual-Big-Data-Executive-Survey-for-2017 (accessed on 30 November 2022).
[13] Yin Y., Shi Y., Zhao Y., Wahab F. "Multi-graph learning-based software defect location," in Wiley Software Evolution and Process, 2023, e2552.
[14] Knilans E. "The 5 V's of Big Data," in Avnet Advantage: The Blog, Solution-Focused Insight for Growth-Minded VARs. http://blogging.avnet.com/ts/advantage/2014/07/the-5-vs-of-big-data/#comment-474 (Last seen 5 April 2015).
[15] Li L., Zhao Y., Li Y., Wahab F., Wang Z. "Searching the most stable community in temporal graph." Knowledge-Based System, 2022;250:109101.
[16] Khalil H., Rahman S. U., Ullah I., Khan I., Alghadhban A. J., Al-Adhaileh M. H., et al. "A UAV-swarm-communication model using a machine-learning approach for search-and-rescue applications." Drones, 2022;6(12):372.
[17] Mazhar T., Irfan H. M., Khan S., Haq I., Ullah I., Iqbal M., et al. "Analysis of cyber security attacks and its solutions for the smart grid using machine learning and blockchain methods." Future Internet, 2023;15(2):83.
[18] Haq I., Mazhar T., Malik M. A., Kamal M. M., Ullah I., Kim T., et al. "Lung nodules localization and report analysis from computerized tomography (CT) scan using a novel machine learning approach." Applied Sciences, 2022;12(24):12614.
[19] Mazhar T., Irfan H. M., Haq I., Ullah I., Ashraf M., Shloul T. A., et al. "Analysis of challenges and solutions of IoT in smart grids using AI and machine learning techniques: A review." Electronics, 2023;12(1):242.
[20] Abu-Salih B., Wongthongtham P., Zhu D., Chan K. Y., Rudra A. "Introduction to big data technology." In Social Big Data Analytics; Springer: New York, NY, 2021; pp. 15–59.

[21] Rasheed Z., Ma Y. K., Ullah I., Al Shloul T., Tufail A. B., Ghadi Y. Y., et al. "Automated classification of brain tumors from magnetic resonance imaging using deep learning." Brain Sciences, 2023;13(4):602.
[22] Yu C., Liu H., Wahab F., Ling Z., Ren T., Ma H., et al. "Global triangle estimation based on first edge sampling in large graph streams." The Journal of Supercomputing, 2023; 79(13):14079–14116.
[23] Ning H., Wang H., Lin Y., Wang W., Dhelim S., Farha F., et al. "A survey on metaverse: The state-of-the-art," Technologies, Applications, and Challenges. arXiv 2021, arXiv:2111.09673.
[24] Chen, M., Mao, S., & Liu, Y. (2014). Big data: A survey. Mobile Networks and Applications, 19(2), 171–209.
[25] Xuze Liu, Yuhai Zhao, Tongze Xu, Fazal Wahab, Yiming Sun, Chen Chen, "Efficient False Positive Control Algorithms in Big Data Mining"; Applied Sciences; Appl. Sci. 2023, 13, 5006.
[26] Chamorro-Premuzic, T. How the Web Distorts Reality and Impairs Our Judgment Skills. The Guardian. 13 May 2014. Available online: https://www.theguardian.com/media-network/media-network-blog/2014/may/13/internet-confirmation-bias (accessed on 30 November 2022).
[27] Davenport, T. H., & Dyche, J. (2013). Big data in big companies. International Institute for Analytics, 1(1), 1–21.
[28] Dean, J., & Ghemawat, S. (2008). MapReduce: Simplified data processing on large clusters. Communications of the ACM, 51(1), 107–113.
[29] Shah A., Ali B., Wahab F., Ullah I., Amesho K. T., Shafiq M. "Entropy-based grid approach for handling outliers: A case study to environmental monitoring data." Environmental Science and Pollution Research, 2023:1–20.
[30] Yousafzai B. K., Khan S. A., Rahman T., Khan I., Ullah I., Ur Rehman A., et al. "Student-performulator: Student academic performance using hybrid deep neural network." Sustainability, 2021;13(17):9775.
[31] Tufail A. B., Ma Y. K., Kaabar M. K., Martinez F., Junejo A. R., Ullah I., et al. "Deep learning in cancer diagnosis and prognosis prediction: A minireview on challenges, recent trends, and future directions." Computational and Mathematical Methods in Medicine, 2021;2021.
[32] Chamorro-Premuzic T. How the web distorts reality and impairs our judgment skills. The Guardian. 13 May 2014. (accessed on 30 November 2022). https://www.theguardian.com/media-network/media-network-blog/2014/may/13/internet-confirmation-bias
[33] Harki N., Ahmed A., Haji L. CPU scheduling techniques: A review on novel approaches strategy and performance assessment. Journal of Applied Science and Technology Trends. 2020;1:48–55.
[34] Zaharia, M., Chowdhury, M., Franklin, M. J., Shenker, S., & Stoica, I. (2010). Spark: Cluster computing with working sets. HotCloud, 10(10–10), 95.
[35] Apache Flink. (n.d.). Retrieved from https://flink.apache.org/.
[36] Ahmad S., Ullah T., Ahmad I., Al-Sharabi A., Ullah K., Khan R. A., et al. "A novel hybrid deep learning model for metastatic cancer detection." Computational Intelligence and Neuroscience, 2022;2022.
[37] Salim N. O., Abdulazeez A. M. "Human diseases detection based on machine learning algorithms: A review," International Journal of Science and Business. 2021; 5:102–113.

[38] Dino H. I., Zeebaree S., Salih A. A., Zebari R. R., Ageed Z. S., Shukur H. M., et al. "Impact of Process Execution and Physical Memory-Spaces on OS Performance," Technology Reports of Kansai University. 2020; 62:2391–2401, 2020.

[39] Jin J., Song A., Gong H., Xue Y., Du M., Dong F., et al. "Distributed storage system for electric power data based on hbase," Big Data Mining and Analytics. 2018;1:324–334.

[40] Ahmed O., Geraldes R., Ahmed A., DeLuca G., Palace J. "Multiple sclerosis and the risk of venous thrombosis: A systematic review," in MULTIPLE SCLEROSIS JOURNAL, 2017;757–758.

[41] Kareem F. Q., Ameen S. Y., Salih A. A., Ahmed D. M., Kak S. F., Yasin H. M., et al. SQL injection attacks prevention system technology. Asian Journal of Research in Computer Science. 2021;13–32.

[42] Shashkina, V. Why Your Medical Organization Can no Longer Do without a Healthcare Data Warehouse. Itrex Archive. 3 December 2021. Available online: https://medium.datadriveninvestor.com/why-your-medical-organization-can-no-longer-do-withouta-healthcare-data-warehouse-774cd374e0a (accessed on 30 November 2022).

[43] Yao, Q., Tian, Y., Li, P.-F., Tian, L.-L., Qian, Y.-M., Li, J.-S. Design and Development of a Medical Big Data Processing System Based on Hadoop. J. Med. Syst. 2015, 39.

[44] Chawla, N. V., Davis, D. Bringing Big Data to Personalized Healthcare: A Patient-Centered Framework. J. Gen. Intern. Med. 2013, 28, 660–665.

[45] Lee, K. M., Park, S. J., & Lee, J. H. (2014). Soft Computing in Big Data Processing.

[46] Shi, Q., Abdel-Aty, M. Big data applications in real-time traffic operation and safety monitoring and improvement on urban express ways. Transp. Res. C Emerg. Technol. 2015, 58, 380–394.

[47] Wu, X., Zhu, X., Wu, G. Q., & Ding, W. (2014). Data mining with big data. Knowledge and Data Engineering, IEEE Transactions on, 26(1), 97–107.

[48] Anju Rathee "survey on decision tree classification algorithms for the evaluation of the student performance", Indian Journal of Chemical Technology. Vol. 4 no. 2 Surjeet Kumar Yadav and Saurabh Pal (2012). "Data mining: A prediction for performance improvement of engineering students using classification", World of science and information technology journal (WCSIT) ISSN: 2221–0741, Vol 2, no. 2.

[49] Wahab F., Khan I., Hussain T., Amir A. An investigation of cyber attack impact on consumers' intention to purchase online. Decision Analytics Journal. 2023 Aug 6:100297.

[50] Veza, I., Afzal A., Mujtaba, M. A., Hoang, A. T., Balasubramanian, D., Sekar, M., Fattah, I. M. R., Soudagar, M. E. M., EL-Seesy, A. I., Djamari, D. W., Hananto, A. L., Putra, N. R., Noreffendy T., Review of artificial neural networks for gasoline, diesel and homogeneous charge compression ignition engine, Alexandria Engineering Journal, Volume 61, Issue 11, 2022, Pages 8363–8391, ISSN 1110-0168, https://doi.org/10.1016/j.aej.2022.01.072.

[51] Mazhar T., Talpur D. B., Shloul T. A., Ghadi Y. Y., Haq I., Ullah I., Ouahada K., Hamam H. Analysis of IoT Security Challenges and Its Solutions Using Artificial Intelligence. Brain Sciences. 2023 Apr 19;13(4):683.

[52] Ismael H. R., Ameen S. Y., Kak S. F., Yasin H. M., Ibrahim I. M., Ahmed A. M., et al. "Reliable communications for vehicular networks," Asian Journal of Research in Computer Science. 2021;33–49.
[53] Rathore, M. M., Ahmad, A., Paul, A., Wan, J., Zhang, D. Real-time Medical Emergency Response System: Exploiting IoT and Big Data for Public Health. J. Med. Syst. 2016, 40, 283.
[54] Mohamed, N., Al-Jaroodi, J. Real-time big data analytics: Applications and challenges. In Proceedings of the 2014 International Conference on High Performance Computing & Simulation (HPCS), Bologna, Italy, 21–25 July 2014; pp. 305–310.
[55] Zhang, J., Wang, F.-Y., Wang, K., Lin, W.-H., Xu, X., Chen, C. Data-Driven Intelligent Transportation Systems: A Survey. IEEE Trans. Intell. Transp. Syst. 2011, 12, 1624–1639.
[56] Poon, C. C. Y., Lo, B. P. L., Yuce, M. R., Alomainy, A., Hao, Y. Body Sensor Networks: In the Era of Big Data and Beyond. IEEE Rev. Biomed. Eng. 2015, 8, 4–16.
[57] Abideen Z. U., Mazhar T., Razzaq A., Haq I., Ullah I., Alasmary H., Mohamed H. G. Analysis of Enrollment Criteria in Secondary Schools Using Machine Learning and Data Mining Approach. Electronics. 2023 Jan 30;12(3):694.
[58] Khan I., Tian Y. B., Ullah I., Kamal M. M., Ullah H., Khan A. Designing of E-shaped microstrip antenna using artificial neural network. Int. J. Comput. Commun. Instrum. Eng. 2018;5(1):23–6.
[59] Khan S., Ullah I., Ali F., Shafiq M., Ghadi Y. Y., Kim T. Deep learning-based marine big data fusion for ocean environment monitoring: Towards shape optimization and salient objects detection. Frontiers in Marine Science. 2023 Jan 11;9:1094915.
[60] Su X., Ullah I., Wang M., Choi C. Blockchain-based system and methods for sensitive data transactions. IEEE Consumer Electronics Magazine. 2021 Apr 30.
[61] Wu, X., Zhu, X., Wu, G. Q., & Ding, W. (2014). Data mining with big data. Knowledge and Data Engineering, IEEE Transactions on, 26(1), 97–107.
[62] Srinivasa, S., & Bhatnagar, V. (Eds.). (2012). Big Data Analytics: First International Conference, BDA 2012, New Delhi, India, December 24–26, 2012: Proceedings (Vol. 7678). Springer.

Chapter 5

Quantum AI

Uniting the future of smart technologies

Dr. Maqsood Muhammad Khan, Dr. Inam Bari, Dr. Omar Usman Khan, Abdullah Akbar, Sanaa Jeehan, and Najeeb Ullah

5.1 INTRODUCTION

In the ever-evolving realm of technology, the convergence of quantum computing and artificial intelligence (AI) signals the beginning of a new era brimming with limitless possibilities [1]. This chapter embarks on a journey to delve into the harmonious interplay between these transformative disciplines, placing a spotlight on the intriguing domain of quantum AI.

5.1.1 Exploring quantum AI

Quantum AI combines the computational power of quantum computing with AI's problem-solving abilities. As we explore this complex relationship, we will unveil the fundamental principles supporting quantum AI. From quantum bits (qubits) to quantum algorithms, this chapter will provide a road map for comprehending the quantum realm and its possible uses in AI.

5.1.2 Significance in modern technology

The significance of quantum AI in contemporary technology is nothing short of transformative. In a world where data-driven decision-making and intricate problem-solving are the standard, quantum AI emerges as a powerful catalyst for innovation. This chapter will take an in-depth look at the pivotal role this emerging field plays in reshaping a myriad of domains, spanning from cryptography and health care to finance and beyond.

By harnessing the distinctive principles of quantum mechanics, quantum AI empowers us to confront once-insurmountable challenges with innovative solutions. Its quantum algorithms and computational prowess promise not only to accelerate tasks that were previously considered computationally intensive but also to unearth entirely new possibilities in fields such as optimization, machine learning (ML), and materials science.

DOI: 10.1201/9781003496410-6

The potential applications are vast. Quantum AI has the capability to revolutionize cybersecurity, enabling us to safeguard sensitive information in an era of ever-increasing digital threats [2]. In health care, it can expedite drug discovery and aid in the development of personalized medicine [3]. The financial sector can benefit from improved risk assessment and portfolio optimization [4]. Quantum AI also extends its reach into emerging areas, such as quantum drones and quantum networks [5].

Quantum drones powered by quantum AI can revolutionize various industries by enabling unprecedented levels of precision and autonomy. These drones can perform tasks such as environmental monitoring, agriculture, and disaster response with unmatched efficiency and accuracy. Quantum networks, on the other hand, have the potential to transform communication and data transfer, offering faster and more secure connections that are resistant to traditional hacking methods.

In short, quantum AI has the potential to reshape industries, redefine what's possible in the digital age, and bring about innovative solutions that were once the realm of science fiction. As we embark on this intellectual journey, we extend a warm invitation to explore the promising convergence of quantum science, AI, quantum drones, and quantum networks with us. Together, we will journey deeper into the heart of quantum AI, where the boundless landscape of innovation awaits.

5.2 QUANTUM COMPUTING ESSENTIALS

The fundamental concepts and principles that underpin quantum computing are presented in this section. Understanding these essentials is crucial for grasping the potential of quantum AI and its applications.

5.2.1 Understanding qubits and quantum gates

In classical computing, the fundamental building block of information is the classical bit, which has the capacity to exist in one of two states: 0 or 1. Quantum computing introduces a revolutionary concept—the quantum bit is a quantum state of a physical entity. In contrast to classical bits, qubits have the unique ability to occupy multiple states simultaneously, a phenomenon referred to as superposition. This means that a qubit can represent both 0 and 1, as well as any arrangement of these states, all at the same time.

Superposition forms the foundation of quantum computing power. This capability empowers quantum computers to perform parallel processing of extensive information, resulting in exponential efficiency gains for specific tasks compared to classical computers [6].

However, when we measure a qubit, it collapses from its superposition state into a classical 0 or 1 state. This collapse is a fundamental aspect of quantum mechanics and distinguishes qubits from classical bits. In the realm of quantum mechanics, the common representation of a qubit state on a unit sphere is known as the Bloch sphere, as illustrated in Figure 5.1.

Quantum gates serve as the quantum counterparts to classical logic gates used in classical computing. They form the foundational elements for constructing quantum circuits, which execute quantum operations on qubits. Quantum gates manipulate qubits, change their states, and enable complex computations.

Here are some essential quantum gates:

Pauli-X Gate (X-Gate): This gate is equivalent to a classical NOT gate. It flips the state of a qubit from $|0\rangle$ to $|1\rangle$ and vice versa.

Pauli-Y Gate (Y-Gate): The Y-gate is analogous to the X-gate but also presents a complex phase shift, resulting in rotations in the Bloch sphere.

Pauli-Z Gate (Z-Gate): The Z-gate leaves the basis states $|0\rangle$ and $|1\rangle$ unchanged but introduces a phase factor. It's analogous to a classical phase inversion.

Hadamard Gate (H-Gate): The Hadamard gate creates superposition by transforming $|0\rangle$ into an equal-weighted superposition of $|0\rangle$ and $|1\rangle$.

CNOT Gate: The Controlled-NOT gate is an entangling gate that operates on two qubits. It flips the second qubit if and only if the first qubit is in the $|1\rangle$ state. It's a fundamental gate for building quantum circuits.

Toffoli Gate: It is a three-qubit gate that performs a classical AND operation, controlled by two qubits, on the third qubit.

SWAP Gate: This gate swaps the states of two qubits. It's essential for rearranging qubit states in quantum algorithms.

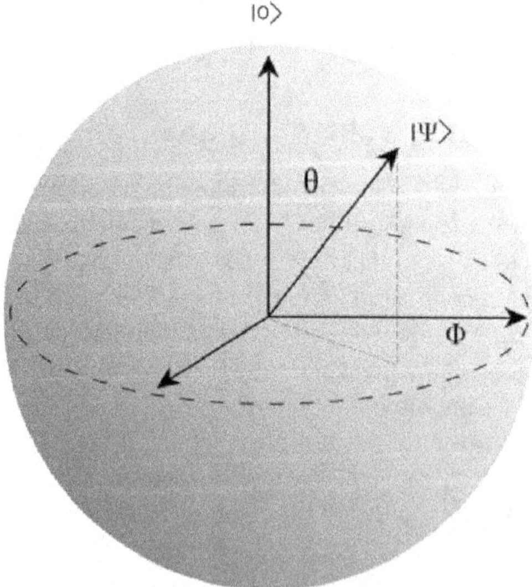

Figure 5.1 Bloch sphere, having a radius of one unit, offers a geometric portrayal of a qubit.

Controlled-U Gate: This is a family of gates that perform controlled rotations or operations on a target qubit built on the state of a control qubit. Controlled gates are crucial for various quantum algorithms.

Quantum gates show a vital role in quantum algorithms and quantum computations. They allow us to manipulate qubits in intricate ways, making it possible to perform complex calculations that classical computers would struggle with. The Clifford+T gate set serves as a comprehensive foundation for describing quantum circuits represented in Figure 5.2.

Understanding qubits and quantum gates is the first step toward grasping the immense potential of quantum computing. These foundational concepts are at the heart of quantum AI and its ability to revolutionize various fields, from cryptography to optimization and beyond. Both the circuit diagram and matrix representations of the Toffoli gate are shown in Figure 5.3.

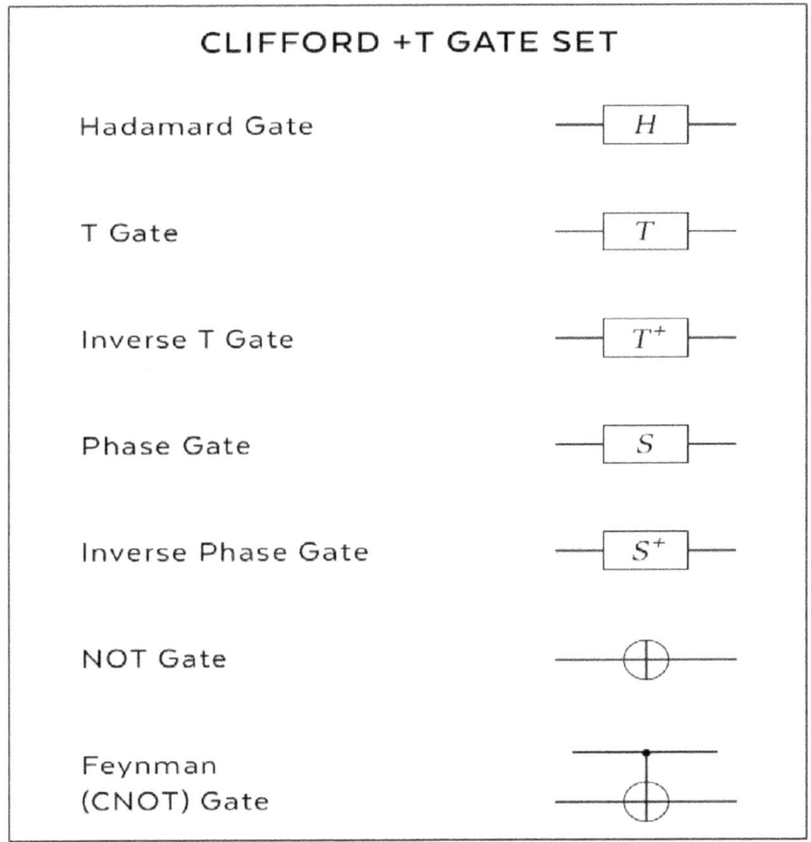

Figure 5.2 Illustration of the universality of the clifford+T gate set in quantum circuits.

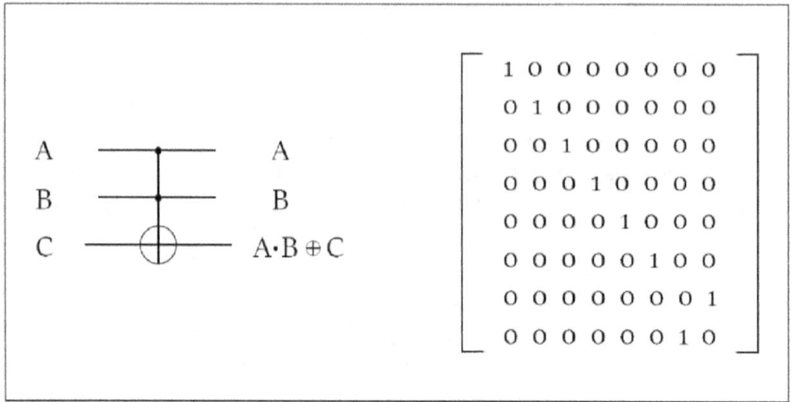

Figure 5.3 Three-qubit toffoli gate circuit diagram alongside its corresponding matrix representation.

Quantum circuits are formed through the consecutive application of quantum gates to qubits. The arrangement and order of these gates dictate how the quantum algorithm behaves. Quantum algorithms harness the innate parallelism of quantum gates to efficiently tackle complex problems.

Quantum gates face several challenges, including issues related to gate fidelity and error correction. Researchers are consistently striving to enhance gate performance and create fault-tolerant quantum computing technologies.

Quantum gates serve as the cornerstone of quantum computation, enabling manipulation and alteration of quantum states. Their distinctive characteristics render them potent instruments for solving problems that classical computers cannot handle, propelling progress in quantum computing and its applications across diverse fields, such as AI and cryptography [7].

5.2.2 The quantum advantage

Quantum computing marks a groundbreaking shift in the world of computation, offering a fascinating look into the future of technology. At its core, quantum computing is incredibly fast and can handle multiple tasks at once, thanks to qubits, the quantum version of regular computer bits. These qubits can exist in several states simultaneously, allowing quantum computers to solve complex problems at incredible speeds. Algorithms that would take regular computers thousands of years to complete can now be done in just seconds or minutes on quantum devices.

One of the most exciting things about the quantum advantage is the development of quantum algorithms that have the potential to revolutionize various industries. For instance, Shor's algorithm can challenge traditional encryption methods by quickly solving complex math problems, while Grover's algorithm can significantly speed up searches in large databases. These innovations have big implications for areas like data analysis and keeping information safe [8].

In the world of keeping information safe, quantum computing is both a challenge and an opportunity. On one hand, it can break traditional encryption methods, which is why we need to develop new ways to keep our information secure. These new methods include lattice and code-based cryptography.

Quantum computing is also game-changing for optimization and machine learning. It's fantastic at solving tough optimization problems that have to do with things like supply chains and financial models [9]. Quantum machine learning uses quantum properties to make computers better at recognizing patterns and making sense of data, which is a big deal for AI.

In terms of communication, quantum computing allows us to create incredibly secure communication channels. This is crucial in today's connected world where there are always threats from hackers trying to access our information. Quantum computing also helps with scientific research. It lets us simulate quantum systems, which is helpful for discovering new materials, understanding how chemicals react, and predicting how things work at the smallest scale. Last, quantum computing can make a positive impact on the environment [10]. It can help us use energy more efficiently, improve how we manage resources, and reduce our impact on the planet. However, while quantum computing has amazing potential, there are also challenges. Things like fixing errors, making it work for bigger problems, and figuring out what kind of resources it needs are all things researchers are actively working on [11].

5.3 QUANTUM MACHINE LEARNING

Machine learning emerges as the promising component of AI. Quantum computing gains popularity due to its fast problem-solving abilities. These advanced systems navigate complex problems through large multidimensional spaces. Various algorithms can work together within these spaces to address these challenges. Quantum machine learning serves as a platform to streamline data mining processes with quantum computing advancements.

Both quantum computing and machine learning are intricate domains. Quantum machine learning strives to rapidly produce solutions within a quantum framework using various algorithms. In contrast, ML employs techniques such as supervised, unsupervised, and semi-supervised learning. ML leverages both labeled and unlabeled data for tasks like classification, clustering, and decision tree analysis for complex challenges. Quantum computing introduces quantum analogs to tackle diverse computational complexities. Quantum machine learning offers comprehensive insights and innovative solutions across diverse domains [12].

5.3.1 Quantum algorithms and their applications

Currently, quantum machine learning algorithms are categorized into three different approaches. These include quantum machine learning algorithms, which are quantum counterparts of the classical ML algorithms, along with algorithms that run on actual quantum computers, such as quantum support vector machine (QSVM), which is inspired by the classical support vector machine (SVM)—a powerful tool for classification and regression tasks in classical ML. QSVM binds the unique features of quantum computing to potentially provide an advantage in solving certain classification problems [13].

A quantum neural network (QNN) represents a fusion of quantum computing principles and artificial neural networks, the core components of classical ML. While classical neural networks rely on classical bits (0s and 1s) for computation, QNNs operate using quantum bits or qubits as their fundamental processing units. This unique aspect enables QNNs to leverage quantum phenomena such as superposition and entanglement. Quantum gates, which manipulate qubit states, take the place of conventional activation functions found in traditional neural networks. Quantum superposition permits qubits to exist in several states concurrently, potentially offering exponential computational advantages for tasks like pattern recognition and optimization. Furthermore, quantum entanglement facilitates rapid convergence by connecting qubit states, even when they are spatially distant [14].

Quantum linear regression (QLR) is a quantum machine learning algorithm specifically designed for addressing linear regression problems with increased computational efficiency compared to classical methods. Linear regression is a fundamental statistical and ML technique used to model associations between a reliant variable and one or more independent variables.

QLR operates by harnessing the exclusive properties of quantum computing to potentially expedite the process of finding optimal parameters in linear regression tasks. In linear regression, the primary objective is to identify the best-fit line or hyperplane that minimizes the differences between predicted and actual values of the dependent variable. This relationship is typically represented by a linear equation involving variables like slope and intercept.

QLR begins by preparing quantum states that represent the dataset and its associated labels, which correspond to the dependent variable values. These quantum states are then manipulated within a quantum circuit. The distinctive advantage of QLR lies in its quantum optimization techniques. Instead of relying on classical optimization methods that can be computationally intensive, QLR exploits quantum parallelism to potentially accelerate the parameter optimization process, particularly for large datasets. Quantum algorithms like the quantum amplitude estimation algorithm (QAE) or the HHL algorithm (Harrow-Hassidim-Lloyd) are employed to perform this optimization more efficiently than classical counterparts. Following the quantum optimization process, measurements are taken to obtain the optimized parameters for the linear equation [15].

While QLR shows promise for faster processing and analysis of extensive datasets due to quantum parallelism, its practical implementation currently faces challenges related to the availability and scalability of quantum hardware and the need to address quantum errors. Quantum-inspired machine learning encompasses a set of approaches that take inspiration from quantum computing principles to enhance and potentially optimize traditional ML algorithms. These methods do not operate on actual quantum computers but rather incorporate quantum concepts to improve the presentation of classical ML models. Here's a breakdown of the quantum-inspired algorithms.

Quantum-Inspired Binary Classifier: This classifier utilizes quantum-inspired techniques to enhance binary classification tasks, exploiting principles like superposition and quantum gates to potentially boost accuracy and efficiency in binary classification [16].

Helstrom Quantum Centroid: The Helstrom quantum centroid is a quantum-inspired binary supervised learning classification algorithm designed to optimize decision boundaries between two classes. It leverages quantum principles, making it especially valuable in situations requiring precise classification boundaries [17].

Quantum-Inspired Support Vector Machines (SVM): Quantum-inspired SVMs infuse quantum concepts into the classical SVM algorithm, a powerful tool for classification and regression tasks. By harnessing quantum parallelism, they expedite SVM decision boundary computations [18].

Quantum Nearest Mean Classifier: This classifier applies quantum-inspired techniques to the nearest mean classifier, commonly used for pattern recognition tasks. Quantum principles have the potential to enhance the identification of the nearest class mean [19].

Quantum-Inspired *k*-Nearest Neighbor (KNN): Quantum-inspired KNN algorithms enhance the traditional KNN, a widely used method for classification and regression. These quantum-inspired approaches may offer improvements in distance calculations and neighbor searches, leading to more accurate predictions [20].

Quantum Algorithms for Ridge Regression: Ridge regression is a regression analysis technique. Quantum algorithms designed for ridge regression aim to utilize quantum computing principles to optimize ridge regression models, potentially delivering faster and more accurate results [21].

Quantum-Inspired Neural Network: Quantum-inspired neural networks incorporate quantum principles into the training and optimization processes of neural networks. These techniques discover the potential of quantum computing for the enhancement of the performance of deep learning models [22].

In conclusion, quantum-inspired machine learning algorithms adapt quantum principles into classical ML methods to potentially achieve performance enhancements, particularly in scenarios where traditional techniques encounter computational challenges or where precision in classification or regression boundaries is paramount. These approaches, although not executed on quantum hardware, demonstrate innovative ways to leverage quantum concepts for advancing the field of ML.

5.3.2 Quantum-enhanced data analysis

Quantum-enhanced data analysis is a powerful blend of quantum computing principles with the field of data analytics. It uses the unique features of quantum computing, like its ability to handle complex tasks and information, to change how we work with complicated datasets. While full-scale quantum computers are still developing, quantum-inspired algorithms are making data analysis much better. These algorithms, though they don't run on quantum machines, use quantum ideas to possibly get faster and more accurate results compared to regular methods. In machine learning, these quantum-inspired methods are improving tasks like sorting data, making predictions, and finding patterns. They also help make big datasets more manageable. Quantum techniques for recognizing patterns are especially useful for tasks like understanding images or text, making it easier to find important details. In the business world, they are helping with tasks like managing supplies and making decisions more efficiently [23]. However, as with any new technology, there are challenges, like the limitations of current quantum hardware and issues with errors. We also have to think about the ethics of handling data and keeping it secure. But despite these challenges, the future of quantum-enhanced data analysis looks bright [24].

5.3.3 Quantum AI in industry

The blend of AI in the quantum domain is triggering a profound transformation across various industries, offering innovative solutions to intricate challenges. In this section, we will explore how quantum AI is revolutionizing businesses and industries, illustrated through real-world applications.

In the realm of finance and investment, quantum AI is redefining portfolio optimization, allowing financial institutions to enhance risk management and maximize returns for their clients [25].

In health care and pharmaceuticals, quantum AI accelerates the process of drug discovery by simulating molecular interactions and forecasting potential drug candidates, with pioneers such as IBM and D-Wave collaborating with pharmaceutical giants to revolutionize drug design [26].

The aerospace and defense sectors are utilizing quantum AI for the design of more fuel-efficient and aerodynamic aircraft. Leading organizations such as NASA are leveraging quantum computing to innovate aerospace systems [27]. In the realm of energy and sustainability, quantum AI is optimizing energy distribution grids for efficient power delivery. Companies like National Grid are collaborating with quantum computing firms to fortify energy networks' sustainability and resilience.

Logistics and supply chain management are benefiting from quantum AI, which optimizes complex routes to streamline product deliveries. Industry players like Volkswagen are employing quantum algorithms to enhance vehicle routing and logistics efficiency [28].

Telecommunications companies are benefiting from quantum AI in network optimization, ensuring seamless connectivity. Different mobile companies are investigating quantum-inspired algorithms to enhance network performance [29].

Environmental conservation efforts are also empowered by quantum AI, which contributes to advanced climate modeling for addressing climate change challenges. Organizations like Google are utilizing quantum computing for high-resolution climate simulations [30].

Cybersecurity is another domain where quantum AI plays a pivotal role. It reinforces cybersecurity measures, with industries reliant on secure communications, such as finance and defense, exploring post-quantum encryption methods to safeguard sensitive data [31].

These real-world applications underscore quantum AI's tangible impact across diverse sectors. While quantum computing hardware is still evolving, collaborations between companies and research institutions are actively harnessing the potential of quantum algorithms and AI. They aim to address complex challenges, foster innovation, and gain a competitive edge in today's rapidly evolving industrial and technological landscape. Quantum AI has transitioned from a futuristic concept to a transformative force shaping the present and paving the way for a future characterized by continually expanding possibilities.

5.4 QUANTUM HARDWARE AND SOFTWARE

Quantum computing represents a revolutionary leap in the world of computing, and at its core, it relies on an interaction between quantum hardware and specialized software. This section explores the sophisticated relationship between quantum hardware and software, shedding light on the key components and tools that enable quantum computing to flourish.

5.4.1 Quantum processors and development tools

Quantum processor architecture is the foundational design and structure of the hardware components that constitute a quantum processor. Unlike classical computers that use bits as their basic units of information, quantum processors rely on qubits (quantum bits). These qubits possess unique quantum properties, which are superposition and entanglement, enabling quantum processors to perform complex calculations in parallel. Quantum gates, akin to classical logic gates, manipulate qubits and form the building blocks for quantum algorithms. Quantum circuits, comprising quantum gates, represent these algorithms. Quantum registers store and manipulate qubits, while quantum interconnects facilitate qubit communication. However, quantum processors are susceptible to decoherence, which necessitates careful management. Quantum processor architecture varies based on physical implementations, including superconducting qubits, trapped ions, and more, each presenting distinct considerations and challenges. Comprehending quantum processor architecture is pivotal for quantum computer scientists and engineers, serving as the cornerstone for developing quantum algorithms and harnessing quantum computing's capabilities to solve currently challenging problems for classical computers [32].

Development tools for quantum hardware constitute an integral part of the quantum computing landscape, facilitating the creation, refinement, and optimization of quantum processors. These tools serve as the backbone for researchers and engineers, simplifying the complex process of quantum hardware development. Quantum circuit simulators stand out as foundational components, allowing for the experiment of quantum circuits and algorithms in a virtual environment, aiding in error detection and algorithm debugging. Quantum compilation tools play a pivotal role by translating high-level quantum code into executable instructions for specific quantum hardware, striving to enhance quantum program efficiency and minimize error rates. Quantum error correction codes [33] and associated tools are indispensable for identifying and rectifying errors that may arise during quantum computations, confirming the reliability and accuracy of quantum hardware [34].

5.4.2 Quantum programming languages

Quantum hardware emulators provide a bridge between simulators and real hardware, mimicking the behavior of quantum processors for benchmarking and evaluation. Quantum development frameworks, such as Qiskit and Microsoft Quantum Development Kit, offer comprehensive libraries and APIs, simplifying quantum programming. Additionally, cloud-based quantum platforms provide remote access to quantum hardware, enabling researchers to experiment with real quantum processors through user-friendly interfaces. These development tools collectively contribute to advancing quantum hardware, accelerating quantum algorithm discovery and unlocking the potential of quantum computing in various domains, including cryptography, optimization, and materials science.

Quantum languages can be categorized into three main groups: imperative, functional, and other quantum programming languages, which may encompass mathematical formalisms not designed for computer execution. Imperative programming languages, often referred to as procedural languages, are primarily rooted in the utilization of statements to modify the overall state of a program or set of variables [35]. These languages have strong ties to Knill's suggested quantum pseudo-code and the quantum computing QRAM model. Functional (or declarative) programming languages do not rely on modifying a global system state. Instead, they perform mathematical transformations by performing mappings that convert inputs into outputs. Recent advancements in quantum programming have predominantly centered on the adoption of functional languages over imperative ones. Freedman, Kitaev, and Wong [36] introduced a significantly distinct quantum programming approach, which revolves around simulating topological quantum field theories (TQFTs) using quantum computers. TQFTs offer a more resilient system of quantum computation by demonstrating quantum states as physical entities that can withstand disturbances.

5.5 ETHICS AND SECURITY IN QUANTUM AI

In the world of quantum AI, there are important considerations when it comes to ethics and security. Imagine that quantum AI has the power to do incredible things, but it also raises important questions.

5.5.1 Ethical considerations

First, we need to think about privacy. Quantum AI can break the codes that protect our private information. This means we must figure out how to keep our information safe while still using these powerful technologies. Regular AI and quantum AI can learn from data that might have biases. This could lead to unfair decisions. We need to make sure quantum AI is fair to everyone. We also need to know who is responsible when things go wrong. Quantum AI can be very complex, making it hard to understand why it makes certain decisions. We need ways to find out who or what is accountable for these decisions.

Another challenge is how we use quantum AI. It can be used for good things, like helping doctors find better treatments, but it can also be used for bad things, like hacking. We need rules and guidelines to make sure we use quantum AI responsibly. Last, we should promote good practices when using quantum AI. This means following ethical rules and using these technologies in ways that don't cause harm. It is essential to tackle ethical issues and set forth guidelines and regulations [37].

5.5.2 Cybersecurity implications

Cybersecurity is all about keeping our data safe. Quantum AI can have a big impact here too. One big concern is how quantum AI could break the encryption that keeps our online communication secure. So we need new ways to protect our data from quantum AI attacks. Researchers are working on what we call "quantum-resistant encryption." This means creating codes that even quantum AI can't break [38].

We also have something called quantum key distribution (QKD). It's like sending secret keys that only the right people can use. This makes our online communication more secure [33]. Quantum AI can also be used to attack computers and networks. So we need to protect our quantum computers from these attacks. Governments and organizations are working on making rules to keep everything safe in the world of quantum AI. So, in a nutshell, we need to think about ethics to make sure we use quantum AI responsibly, and we need to be smart about cybersecurity to keep our data safe from the power of quantum AI.

5.6 THE FUTURE OF QUANTUM AI AND CONCLUSION

The future of quantum AI holds tremendous promise and potential. Imagine a world where quantum AI becomes an indispensable part of our lives, transforming various aspects of our society.

5.6.1 Envisioning quantum AI's role

As quantum AI stands poised to revolutionize various industries, its potential benefits are far-reaching. From health care to finance, transportation, and environmental sustainability, the applications are promising. However, along with these opportunities come important challenges, notably in the realms of ethics, accessibility, and fairness. Balancing the transformative power of quantum AI with responsible and inclusive implementation is crucial for shaping a brighter future across diverse sectors.

In health care, quantum AI could revolutionize drug discovery and medical diagnostics. Researchers and doctors might use quantum AI to analyze vast amounts of medical data, identifying novel treatments and predicting diseases more accurately. Personalized medicine could become a reality, tailoring treatments to individual genetic profiles. In finance, quantum AI could lead to more precise risk assessment, optimized investment strategies, and enhanced fraud detection. Financial institutions might employ quantum AI to navigate complex markets and make real-time decisions, ultimately benefiting both investors and consumers. Transportation and logistics could see vast improvements as well. Quantum AI–powered systems might optimize traffic flow, reduce energy consumption, and enhance the safety of autonomous vehicles. Supply chains could become more efficient and resilient, ensuring the timely delivery of goods worldwide. Environmental sustainability could be another area where quantum AI plays a pivotal role. Quantum AI algorithms could contribute to the discovery of new materials for renewable energy sources, helping address climate change and reducing our carbon footprint.

5.6.2 Ongoing developments and challenges

Quantum AI's path forward is not without obstacles. Quantum computing hardware is still evolving, and practical quantum computers that can handle complex AI tasks are a work in progress. Researchers face the challenge of making quantum processors more stable and scalable. Furthermore, the interdisciplinary nature of quantum AI means that experts from various fields, including quantum physics, computer science, and ethics, must collaborate closely. Ensuring that quantum AI aligns with ethical and societal norms will require ongoing dialogue and regulation. Additionally, cybersecurity concerns persist. Quantum AI has the potential to break current encryption methods, highlighting the need to create encryption techniques resistant to quantum attacks to protect sensitive data. Overall, the future of quantum AI holds the promise of groundbreaking advancements across multiple domains, from health care to finance and transportation. However, realizing this future requires addressing ethical considerations, fostering collaboration, and overcoming technological challenges. As quantum AI continues to evolve, its role in shaping the future of smart technologies is one of great anticipation and responsibility.

5.6.3 Embracing the quantum AI fusion

The profound significance of integrating quantum computing and AI, this synergy unlocks unprecedented potential, allowing us to tackle complex problems that were previously insurmountable with classical methods. By combining the computational power of quantum systems with the data-driven decision-making capabilities of AI, we gain a formidable tool for addressing real-world challenges. Interdisciplinary collaboration becomes paramount in this endeavor, as quantum physicists, computer scientists, and data experts need to work together to harness the full power of quantum AI. Industries such as finance, health care, logistics, and materials science stand to benefit significantly from this fusion, driving innovation and creating competitive advantages.

Moreover, educational initiatives are emerging to train the next generation of quantum AI experts, ensuring a steady influx of talent into this cutting-edge field. Global collaboration is also crucial, as international efforts and partnerships accelerate research and development. At the same time, we must consider the broader societal implications of quantum AI, recognizing its potential to address pressing global challenges like climate change and health care crises. Responsible development and ethical considerations are essential as we navigate this transformative territory. It is vital to acknowledge and address challenges and risks, including the need for robust security measures and ethical frameworks. In conclusion, embracing the quantum AI fusion is not just an opportunity; it's a necessity for a smarter, more innovative, and technologically advanced future. Stay engaged with ongoing developments in this exciting field to be a part of this transformative journey [39].

5.6.4 Conclusion

In this chapter, we delved into quantum AI, a convergence of quantum computing and AI poised to revolutionize modern technology. We discovered the fundamentals of quantum computing, including qubits and quantum gates and underscored its advantages over classical computing. The chapter delved into quantum AI's role in machine learning and data analysis. It also addressed the hardware and software aspects, ethical considerations, and cybersecurity implications. Concluding with a glimpse into the future, we envisioned quantum AI's evolving role, ongoing developments, and the crucial importance of embracing this fusion for a smarter technological future.

REFERENCES

[1] Bhowmik, B. R., & Manjunath, T. D. (2023). Quantum Learning and Its Related Applications for the Future. In Handbook of Research on Quantum Computing for Smart Environments (pp. 25–47). IGI Global.

[2] Geluvaraj, B., Satwik, P. M., & Ashok Kumar, T. A. (2019). The Future of Cybersecurity: Major Role of Artificial Intelligence, Machine Learning, and Deep Learning in Cyberspace. In International Conference on Computer Networks and Communication Technologies: ICCNCT 2018 (pp. 739–747). Springer Singapore.

[3] Rani, K. S. K., Priyadharsheni, J. M., Karthikeyan, B., & Pugalendhi, G. S. (2023). Applications of quantum AI for healthcare. Quantum Computing and Artificial Intelligence: Training Machine and Deep Learning Algorithms on Quantum Computers, 271.

[4] Lee, R. S. (2020). Quantum Finance. Springer Singapore.

[5] Kumar, A., Bhatia, S., Kaushik, K., Gandhi, S. M., Devi, S. G., Diego, A. D. J., & Mashat, A. (2021). Survey of promising technologies for quantum drones and networks. IEEE Access, 9, 125868–125911.

[6] Humble, T. S., Thapliyal, H., Munoz-Coreas, E., Mohiyaddin, F. A., & Bennink, R. S. (2019). Quantum computing circuits and devices. IEEE Design & Test, 36(3), 69–94.

[7] Fisher, M. P., Khemani, V., Nahum, A., & Vijay, S. (2023). Random quantum circuits. Annual Review of Condensed Matter Physics, 14, 335–379.

[8] Lu, Y., Sigov, A., Ratkin, L., Ivanov, L. A., & Zuo, M. (2023). Quantum computing and industrial information integration: A review. Journal of Industrial Information Integration, 100511.

[9] Herman, D., Googin, C., Liu, X., Galda, A., Safro, I., Sun, Y., . . . Alexeev, Y. (2022). A survey of quantum computing for finance. arXiv preprint arXiv:2201.02773.

[10] Ajagekar, A., & You, F. (2019). Quantum computing for energy systems optimization: Challenges and opportunities. Energy, 179, 76–89.

[11] Upama, P. B., Faruk, M. J. H., Nazim, M., Masum, M., Shahriar, H., Uddin, G., . . . Rahman, A. (2022, June). Evolution of Quantum Computing: A Systematic Survey on the Use of Quantum Computing Tools. In 2022 IEEE 46th Annual Computers, Software, and Applications Conference (COMPSAC) (pp. 520–529). IEEE.

[12] Prajapati, J. B., Paliwal, H., Prajapati, B. G., Saikia, S., & Pandey, R. (2023). Quantum machine learning in prediction of breast cancer. In Quantum Computing: A Shift from Bits to Qubits (pp. 351–382). Springer Nature Singapore.

[13] Rebentrost, P., Mohseni, M., & Lloyd, S. (2014). Quantum support vector machine for big data classification. Physical Review Letters, 113(13), 130503.

[14] da Silva, A. J., Ludermir, T. B., & de Oliveira, W. R. (2016). Quantum perceptron over a field and neural network architecture selection in a quantum computer. Neural Network, 76, 55–64.

[15] Schuld, M., Sinayskiy, I., & Petruccione, F. (2016). Prediction by linear regression on a quantum computer. Physical Review A, 94, 022342.
[16] Tiwari, P., & Melucci, M. (2019). Towards a quantum-inspired binary classifier. IEEE Access, 7, 42354–42372.
[17] Sergioli, G., Giuntini, R., & Freytes, H. (2019). A new quantum approach to binary classification. PLoS One, 14, e0216224.
[18] Ding, C., Bao, T. Y., & Huang, H. L. (2021). Quantum-inspired support vector machine. IEEE Transactions on Neural Networks and Learning Systems, 33, 7210–7222.
[19] Sergioli, G., Russo, G., Santucci, E., Stefano, A., Torrisi, S. E., Palmucci, S., . . . Giuntini, R. (2018). Quantum-inspired minimum distance classification in a biomedical context. International Journal of Quantum Information, 16, 18400117.
[20] Chen, H., Gao, Y., & Zhang, J. (2015). Quantum k-nearest neighbor algorithm. Dongnan Daxue Xuebao, 45, 647–651.
[21] Yu, C. H., Gao, F., & Wen, Q. Y. (2019). An improved quantum algorithm for ridge regression. IEEE Transactions on Knowledge and Data Engineering, 33, 858–866.
[22] Sagheer, A., Zidan, M., & Abdelsamea, M. M. (2019). A novel autonomous perceptron model for pattern classification applications. Entropy, 21, 763.
[23] Stein, J., Christ, I., Kraus, N., Mansky, M. B., Müller, R., & Linnhof-Popien, C. (2023). Applying QNLP to sentiment analysis in finance. arXiv preprint arXiv:2307.11788.
[24] Padha, A., & Sahoo, A. (2022, August). Quantum enhanced machine learning for unobtrusive stress monitoring. In Proceedings of the 2022 Fourteenth International Conference on Contemporary Computing (pp. 476–483). ACM.
[25] Sodhi, M. S., & Tayur, S. R. (2022). Make your business quantum-ready today. Forthcoming: Management and Business Review. doi: 10.2139/ssrn.4081201
[26] Zinner, M., Dahlhausen, F., Boehme, P., Ehlers, J., Bieske, L., & Fehring, L. (2022). Toward the institutionalization of quantum computing in pharmaceutical research. Drug Discovery Today, 27(2), 378–383.
[27] Broderick, D., & Serapiglia, G. B. (2023, March). The emergence of quantum computing: Intellectual property, partnerships, and the aerospace sector. In 2023 IEEE Aerospace Conference (pp. 1–12). IEEE.
[28] Rietsche, R., Dremel, C., Bosch, S., Steinacker, L., Meckel, M., & Leimeister, J. M. (2022). Quantum computing. Electronic Markets, 32(4), 2525–2536.
[29] Boev, A. S., Usmanov, S. R., Semenov, A. M., Ushakova, M. M., Salahov, G. V., Mastiukova, A. S., . . . Fedorov, A. K. (2023). Quantum-inspired optimization for wavelength assignment. Frontiers in Physics, 10, 1092065.
[30] Singh, M., Dhara, C., Kumar, A., Gill, S. S., & Uhlig, S. (2022). Quantum artificial intelligence for the science of climate change. In Artificial Intelligence, Machine Learning and Blockchain in Quantum Satellite, Drone and Network (pp. 199–207). CRC Press.
[31] Malhotra, Y. (2022). Existing & Emerging Sectors of AI, Cybersecurity, Defense and Law Enforcement. ConnectAI 2022 Digital Masterclass.
[32] Li, G., Ding, Y., & Xie, Y. (2020, March). Towards efficient superconducting quantum processor architecture design. In Proceedings of the Twenty-Fifth International Conference on Architectural Support for Programming Languages and Operating Systems (pp. 1031–1045). ACM.

[33] Khan, M. M., Bari, I., Khan, O., Ullah, N., Mondin, M., & Daneshgaran, F. (2021). Soft decoding of short/medium length codes using ordered statistics for quantum key distribution. International Journal of Quantum Information, 19(6), 2150025.

[34] Chong, F. T., Franklin, D., & Martonosi, M. (2017). Programming languages and compiler design for realistic quantum hardware. Nature, 549(7671), 180–187.

[35] Sofge, D. A. (2008, February). A survey of quantum programming languages: History, methods, and tools. In Second International Conference on Quantum, Nano and Micro Technologies (ICQNM 2008) (pp. 66–71). IEEE.

[36] Freedman, M., Kitaev, A., & Wong, Z. (2000). Simulation of topological field theories by quantum computers. arXiv:quant-ph/0001071/v3.

[37] Sihare, S., & Khang, A. (2023). Effects of quantum technology on the metaverse. In Handbook of Research on AI-Based Technologies and Applications in the Era of the Metaverse (pp. 174–203). IGI Global.

[38] Soni, L., Chandra, H., Gupta, D. S., & Keval, R. (2022). Quantum-resistant public-key encryption and signature schemes with smaller key sizes. Cluster Computing, 1–13.

[39] Aithal, P. S. (2023). Advances and new research opportunities in quantum computing technology by integrating it with other ICCT underlying technologies. International Journal of Case Studies in Business, IT and Education (IJCSBE), 7(3), 314–358.

Part II

Secure artificial intelligence in computing systems

Chapter 6

The significance of artificial intelligence in cybersecurity

Fazal Wahab, Anwar Shah, Inam Ullah, Habib Khan, and Deepak Adhikari

6.1 INTRODUCTION

The quality known as "intelligence" is the only trait that sets humans apart from all other life on our planet. Even if artificial computers cannot inherit intellect like humans do, it is still fascinating to think that they might one day have intelligence like that. In place of the intelligence that comes naturally to humans, members of the philosophical, scientific, and other groups working toward improved knowledge of the human mind began to consider "the reason why the machines are unable to think." Researchers worldwide began to pay attention to this idea of creating "artificial intelligence (AI)" because of multidisciplinary investigations in the areas of computer science, cognitive science, and neurology. Researchers started to have very high aspirations for AI research in the 1960s and 1970s, but their efforts were ineffective, primarily because no notable advancements were produced.

These days, people spend a great deal of their time in online communities that offer both private and public social platforms and services. Therefore, these types of situations require protection from hackers who could steal data or disrupt operations [1, 2]. Cybersecurity refers to the practice of safeguarding digital information and communication systems from unauthorized access and misuse for the purposes of ensuring the continuation of their operations, protecting the privacy and security of sensitive data, and protecting people from hazards and hacking attempts. Considering this, this chapter discusses both the role AI plays in cybersecurity and the methods used to protect computer systems from attacks, breaches in security, and other forms of cybercrime.

The internet has made the globe into a very tiny village in which individuals can share their information and their cultures with one another. This has made many things easier for us in our daily lives. Networks are the fundamental building blocks upon which mobile phones, computers, and the internet are all constructed. These devices lose practically all their value once they are disconnected from the internet. Using wires and electromagnetic waves, networks connect computers so that they can share and exchange data, information, and applications [3]. The transmission of data that includes personally identifiable information is among the most important types of data transmission that can occur across

DOI: 10.1201/9781003496410-8

computer networks. Because this data could be used by hackers to steal identities or establish fake profiles on social media platforms, networks are implementing safeguards to prevent unauthorized access to it. As a result of the COVID-19 pandemic, the practice of conducting business exclusively online, with as little interaction with other people as possible, has become widespread across numerous industries [4, 5]. Physical violence is more likely now because internet-connected services are more common, and working from home is becoming more popular, potentially leading to increased interpersonal tensions and conflicts without the usual physical boundaries of traditional workplace environments. Using signatures or other outmoded tactics to defend against these assaults is becoming an increasingly dated tactic.

Organizations are exposed to attack because it takes too long to identify a threat, stop it once it is identified, or wait for someone else to report it. Progress in AI happens throughout this period [6]. Defense systems must be able to recognize and react to dangers as they arise, to overcome the numerous, intricate, and constantly changing threats. Additionally, it is possible to take advantage of the system's terminus, or "edge." At this point, eliminating humans speeds up the process because AI can quickly sift through millions of datasets in search of any potential threat. It might be able to identify behaviors linked to malware, phishing, and "crypto-jacking"—a form of cybercrime in which compromised computers are instructed to mine the digital currency for the hacker—even without explicitly searching for software fingerprints linked to known threats. If any of these vulnerabilities are found, AI may use this knowledge to fortify its defenses and ensure the security of humanity [7]. Figure 6.1 illustrates AI in cybersecurity.

The use of AI to combat cybercrime has become increasingly common. Solutions for cybersecurity that are driven by AI can identify, evaluate, and respond to hostile assaults more quickly. Solutions that are powered by AI can sort through vast volumes of information to help and rapidly discover malicious activities, such as a new zero-day attack [8, 9]. You will find it much simpler to meet your organization's cybersecurity requirements if you make use of AI's ability to automate a variety of security operations, such as handling patches. Automating some actions, such as redirecting information out of an exposed system for reporting possible vulnerabilities to the IT staff, can help you respond more quickly to assaults and improve the security of your network [10–12].

This chapter makes a significant contribution to the field of cybersecurity by examining the crucial function that machine learning (ML) and deep learning (DL) approaches serve in the safety and security field. It does so by illustrating how these techniques contribute to lowering computer intrusions and incidents and by highlighting the different applications that make use of these techniques. This chapter also provides a concise summary of the most significant research that has used methods involving DL and ML in the subject of cybersecurity, analyzes the findings drawn from those studies, and investigates how those studies contribute

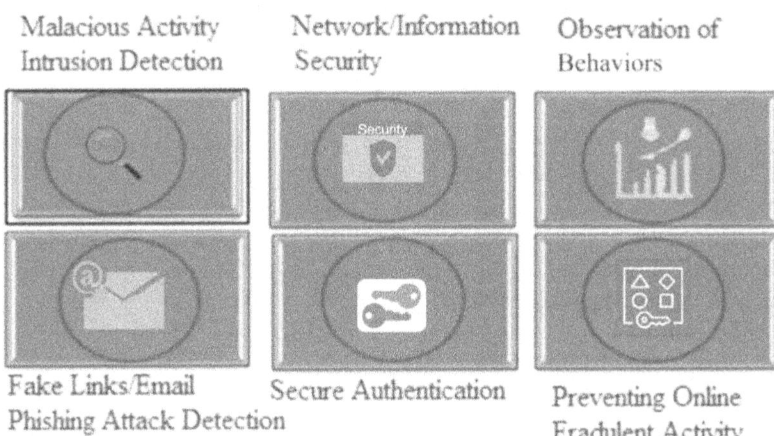

Figure 6.1 AI in cybersecurity.

to the process of decision-making. Researchers who are interested in cybersecurity can save time and effort by using the data that is used in this chapter, which has been gathered from numerous web sources and published literature.

6.2 AI APPLICATIONS IN CYBERSECURITY

AI cybersecurity systems can deliver superior real-time detection and response for recognized and new threats, reduce time in cyber defense, and produce novel solutions beyond the capability of present humans. These capabilities are made possible through automation and machine learning. Tools based on generative AI can alter the method in which cyber threats are designed and carried out. Because these models can generate text and speech that sounds and looks human, they can be used to automate the production of malicious content, such as social engineering attacks, phishing emails, and other forms of harmful content.

6.2.1 Identifying and preventing threats through AI

By utilizing ML approaches to scan massive volumes of data and find anomalies, patterns, and indicators of breach, AI introduces a paradigm shift in attack identification and mitigation. Contrary to conventional methods, AI systems can adapt to changing attack methodologies, learn from prior attack data, and continuously improve their detection abilities. With this proactive strategy, threat detection accuracy is increased, false positives are decreased, and early identification and reaction to developing threats are made possible [13–15].

6.2.2 The detection and prevention of advance threats

Using AI, computers can sift through huge amounts of data in search of patterns and outliers that can indicate a security breach. AI systems may continuously learn from previous attacks, adjust to new attack vectors, and develop their defense mechanisms by utilizing ML and DL techniques. With this feature, threat detection accuracy is improved, false positives are decreased, and a proactive response to new threats is made possible [16, 17].

6.2.3 Automated and intelligent response to incidents

By automating numerous processes, including notification setting priorities and event evaluation and restoration, AI may significantly improve incident response capabilities. AI systems can quickly detect and respond to security issues, lowering response times and the impact of attacks, because they can process and analyze enormous volumes of security event data [18]. Additionally, AI speeds up reaction times and also lessens the workload on human analysts, so they may concentrate on more difficult and strategic jobs.

6.2.4 Malware detection and analysis improvements

Malware continues to be one of the foremost hazards in cyberspace, and it is continuously changing to find new ways to circumvent existing security measures. Methods that are powered by AI, such as behavioral analysis, anomaly detection, and DL, make it possible to detect and analyze malware more effectively. AI models can identify novel and zero-day malware by identifying coding anomalies, patterns of behavior, or destructive command and control connections. These can be identified by learning from large-scale datasets [19, 20]. Organizations can be one step ahead of their adversaries and better defend their systems and networks when they have the capability to detect and analyze malware that has not been seen before.

6.2.5 Behavioral analytics and user monitoring

Significant hazards to corporations come from hacked user accounts and insider attacks. User behavior analytics (UBA) systems powered by AI can track user activity, define baseline behavior, and spot variances that can point to account compromise or malicious intent. UBA systems can identify aberrant behavior and issue alarms for further inquiry by examining user activities, access patterns, and contextual data [21]. This proactive strategy strengthens the overall security posture by enabling early identification of insider threats, privilege abuse, and credential theft.

6.2.6 Proactive vulnerability management

A crucial component of cybersecurity is the process of locating and fixing flaws in software or hardware before a threat may take advantage of them. AI can assist in proactive vulnerability management by evaluating configurations of systems and communication over the network, as well as application behavior to discover potential vulnerabilities. This can help to identify potential threats. AI algorithms may prioritize vulnerabilities based on their severity, possible effect, and exploitability [22]. This enables businesses to efficiently deploy resources and focus on addressing the most serious vulnerabilities first. This effective vulnerability management helps lower the attack surface and strengthens the overall security posture of the organization.

6.2.7 Collaborative threat intelligence

Aggregating, assessing, and distributing threat intelligence across enterprises and industries are critical tasks for AI. AI systems can process enormous amounts of security data from many sources, spot trends, and produce valuable insights by utilizing natural language processing and ML approaches [23]. Organizations may make use of their common knowledge, spot emerging hazards, and react to attacks more skillfully by sharing collaborative threat intelligence.

6.3 ARTIFICIAL INTELLIGENCE

The development of computers, automated robots, and software programs that can simulate human intelligence is made possible by AI. Studying human cognition, which includes topics such as how people learn, approach issues, and make decisions, is how AI is produced. This study will help to shape the development of software and devices powered by AI [24]. It is common practice to describe intelligence as the capacity to assimilate new information and use that information for the resolution of difficult situations. In the future, it is possible that intelligent robots may be able to handle many of the jobs that are currently performed by people. The field of study known as AI focuses on the creation of intelligent technologies and computer programs that have cognitive capabilities comparable to those of humans as well as their subsequent application. AI is distinct from both the field of psychology and the field of computer science. Computer science, on one hand, concentrates on perception, the mind, and action, whereas psychology, on the other hand, focuses on thinking, action, and perception [24]. This is how computer science varies from psychology. This promotes machine learning, which ultimately results in enhanced capabilities.

6.3.1 Machine learning in cybersecurity

For network intrusion detection, several methods have been suggested. Approaches based on ML offer an advantage over all old approaches since they are reliable, highly functional, and refined. The widespread adoption of AI in areas like big data and intrusion detection has inspired us to create an elegant and inexpensive method for preventing cyberattacks in critical scenarios [25–27]. A network intrusion detection system's primary objective is to detect malicious network activity. The essential task is to sort data packets and network events into two categories: normal and abnormal or anomaly. The categorization job in supervised machine learning is a natural application. This is a specific case of a binary classification problem, with normal and anomalous serving as the classes in consideration [28, 29]. To represent this as a machine learning problem, we need access to a dataset containing examples of network activity. Individuals have already purchased such datasets as those made available by the Cyber Range Lab of the Australian Cyber Security Centre (ACSC) and similar institutions. This data collection is a composite of real-world, modern, normal behaviors and aggressive ones. The over two million data points in this dataset represent 47 different qualities. We only select features that are associated with the target variable (also known as the label variable), as not all features are equally important. New data patterns or practices can be discovered and predicted using several ML principles and methods [30–32]. Cybersecurity can benefit from these techniques, which can be categorized as either supervised or unsupervised. Figure 6.2 presents the ML uses in cybersecurity.

6.3.2 Supervised learning

Supervised learning works in a systematic manner by establishing precise goals and achieving them via a given set of inputs [33, 34]. Supervised learning methods, which have become popular in a variety of domains, are simple to implement and track. Methods that classify security data or make predictions about future security issues are categorized as either classification or regression methods. A company's ability to function could be jeopardized if it does not take precautions to prevent an expected attack on its IT infrastructure. Therefore, it is recommended that all industries implement approaches based on ML in cybersecurity to better safeguard their data and the data of their users. Some of the most widely used supervised learning approaches for classification include decision trees, support vector machines, *k*-nearest neighbors, logistic regression, and so on [35]. Because of their ability to construct a data-driven predictive model, these techniques are also used in prediction. For example, the behaviors of users in particular networks inside a private or public organization could be anticipated by following all their conducted processes, gathering information continuously, and figuring out who or what is releasing processes that have an impact on the web. In the meantime, the most employed regression approaches in supervised learning are support vector

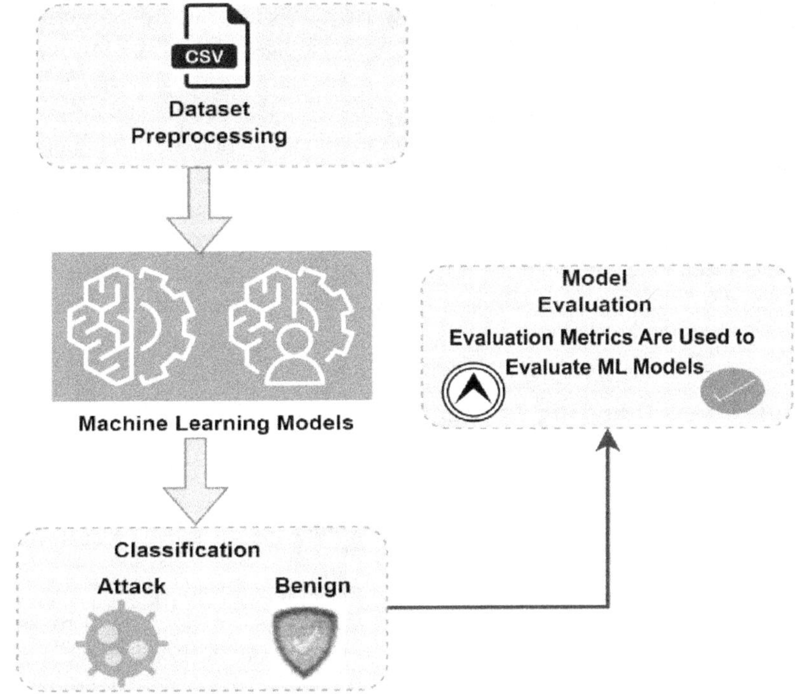

Figure 6.2 ML in cybersecurity.

regression and linear regression, which are used to uncover the underlying causes of severe cybercrimes that have a significant impact on the lives of many people and discover appropriate remedies [36]. The findings of classification and regression approaches can be used to understand the differences between them. The results of a classification analysis are categorical or discrete, while those of a regression analysis are quantitative or continuous.

6.3.3 Unsupervised learning

Unsupervised learning techniques are used to extract knowledge, patterns, or structures from unlabeled data. Malware operates dynamically in cybersecurity but remains hidden to avoid detection and remediation [37, 38]. Unsupervised learning approaches for clustering, such as *k*-means, *k*-medoids, and single linkage, are used to identify and learn about sophisticated and covert attacks by looking for previously unknown patterns and structures in a vast dataset. Additionally, these methods locate and notify users or programmers of system irregularities, privacy law infractions, and illegal access to data. The engineering activities utilized in these technologies, such as optimizing a dataset's properties or identifying the relevant aspects of a specific security issue in a system, are categorized as functional tasks for conducting further study, regardless of the size of the dataset. Additionally, security elements are chosen based on their importance. To handle security risks, discover hidden programs, and stop such breaches and information vandalism, approaches based on ML and additional techniques, including principal component analysis, linear discriminant analysis, and Pearson correlation analysis, can be helpful. Expert systems use rules that are manually created by an experienced engineer working with a data security specialist. For the purpose of extracting the available security features or qualities, association rules learning seeks to discover the available rules or relationships among datasets. The strength of the association between datasets is assessed by correlation analysis.

6.3.4 Deep learning in cybersecurity

Different approaches are utilized to address complicated issues in cybersecurity depending on specific factors, including issue nature, data volume, issue severity, and decision intolerance in the outcome. Researchers who lack expert topic knowledge can use convolutional neural networks (CNNs) to extract valuable characteristics from data. CNNs have made some headway in the field of DL-based software bug detection in recent years [39–41]. DL methods based on parallel processing are beneficial for massive data and call for intricate procedures. This section analyzes the research on the application of DL algorithms for malware, intrusion, and assault detection. To guarantee accuracy of data, privacy, and dependability, as well as guarantee that only authorized people can access the system, instead of being created on a personal computer, DL architectures are developed on server-based systems. A DL model must go through two steps to function correctly in cybersecurity. The first step is to encrypt the local environment region where data is sent to the server [42]. Sending these data to the server for processing, classification, and type determination constitutes the second stage. For instance, the first stage of character recognition from images entails character encoding and transmission; data categorization relies on the second phase, which entails evaluating the collected information to see if a man-in-the-middle attack occurred between the server and the local system. By doing so, sensitive information is transmitted to the intended recipients without risk of interception.

6.5 CYBERSECURITY AND DIFFERENT ATTACK TYPES

Cybersecurity is the protection of interconnected networks, information, hardware, and software from cyberattacks, often known as illegal access and adverse effects. There are many different methods and strategies that fall under this category. When seen from the perspective of computing, the safety of both cyberspace and physical space should be considered of equal importance. Some of the components of cybersecurity, such as operational security, application security, network security, information security, and so on, are what integrate the complete information system [43]. This configuration appears to employ multiple layers of security across the system and its many parts. It is one of the modern defenses that works.

6.5.1 Physical security attack

Physical attacks, in which the victim's device's physical properties are exploited, are a common type of low-tech cyberattack. There are many kinds of physical attack. Physical damage, in which machines or their parts get damaged to prevent appropriate performance; malicious code injection, where an attacker plugs a USB that contains a malicious program into the intended device; and object jamming, where signal blockers are utilized to disable and control the signals that are emitted by the objects. Outage attacks are another type of cyberattack; in this kind of attack, the network to which the devices are linked is intentionally disabled to interfere with their operations [19, 26, 37].

6.5.2 Man-in-the-middle

Man-in-the-middle (MITM) attacks are among the most common ones against network devices. In terms of computers, in general, an MITM attack allows the attacker to act as a proxy by intercepting communication between two nodes. Connections between computers and routers; mobile devices; and, most commonly, servers and clients are all vulnerable to attack. When it comes to the internet of things (IoT), the attacker typically conducts MITM incidents against a device in the IoT and the application it interfaces with [42]. IoT is especially vulnerable to MITM attacks because it lacks the standard methods to guard against the attacks. Figure 6.3 illustrates MITM.

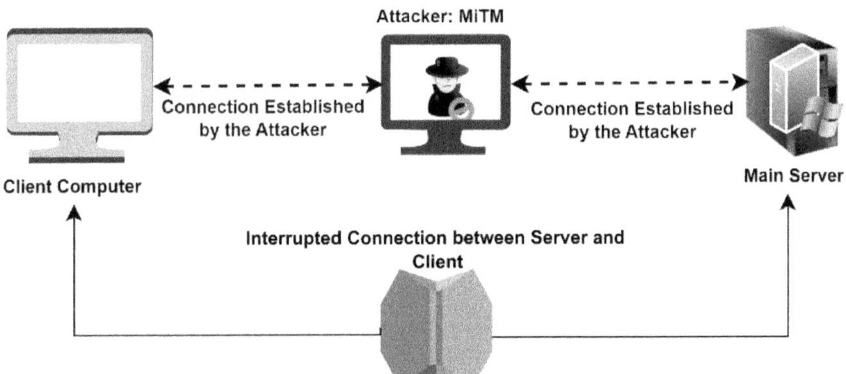

Figure 6.3 Man-in-the-middle attack.

6.5.3 Bluetooth man-in-the-middle

Bluetooth connections are often the target of MITM attacks directed against IoT devices. The Bluetooth Low Energy (BLE) standard, which is used by many IoT devices, was developed specifically for IoT devices with the goal of being more compact, affordable, and power efficient [44–47].

However, MITM attacks are possible with BLE connections. BLE employs the AES-CCM encryption standard. Although AES encryption is thought to be secure, the way encryption keys are often shared is frequently not secure. The technique of pairing that is used to exchange temporary keys between the devices is what determines the level of security that may be achieved [18, 22]. The BLE basically includes three-phase pairing processes: in the initial stages, the requesting machine transmits a connection request, and the gadgets exchange connecting functions through a vulnerable channel; second, the machines transfer short-term credentials and confirm they genuinely share the same temporary key, and that is subsequently utilized to produce a temporary key; and third, data can be encrypted using the newly created key, which is transferred through a secure link [37].

6.5.4 False data injection attack

An example of an attack that an adversary could carry out after getting access to a few or all the electronic devices that make up an IoT network by means of an MITM attack is referred to as a false data injection (FDI) assault. In FDI attacks, measurements from IoT sensors are slightly altered by the attacker to evade detection, and the tampered data is then output [48–50]. There are other techniques to carry out FDI attacks; however in reality, MITM attacks are the most useful. Sensors that send information to a machine that does some sort of prediction or analysis based on that information are common targets of FDI attacks. These systems, also known as predictive maintenance systems, are widely employed to monitor the health of mechanical equipment and foretell when servicing or fine-tuning would be necessary.

6.5.5 Botnets

Attacks on network gadgets frequently take the form of coordinated distributed denial of service (DDoS) attacks launched from a network of compromised devices known as a botnet. Denial of service (DoS) attacks, in which many parties work together to disrupt a service's availability for legitimate users, are one type of DDoS attack. DDoS attacks are intended to overburden the intended service's infrastructure and interrupt usual information transmission [43, 51]. DDoS attacks typically consist of different phases: recruitment, whereby the intruder searches for machines that are susceptible to use in the DDoS attack to harm the target; communication, in which the attacker assesses the affected machines, interprets those that are available online, and chooses when to plan the attempts or update a device; exploitation and infection, in which the compromised systems are exploited and malicious program is inserted. The botnet hierarchy is shown in Figure 6.4.

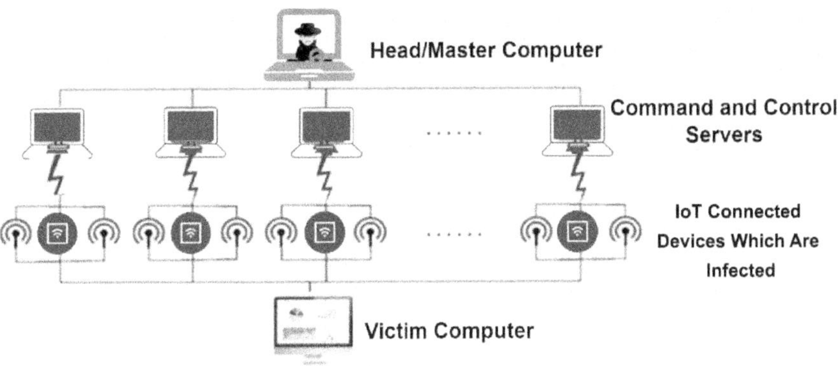

Figure 6.4 Botnet hierarchy.

6.6 FUTURE OF CYBERSECURITY

There are several important things to consider despite the positive predictions made regarding the future of this type of safety. Datasets are essential to the ML process, yet compliance with privacy regulations can be challenging.

The "right to be forgotten" is in direct conflict with the requirement that software systems have access to massive datasets to generate reliable forecasts. Because of the possibility of data breaches, including personal identifiers, it will be necessary to take corrective actions. Increasing the difficulty of gaining access to the raw data that is required to train the algorithm is one potential solution to the problem. Anonymizing data points is yet another approach; however, this method involves additional work to guarantee that it does not disrupt the logic of the program in any manner. There is a significant need for specialists in AI and ML security who are also familiar with this type of programming. If people oversaw making updates and alterations, then network security that relies on machine learning would be substantially more efficacious [26, 52]. However, there is a significant gap between the global demand for people who can provide these answers and the current supply of such people. Even in the far-off future, collaborative efforts from individuals must be maintained. The capacity to think critically as well as creatively will be beneficial when it comes to forming judgments [53].

6.7 CONCLUSION

One of the Fourth Industrial Revolution's most significant achievements is AI. Given its benefits to humanity, AI is predicted to continue to evolve and develop on a vast scale. The issues surrounding AI and cybersecurity merit more research to strike a balance between the advancement of technological equipment and core human values. As the landscape of cyber threats continues to evolve, organizations confront significant hurdles in securing their sensitive data, vital infrastructure, and digital assets. These challenges stem from the fact that the environment of cyber threats is constantly shifting. The conventional methods of cybersecurity, such as rule-based systems and signature-based detection, are no longer adequate to tackle the complexity and breadth of new threats since they are reliant on patterns and signatures. As a direct consequence of this, there is an increasing awareness of the significance of utilizing AI in the process of strengthening cybersecurity defenses. This chapter explored the many facets of AI's role in cybersecurity and emphasized its potential to revolutionize the way we protect our digital environment.

REFERENCES

[1] N. Bhalaji, "Reliable data transmission with heightened confidentiality and integrity in IoT empowered mobile networks," Journal of IoT in Social, Mobile, Analytics, and Cloud, vol. 2, no. 2, pp. 106–117, May 2020.

[2] D. Ghillani, Deep Learning and Artificial Intelligence Framework to Improve the Cyber Security. Authorea Preprints, 2022.

[3] J. Budd, B. S. Miller, E. M. Manning, V. Lampos, M. Zhuang, M. Edelstein, G. Rees, V. C. Emery, M. M. Stevens, N. Keegan, M. J. Short, D. Pillay, E. Manley, I. J. Cox, D. Heymann, A. M. Johnson Rachel, A. McKendry, "Digital technologies in the public health response to COVID-19," Nature Medicine, vol. 26, pp. 1183–1192, August 2020. https://doi.org/10.1038/s41591-020-1011-4.

[4] M. G. Tolani, H. G. Tolani, "The use of artificial intelligence in cyber defense," International Research Journal of Engineering and Technology (IRJET), vol. 6, no. 7, pp. 3084–3087, 2019.

[5] R. Calderon, "The benefits of artificial intelligence in cybersecurity," Economic Crime Forensics Capstones, vol. 36, 2019.

[6] Google, "Artificial intelligence in cyber security," 10–20, 6; 2023, https://terranovasecurity.com/ai-in-cyber-security/.

[7] J. Yang, J. Li, and X. Wei, "An artificial intelligence-based network intrusion detection system using improved convolutional neural network," Neural Computing and Applications, vol. 33, no. 10, pp. 4589–4599, 2021.

[8] E. Hemberg and U. M. O'Reilly, "Using a collated cybersecurity dataset for machine learning and artificial intelligence," Arxiv Preprint Arxiv:2108.02618, 2021.

[9] E. Proko, A. Hyso, and D. Gjylapi, "Machine learning algorithms in cyber security." In RTA-CSIT (pp. 203–207), CEUR Workshop Proceedings, Germany, 2018.

[10] I. Ahmad, T. Rahman, A. Zeb, I. Khan, I. Ullah, H. Hamam, and O. Cheikhrouhou, "Analysis of security attacks and taxonomy in underwater wireless sensor networks," Wireless Communications and Mobile Computing, vol. 2021, pp. 1–5, 2021.

[11] S. Khan, I. Ullah, F. Ali, M. Shafiq, Y. Y. Ghadi, and T. Kim, "Deep learning-based marine big data fusion for ocean environment monitoring: Towards shape optimization and salient objects detection," Frontiers in Marine Science, vol. 9, p. 1094915, 2023.

[12] X. Su, I. Ullah, M. Wang, and C. Choi, "Blockchain-based system and methods for sensitive data transactions," IEEE Consumer Electronics Magazine, vol. 13, no. 2, pp. 87–96, 2021.

[13] A. Alshamrani, S. Cranefield, and A. Gravell, "Artificial intelligence techniques for cybersecurity vulnerability analysis: Survey and challenges," Future Generation Computer Systems, vol. 92, pp. 546–563, 2019.

[14] A. Shah, B. Ali, F. Wahab, I. Ullah, K. T. Amesho, and M. Shafiq, "Entropy-based grid approach for handling outliers: A case study to environmental monitoring data," Environmental Science and Pollution Research, pp. 1–20, 2023.

[15] B. K. Yousafzai, S. A. Khan, T. Rahman, I. Khan, I. Ullah, A. Ur Rehman, M. Baz, H. Hamam, and O. Cheikhrouhou, "Student-performulator: Student academic performance using hybrid deep neural network," Sustainability, vol. 13, no. 17, p. 9775, 2021.

[16] B. Geluvaraj, P. M. Satwik, and T. A. Ashok Kumar, "The future of cybersecurity: Major role of artificial intelligence, machine learning, and deep learning in cyberspace." In International Conference on Computer Networks and Communication Technologies (pp. 739–747). Springer, Singapore, 2019.

[17] A. Alshamrani, S. Cranefield, and A. Gravell, "Artificial intelligence techniques for cybersecurity vulnerability analysis: Survey and challenges," Future Generation Computer Systems, vol. 92, pp. 546–563, 2019.

[18] K. Leung., J. T. Wu, and G. M. Leung, "Real-time tracking and prediction of COVID-19 infection using digital proxies of population mobility and mixing," Nature Communications, vol. 12, no. 1501, pp. 1–8, March 2021. https://doi.org/10.1038/s41467-021-21776-2.

[19] M. Ssenyonga, "Imperatives for post COVID-19 recovery of Indonesia's education, labor, and SME sectors," Cogent Economics & Finance, vol. 9, no. 1, pp. 1–51, April 2021.

[20] H. S. Lallie, et al., "Cyber security in the age of COVID-19: A timeline and analysis of cyber-crime and cyber-attacks during the pandemic," Computers & Security, vol. 105, p. 102248, June 2021. https://doi.org/10.1016/j.cose.2021.102248.

[21] M. M. Mijwil, "Implementation of machine learning techniques for the classification of lung X-ray images used to detect COVID-19 in humans," Iraqi Journal of Science, vol. 62, no. 6, pp. 2099–2109, July 2021. https://doi.org/10.24996/ijs.2021.62.6.35.

[22] I. Santos, R. Breinlinger, and S. Kounev, "AI-based malware detection: A systematic review," ACM Computing Surveys, vol. 52, no. 5, article no. 108, 2019.

[23] F. Wahab, Y. Haizhao, D. Javeed, M. Hmoud, S. Ahmad, W. Khan, M. Shahid, and R. Kumar, "An AI-driven hybrid-framework for intrusion detection in IoT-Enabled E-health", Computational Intelligence and Neuroscience, vol. 2022, p. 11, 2022.

[24] M. Rouse, "What is IoT (Internet of Things) and how does it work?" *IoT Agenda, TechTarget.* www.internetofthingsagenda.techtarget.com/definition/Internet-of-Things-IoT. Accessed 11 Feb 2020.

[25] A Lakhani, "The role of artificial intelligence in IoT and OT security," www.csoonline.com/article/3317836/the-role-of-artificial-intelligence-in-iot-and-ot-security.html. Accessed 11 Feb 2020.

[26] T. Liu, J. Yang, and J. Zhang, "Deep learning for zero-day flash malware detection," IEEE Transactions on Information Forensics and Security, vol. 13, no. 9, pp. 2331–2342, 2018.

[27] F. Wahab, I. Ullah, A. Shah, R. A. Khan, A. Choi, and M. S. Anwar, "Design and implementation of real time object detection system based on single shot detector and OpenCV," Frontiers in Psychology, vol. 10, pp. 1–17, 2022.

[28] T. Ghosh, et al., "Artificial intelligence and internet of things in screening and management of autism spectrum disorder," Sustainable Cities and Society, vol. 74, p. 103189, November 2021. https://doi.org/10.1016/j.scs.2021.103189.

[29] S. Almashaqbeh, H. Al-Aqrabawi, and M. Alsmadi, "Deep learning approaches for user behavior analytics in cybersecurity: A survey," IEEE Access, vol. 9, pp. 4974–4997, 2021.

[30] M. Abdullahi, et al., "Detecting cybersecurity attacks in Internet of Things using artificial intelligence methods: A systematic literature review," Electronics, vol. 11, no. 2, pp. 1–27, January 2022. https://doi.org/10.3390/electronics11020198.

[31] Y. Ying, Y. Shi, Y. Zhao, and F. Wahab, "Multi-graph learning-based software defect location", special issue—technology paper. Wiley Software Evolution and Process, 2023.

[32] N. Rawindaran, A. Jayal, E. Prakash, and C. Hewage, "Cost benefits of using machine learning features in NIDS for cyber security in UK small medium enterprises (SME)," Future Internet, vol. 13, no. 8, pp. 1–36, July 2021.

[33] H. Khalil, S. U. Rahman, I. Ullah, I. Khan, A.J. Alghadhban, M. H. Al-Adhaileh, G. Ali, and M. ElAffendi, "A UAV-swarm-communication model using a machine-learning approach for search-and-rescue applications," Drones, vol. 6, no. 12, p. 372, 2022.

[34] I. Haq, T. Mazhar, M. A. Malik, M. M. Kamal, I. Ullah, T. Kim, M. Hamdi, and H. Hamam, "Lung nodules localization and report analysis from computerized tomography (CT) scan using a novel machine learning approach," Applied Sciences, vol. 12, no. 24, p. 12614, 2022.

[35] L. Li, Y. Zhao, Y. Li, F. Wahab, and Z. Wang, "Searching the most stable community in temporal graph," Knowledge-Based System, vol. 250, p. 109101, 2022.

[36] T. Mazhar, H. M. Irfan, I. Haq, I. Ullah, M. Ashraf, T. A. Shloul, Y. Y. Ghadi, Imran, and D. H. Elkamchouchi, "Analysis of challenges and solutions of IoT in smart grids using AI and machine learning techniques: A review," Electronics, vol. 12, no. 1, p. 242, 2023.

[37] M. Tehseen, H. M. Irfan, S. Khan, I. Haq, I. Ullah, M. Iqbal, and H. Hamam, "Analysis of cyber security attacks and its solutions for the smart grid using machine learning and blockchain methods," Future Internet, vol. 15, no. 2, p. 83, 2023.

[38] C. Yu, H. Liu, F. Wahab, Z. Ling, T. Ren, H. Ma, and Y. Zhao, "Global triangle estimation based on first edge sampling in large graph streams," The Journal of Supercomputing, vol. 79, no 13, pp. 14079–14116, 2023.

[39] X. Liu, Y. Zhao, T. Xu, F. Wahab, Y. Sun, and C. Chen, "Efficient false positive control algorithms in big data mining," Applied Sciences, vol. 13, p. 5006, 2023.

[40] S. B. Atiku, A. U. Aaron, G. K. Job, F. Shittu, and I. Z. Yakubu, "Survey on the applications of artificial intelligence in cyber security," International Journal of Scientistic and Technology Research, vol. 9, no. 10, pp. 165–170, 2020.

[41] M. M. Mijwil, I. E. Salem, and M. M. Ismaeel, "The significance of machine learning and deep learning techniques in cybersecurity: A comprehensive review," Iraqi Journal for Computer Science and Mathematics, vol. 4, no 1, pp. 87–101, 2023.

[42] M. Kuzlu, C. Fair, and O. Guler, "Role of artificial intelligence in the Internet of Things (IoT) cybersecurity," Discover Internet of Things, vol. 1, no 1, p. 7, 2021.

[43] T. C. Truong, I. Zelinka, J. Plucar, M. Čandík, and V. Šulc, "Artificial intelligence and cybersecurity: Past, presence, and future." In Artificial Intelligence and

Evolutionary Computations in Engineering Systems (pp. 351–363). Springer, Singapore, 2020.

[44] I. H. Sarker, H. Furhad, and R. Nowrozy, "AI-driven cybersecurity: An overview, security intelligence modeling and research directions," SN Computer Science, vol. 2, no. 173, March 2021. https://doi.org/10.1007/s42979-021-00557-0.

[45] I. Ullah, A. Noor, S. Nazir, F. Ali, Y. Y. Ghadi, and N. Aslam, "Protecting IoT devices from security attacks using effective decision-making strategy of appropriate features," The Journal of Supercomputing, pp. 1–30, 2023.

[46] Z. U. Abideen, T. Mazhar, A. Razzaq, I. Haq, I. Ullah, H. Alasmary, and H. G. Mohamed, "Analysis of Enrollment criteria in secondary schools using machine learning and data mining approach," Electronics, vol. 12, no. 3, p. 694, 2023.

[47] H. U. Khan, M. Sohail, F. Ali, S. Nazir, Y. Y. Ghadi, and I. Ullah, "Prioritizing the multi-criterial features based on comparative approaches for enhancing security of IoT devices," Physical Communication, vol. 59, p. 102084, 2023.

[48] F. Farivar, M. S. Haghighi, A. Jolfaei, and M. Alazab, "Artificial intelligence for detection, estimation, and compensation of malicious attacks in nonlinear cyber-physical systems and industrial IoT," IEEE Transactions on Industrial Informatics, vol. 6, no. 4, pp. 2716–2725, 2020. https://doi.org/10.1109/TII.2019.29564 74.

[49] I. Khan, Y. B. Tian, I. Ullah, M. M. Kamal, H. Ullah, and A. Khan, "Designing of E-shaped microstrip antenna using artificial neural network," International Journal of Computing, Communication and Instrumentation Engineering, vol. 5, no. 1, pp. 23–26, 2018.

[50] T. Mazhar, D. B. Talpur, T. A. Shloul, Y. Y. Ghadi, I. Haq, I. Ullah, K. Ouahada, and H. Hamam, "Analysis of IoT security challenges and its solutions using artificial intelligence," Brain Sciences, vol. 13, no. 4, p. 683, 2023.

[51] F. Wahab, I. Khan, T. Hussain, and A. Amir, "An investigation of cyber attack impact on consumers' intention to purchase online," Decision Analytics Journal, p. 100297, 2023.

[52] R. Dalal, R. Varahamurthy, and R. Talegaon, "A to I of artificial intelligence," Work, vol. 7, p. 8, 2020.

[53] F. Meneghello, M. Calore, D. Zucchetto, M. Polese, and A. Zanella, "IoT: Internet of threats? A survey of practical security vulnerabilities in real IoT devices," IEEE Internet of Things Journal, vol. 6, no. 5, pp. 8182–8201, 2019.

Chapter 7

Securing the internet of things with blockchain

Hamed Taherdoost

7.1 INTRODUCTION

The internet of things (IoT) is a network of linked gadgets that can exchange data and communicate without human involvement [1]. It includes sensors and networking capabilities, from commonplace items to industrial machines. The importance of IoT resides in its capacity to alter various sectors and facets of daily life. Automating procedures and streamlining operations increase productivity and efficiency [2]. IoT devices generate enormous amounts of data, which allows for data-driven decision-making and better results [3]. IoT significantly impacts the quality of life through smart homes and wearable technology, enhancing safety, convenience, and well-being [4]. IoT in health care makes it possible for customized health care solutions and remote monitoring [5, 6]. IoT supports sustainability initiatives by encouraging renewable energy sources and improving resource management [4]. With the integration of technologies like artificial intelligence (AI) and big data analytics made possible by the connectivity and interoperability fostered by the IoT, new opportunities and innovation are made possible [3]. However, there are now many more IoT gadgets on the market that pose serious security risks.

Cybercriminals are attracted to the IoT due to its unique characteristics, such as large-scale deployment, diverse device ecosystems, and inherent vulnerabilities. Due to IoT systems' centralized nature and reliance on insecure communication channels, traditional security measures fail to protect them [7]. Consequently, protecting IoT infrastructure and data has become a top priority. Initially introduced as the underlying technology for cryptocurrencies such as bitcoin, blockchain technology has emerged as a powerful solution for addressing security and trust issues in various domains [8]. Blockchain is a decentralized and distributed ledger that facilitates immutable and transparent transaction recording. Its decentralized consensus mechanism offers high security, data integrity, and trust [8, 9].

Transparency, immutability, and tamper resistance are three core characteristics of blockchain that make it an interesting platform for boosting IoT security [10]. IoT systems can provide improved data integrity, secure device identity management, strong access control, and privacy preservation by utilizing blockchain.

DOI: 10.1201/9781003496410-9

Since there is no need for a centralized authority thanks to blockchain, there are fewer vulnerabilities and single points of failure [11]. Despite the potential advantages of blockchain for IoT security, there is a need to investigate and comprehend its real-world uses, difficulties, and limitations. This chapter's goals are to examine the fundamental ideas and principles of IoT and blockchain technology; to look into the specific security issues that IoT systems face and the limitations of current security measures; to examine the potential advantages and applications of blockchain technology in addressing IoT security issues; to identify the key considerations, challenges, and limitations in implementing blockchain for IoT security; and to explore the potential benefits and applications of blockchain technology in addressing these issues.

7.2 IOT SECURITY CHALLENGES AND VULNERABILITIES

Due to IoT ecosystems' distinctive qualities and complexity, securing IoT poses substantial challenges and vulnerabilities. Inconsistencies and compatibility problems are caused by the absence of standardization in security protocols and frameworks across IoT platforms and devices, leaving vulnerabilities that attackers might exploit [12]. IoT devices are vulnerable to unauthorized access due to inadequate authentication and authorization systems and the usage of default or weak credentials. Due to resource limitations and occasional security updates, firmware and software flaws expose devices to known hazards. Sensitive information is susceptible to interception and unauthorized access due to insecure communication methods and poor encryption [13]. IoT devices' physical security should be more frequently addressed, to avoid leaving them open to theft or tampering [14, 15].

Due to the large volumes of personal data that IoT devices collect and handle, data privacy and permission are major obstacles to IoT security. Device inventory, security updates, and monitoring are difficult because of scalability and administrative complexity [16]. Weak passwords and insecure device usage result from user ignorance about IoT security best practices and hazards. Malicious actors who target weak points in the manufacturing or distribution processes might put the supply chain at risk [17].

A comprehensive strategy is necessary to overcome these obstacles. Security-by-design concepts, standardization initiatives, and industry cooperation are crucial. To improve IoT security, robust authentication procedures, secure communication protocols, and frequent firmware and software updates are required. Campaigns that promote user education and awareness are essential for encouraging appropriate gadget use. Additionally, it is essential to undertake rigorous security audits and guarantee the supply chain's integrity to reduce supply chain risks.

7.2.1 Risks associated with compromised IoT devices

IoT device security risks can seriously impact user privacy, data security, and overall system integrity. First of all, since hackers can use IoT device vulnerabilities to take over and access critical data, unauthorized access becomes a serious problem that threatens both business and personal data. Furthermore, infected IoT devices can act as entry points for data breaches, enabling hackers to enter networks and access important data, posing a serious risk to individuals and organizations.

Attackers frequently use infected IoT devices to perform massive distributed denial of service (DDoS) attacks. Attackers overpower targeted systems or networks via a botnet of infected devices, resulting in service interruptions and potential financial losses [18]. In addition, hacked IoT devices in crucial infrastructure, such as industrial control systems, present dangers to physical safety. Attackers may use these devices to harm people, damage property, or even jeopardize their lives [19].

Malware spreads through the use of compromised IoT devices. Once a device has been infected, the malware can propagate swiftly throughout networks, contaminating further exposed devices and enlarging the scope of the attack [20]. In addition to compromising data security, this makes it difficult to patch and update these devices. The risk of unauthorized control, privacy invasion, and degraded device performance is increased when devices are not regularly updated with security patches or upgrades [21].

7.2.2 Existing security measures

To reduce risks and defend against potential threats, IoT device security mechanisms already have been put in place. Password-based authentication is one of the traditional authentication methods but is vulnerable to brute-force attacks and password-cracking methods [22]. While offering a secure authentication method, public key infrastructure (PKI) and digital certificates can be difficult to set up and manage, and a compromised private key presents a single point of failure. Transport layer security (TLS) encryption provides security but might only partially guard against sophisticated attacks like man-in-the-middle attacks. TLS installation in IoT devices could also result in security flaws [12].

Firewalls and network segmentation separate IoT devices and filter network traffic. Although this strategy aids in minimizing breaches, it is ineffective if an intruder gains access to the internal network. Complex network segmentation can be difficult to manage and maintain [23]. The ability of intrusion detection systems (IDS) and intrusion prevention systems (IPS) to detect new and undiscovered threats is constrained by their reliance on signature-based detection. Additionally, they could produce many false positives, requiring much manual analysis and verification [24].

Updates to software and device firmware are essential for fixing security flaws. However, the lack of centralized management and coordination makes updating many remote IoT devices challenging. Furthermore, IoT devices' processing and memory capabilities are frequently constrained, affecting how well they can handle updates. Centralized security systems provide management and monitoring capabilities, but exploiting them could create a single point of failure. It can be challenging to scale and guarantee performance when dealing with many IoT devices. Dependence on outside service providers for security could raise issues with privacy and trust.

Physical barriers and access controls are part of security measures to safeguard IoT devices. However, they have difficulty preventing physical manipulation or theft, especially in distributed IoT deployments. The heterogeneous IoT ecosystem makes it difficult to deploy uniform security measures due to standardization and compliance. Security initiatives are made more difficult by the fragmented nature of IoT standards and compliance requirements. Since human weaknesses can be exploited, social engineering and human variables like user awareness and training are important factors to consider. IoT devices capture, retain, and exchange sensitive data but frequently need clearer consent processes and user control, posing problems to data privacy and consent. IoT networks lack transparency and auditability, which makes it challenging to spot and look into security lapses or illegal access. Figure 7.1 illustrates the challenges in physical security, standardization, user awareness, data privacy, and network transparency in IoT deployments.

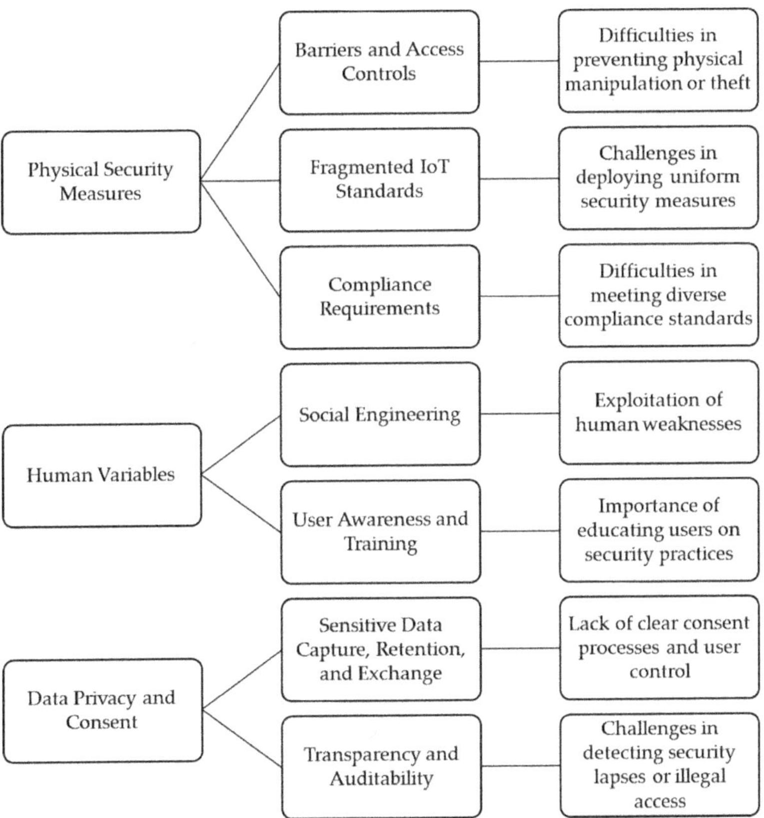

Figure 7.1 IoT security challenges and considerations.

7.3 BLOCKCHAIN TECHNOLOGY FOR IOT SECURITY

Blockchain technology is a possible solution for addressing IoT security issues since it has built-in security features, including immutability, transparency, and decentralization. It secures device authentication and identity management, guarantees data integrity and provenance in IoT data streams, enables access control and authorization frameworks for IoT networks, and protects privacy and confidentiality in IoT communications. However, considering elements like scalability, resource efficiency, and interoperability is necessary when choosing the best blockchain implementation for IoT security. Optimizing blockchain integration in IoT systems requires scalability solutions, including sharding and off-chain processing, resource-efficient consensus methods, and interoperability frameworks. Blockchain can greatly improve the security and dependability of IoT systems by utilizing these characteristics.

7.3.1 Blockchain's security features

A variety of security characteristics offered by blockchain technology make it an appealing option for boosting security in various applications, including the IoT. Immutability and tamper resistance are important security features that make it nearly impossible to change or tamper with a transaction once it has been added to the blockchain. Because of the decentralized architecture of the blockchain network and cryptographic hashing techniques, it is very challenging for hostile parties to alter transaction records. This immutability provides a strong foundation for protecting IoT data and transactions by ensuring data integrity and preventing unauthorized alterations.

Transparency and auditability are two additional key security features of blockchain technology. All network users can see and verify the entire history of transactions because of the blockchain's transparency. The transparency and accountability of every transaction recorded on the blockchain provide for improved auditability and accountability. This openness makes the environment for IoT applications more secure and reliable, fostering players' confidence and decreasing dependency on centralized intermediaries or authorities.

As blockchain technology relies on a decentralized network of nodes rather than a central authority or single point of failure, both central and single points are optional. The blockchain's consensus mechanism ensures that many users verify and concur that a transaction is correct. IoT systems are more secure because there are fewer points of vulnerability to intrusion or system failure due to decentralization. The decentralized nature of blockchain protects the integrity and availability of the overall system, even if some nodes or devices are compromised.

7.3.2 Key applications of blockchain in IoT security

Numerous crucial applications of blockchain technology are available to improve the security of IoT deployments. These applications handle numerous security issues and offer effective solutions for protecting IoT interactions, data, and devices.

7.3.2.1 Management of device identity and authentication

Blockchain technology is used to improve authentication and device identity management in the IoT. Blockchain-based authentication offers a secure and open mechanism for confirming IoT devices' authenticity by storing identity information on an immutable and decentralized blockchain ledger. Each device is given a distinct cryptographic key or identity, verified via cryptography compared to the identity record on the blockchain [25]. This decentralized strategy removes the flaws in centralized authentication systems, guarding against device spoofing and unlawful entry. Blockchain's openness improves accountability and makes it possible to trace device IDs over their entire life cycle, guaranteeing the integrity and dependability of IoT devices in a tamperproof and auditable manner [26].

7.3.2.2 Secure data transmission and communication

The IoT can substantially improve secure data transport and communication by incorporating blockchain technology. Blockchain preserves the confidentiality and integrity of data sent between IoT devices by combining encryption mechanisms. Direct and secure connections can be made through peer-to-peer communication made possible by blockchain, reducing the need for centralized intermediaries [27]. Blockchain consensus mechanisms validate and authenticate data exchanges, guaranteeing their veracity and avoiding tampering. In addition to enhancing the overall security and privacy of IoT connectivity, the immutable recordkeeping on the blockchain enables the creation of an auditable and visible history of communication. In IoT environments, blockchain's trust and transparency facilitate a strong framework for secure and trustworthy data movement and communication.

7.3.2.3 Storage of tamperproof data and auditability

A reliable solution for guaranteeing the accuracy and auditability of IoT data is provided by blockchain technology. Blockchain offers tamperproof storage of IoT data by enabling immutability, decentralization, and distributed consensus methods. This makes it nearly impossible for unauthorized parties to manipulate or alter the data. Data can be easily verified and traced because of the transparent and auditable nature of blockchain transactions and secure hash algorithms, giving stakeholders a trustworthy and reliable record of the data history [28]. The quality and dependability of the information shared within the IoT ecosystem are improved by blockchain's use of these qualities to provide IoT deployments with a safe and verifiable data storage solution.

7.3.2.4 Access management and permissions control

Blockchain technology can dramatically improve access control and permissions management in IoT ecosystems. IoT deployments can use blockchain to construct decentralized access control systems, ensuring safe and precise permissions management. Access rights can be properly established and enforced by smart contracts, eliminating the need for centralized authorities and lowering the likelihood of unwanted access [29]. For the sake of accountability and compliance, the openness and immutability of the blockchain offer an auditable record of access actions and create a tamperproof audit trail. Blockchain-based access control enables frictionless and secure interactions between IoT devices and stakeholders thanks to conditional access triggers and trustless interactions.

7.3.2.5 Compliance with regulations and standards

Regulatory compliance and standards are essential for IoT installations to be secure and trustworthy, and blockchain technology has significant advantages. The transparent and auditable nature of blockchain enables enterprises to automate compliance using smart contracts, boost supply chain traceability, preserve immutable audit trails, and accomplish data protection and privacy compliance. Organizations may build a solid foundation for regulatory compliance by utilizing blockchain, which ensures data integrity, consent management, and adherence to rules relevant to the industry [3]. While industry collaborations strive toward standardization and best practices, further encouraging secure and compliant IoT ecosystems, blockchain's transparent and decentralized nature builds confidence between enterprises and regulatory agencies.

7.3.2.6 Security and traceability of the supply chain

Blockchain technology has much to offer for improving the security and traceability of supply chains [30]. Stakeholders can ensure product authenticity, improve transparency, streamline tracking and traceability, enforce quality control and compliance, facilitate quick issue resolution, improve supplier management and accountability, secure payments through smart contracts, simplify regulatory compliance and audits, and ultimately, build reliable and trustworthy supply chain ecosystems by taking advantage of blockchain's decentralized and immutable nature. Blockchain technology's transparency and auditability make real-time tracking and verification of commodities possible, lowering the danger of fake goods and improving supply chain participant communication.

7.3.2.7 Enhanced collaboration on cybersecurity

Blockchain technology provides safe threat intelligence sharing, coordinated incident response, decentralized security operations centers (SOCs), and trustworthy threat data exchange platforms, improving cybersecurity collaboration [31]. Blockchain's immutability guarantees tamperproof forensics and audit trails, promoting compliance and accountability. Blockchain-based trust and verification methods improve stakeholder engagement, and adopting standards and best practices fosters consistency and interoperability. Organizations may build a strong cybersecurity ecosystem by utilizing blockchain, allowing for proactive defensive tactics, real-time information sharing, and coordinated responses to emerging cyber threats in our linked digital world.

7.3.2.8 Micropayments and secure transactions

The IoT can benefit from blockchain technology's secure and effective payment solutions by utilizing decentralized consensus and cryptographic algorithms. Blockchain enables direct peer-to-peer transactions between IoT devices, avoiding intermediaries and cutting expenses. Blockchain's openness and immutability guarantee transaction integrity and allow for real-time visibility and auditability. Smart contracts automate payments by predetermined criteria, enabling quick and easy settlements. Blockchain also supports new business models and monetization options in the IoT by enabling micropayments and machine-to-machine transactions [32]. Cross-border transactions are made possible by incorporating cryptocurrencies or tokens in IoT payments, improving security. Blockchain privacy features provide discretion while upholding payment transparency and confidence.

7.3.2.9 Updates to firmware and software

IoT device security and operation depend on regular firmware and software updates, and blockchain technology has many benefits in this area. Through cryptographic verification and consensus techniques, blockchain assures the secure dissemination of updates while prohibiting tampering and unauthorized modifications [33]. Storing modifications as immutable transactions, confirming their provenance, and avoiding compromised versions ensures their validity and integrity. Along with enabling reliable source verification, auditability, and transparency of the update process, blockchain also lowers risks and makes compliance easier. Blockchain improves IoT devices' general security, dependability, and longevity by guaranteeing they receive secure and effective upgrades. Blockchain features decentralized management, rollback mechanisms, and version control.

7.3.2.10 Management of consent and data privacy

Protecting the personal data gathered by IoT devices and creating clear consent methods are key components of data privacy and consent management in the IoT space. With its decentralized, immutable ledger and transparent, auditable data processing, blockchain technology offers healthy options to improve data privacy. It enables self-sovereign identity, allowing people to alter their consent preferences and govern their data. Blockchain-based smart contracts simplify consent management, guaranteeing that data usage follows predetermined guidelines. Blockchain can be coupled with other privacy-enhancing technologies like anonymization and data minimization strategies while still adhering to data protection laws [34]. Although there are still difficulties, blockchain for IoT data privacy and permission management empowers people, enhances privacy protections, and fosters confidence in the developing IoT ecosystem.

7.3.3 Comparative analysis of blockchain implementations for IoT security

It is critical to compare various blockchain designs and consensus processes when contemplating the use of blockchain for IoT security. This research aids in selecting the best blockchain options for resolving the unique needs and difficulties associated with IoT security. The comparison study should take into account several important elements.

Organizations may choose the best blockchain implementation for their IoT security needs by comparing scalability, performance, resource efficiency, and interoperability. They can choose a solution that balances the unique needs of IoT installations, providing smooth integration and efficient IoT security enhancement. Many elements must be considered when choosing a blockchain implementation for IoT security. Table 7.1 thoroughly analyzes and compares these factors and their implications, integration issues, and regulatory compliance concerns.

7.3.3.1 Comparing several blockchain systems and consensus methods

It is crucial to consider both the advantages and disadvantages of blockchain topologies and consensus processes when assessing their suitability for IoT security. Although public blockchains provide great security and decentralization, they may not be appropriate for IoT devices with limited resources due to scalability and privacy issues. Private blockchains offer more privacy and control, making them suitable for IoT deployments in business, but they also pose a risk of centralization. Blockchain consortiums are well suited for cooperative IoT applications because they achieve a compromise between centralization and decentralization. Hybrid blockchains enable cross-domain interoperability by combining the advantages of public and private/consortium blockchains, but their complexity necessitates careful design considerations.

Proof of work (PoW), one of the consensus processes, offers excellent security but is computationally and energetically demanding, making it less suitable for IoT devices. By choosing validators based on the amount of cryptocurrency they own, proof of stake (PoS) enables scalability and energy efficiency, but centralization has its hazards. Fast consensus is ensured by practical byzantine fault tolerance (PBFT), but scaling is constrained by validators' communication overhead. The high throughput and scalability of delegated proof of stake (DPoS) make it appropriate for large-scale IoT applications. However, the fact that it depends on a select group of reliable delegates raises questions about centralization [35].

Table 7.1 Factors, concerns, integration issues, and regulatory compliance in a comparative examination of blockchain solutions for IoT security

Factors to consider	*Considerations*	*Integration challenges*	*Regulatory compliance*
Scalability	Transaction throughput, sharding, off-chain processing	Network congestion, consensus algorithm selection	Compliance with data protection regulations, data privacy laws
Performance	Latency, response time, network bandwidth	Resource constraints, synchronization delay	Compliance with industry-specific regulations, data integrity requirements
Resource efficiency	Lightweight consensus algorithms, data compression, selective data storage	Limited computational power, storage capacity	Compliance with security standards, access control regulations
Interoperability	Compatibility with IoT systems, cross-chain communication protocols, standardization efforts	Protocol mismatches, data format conversion	Compliance with data sharing agreements, interoperability standards
Security	Cryptographic algorithms, consensus protocols, identity management systems	Vulnerability to 51% attacks, private key management	Compliance with data protection laws, secure data transmission
Governance and models	Governance framework, consensus mechanisms, decision-making processes, upgrade mechanisms	Decision-making conflicts, community governance	Compliance with regulatory reporting requirements, audit trails

Note: IoT, internet of things.

The choice of blockchain topologies and consensus methods for IoT security ultimately depends on several criteria, including scalability, privacy, resource efficiency, governance, and the IoT devices' characteristics. Organizations should carefully evaluate these considerations to balance security, scalability, and operational constraints. To meet the specific requirements of IoT environments, hybrid architectures that use private or consortium blockchains coupled with the proper consensus mechanisms may present a flexible solution. These architectures can guarantee data integrity, privacy, and regulated data exchange. Table 7.2

summarizes several blockchain designs, consensus processes, their evaluation for IoT security, and their applicability for various use cases, assisting in selecting the best blockchain solution for IoT security applications.

Table 7.2 Blockchain architectures and consensus processes compared for IoT security

Blockchain architecture	*Description*	*Evaluation for IoT security*	*Use case suitability*
Public blockchain	Open and permissionless network	• Not suitable for resource-constrained IoT devices • Transparency may raise privacy concerns for certain IoT applications	• IoT applications with less sensitivity to privacy and resource constraints
Private blockchain	Restricted participation and controlled by a central authority or consortium	• Improved scalability and privacy • Suitable for enterprise IoT deployments • Centralization risks and may not align with the decentralized nature of blockchain	• Enterprise IoT deployments where privacy and control are critical
Consortium blockchain	Governed by a consortium of organizations	• Balance between decentralization and control • Suitable for collaborative IoT applications involving multiple stakeholders • Governance model needs careful consideration to ensure fairness and security	• Collaborative IoT applications involving multiple stakeholders
Hybrid blockchain	Combination of public and private/ consortium blockchains	• Flexibility and interoperability in multi-domain IoT environments • Ensures data integrity and privacy while allowing controlled data sharing with external entities • Complexity requires careful design and implementation considerations	• IoT applications requiring interoperability across multiple domains

(Continued)

Table 7.2 (Continued)

Blockchain architecture	*Description*	*Evaluation for IoT security*	*Use case suitability*
Consensus Mechanisms Proof of work (PoW)	Validators solve computationally intensive puzzles	• High security but consumes significant computational resources and energy • Unsuitable for resource-constrained IoT devices	• IoT applications prioritizing high security and where computational resources are available
Proof of stake (PoS)	Validators selected based on ownership of cryptocurrency	• Energy efficient and scalable • Suitable for IoT deployments • Potential centralization risks if ownership concentration becomes an issue	• IoT applications prioritizing energy efficiency and scalability
Practical byzantine fault tolerance (PBFT)	Validators reach an agreement through a voting process	• Fast consensus and fault tolerance • Limited scalability due to communication overhead among validators	• IoT applications requiring fast consensus and fault tolerance
Delegated proof of stake (DPoS)	Trusted delegates validate transactions and produce blocks	• Scalable and high transaction throughput • Suitable for large-scale IoT applications • Relies on a small number of trusted delegates, raising concerns about centralization and potential collusion	• Large-scale IoT applications prioritizing high transaction throughput and trusted delegate–based consensus

Note: IoT, internet of things.

7.3.4 Scalability, performance, and resource efficiency considerations

Concerns are critical when integrating blockchain technology with IoT systems for improved security, scalability, performance, and resource efficiency. Supporting the enormous number of IoT devices producing data and processing transactions becomes crucial in blockchain scalability. Layer-two solutions, off-chain processing, and sharding are being investigated to increase scalability. Off-chain processing includes moving some transactions or calculations off the main blockchain, whereas sharding involves splitting the blockchain network into smaller divisions for parallel transaction processing. Layer-two solutions use side chains or extra layers to manage large transactions independently.

Blockchain performance—which includes transaction throughput and confirmation times—is essential for IoT applications. PoW, a common consensus mechanism, can be computationally expensive and slow down transaction speeds. Faster validation improves performance in IoT applications. Alternative consensus algorithms include PoS, DPoS, and PBFT [36]. Efficient data structures, optimal data networks, and optimized protocols improve blockchain performance.

Resource efficiency is a crucial factor for IoT devices with limited resources. IoT blockchain systems need to balance security and resource efficiency. Computing needs and energy usage can be decreased by using lightweight consensus techniques like proof of authority (PoA) or directed acyclic graph (DAG) architectures [37]. Additionally, while maintaining the integrity of the blockchain, data compression methods like Merkle tree architectures or data pruning can lower storage requirements [38].

IoT blockchain integration is significantly influenced by network and hardware factors. IoT devices frequently have little storage, memory, and computing capabilities. Blockchain technologies should be created with these limitations in mind. Optimized communication protocols and network bandwidth allocation are crucial to handle increasing data traffic brought on by blockchain integration without creating bottlenecks. Cryptographic techniques and hardware-accelerated encryption can guarantee data security while reducing the computing load on IoT devices [39].

It might be challenging to balance scalability, performance, and resource efficiency when combining blockchain with IoT for security reasons. For blockchain solutions to successfully satisfy the requirements of IoT contexts, careful architecture design, consensus mechanism selection, protocol improvements, and consideration of hardware and network restrictions are required. By considering these factors, blockchain and IoT can increase security without sacrificing operational effectiveness. The factors for integrating blockchain technology with IoT security are shown in Figure 7.2, focusing on scalability, performance, resource efficiency, and hardware/network limitations.

7.3.5 Use of smart contracts for secure and automated transactions in IoT

Blockchain technology–enabled smart contracts provide improved security and automation for transactions within the IoT [40]. Smart contracts offer a trustless and tamperproof environment utilizing blockchain's immutability and decentralized characteristics. They do away with intermediaries, lowering the possibility of fraud and guaranteeing the accuracy of transactional data. Smart contracts allow for smooth and independent transactions between IoT devices thanks to automated execution based on established circumstances [41, 42]. Automation improves efficiency and lowers costs by removing the need for intermediaries and streamlining processes [43].

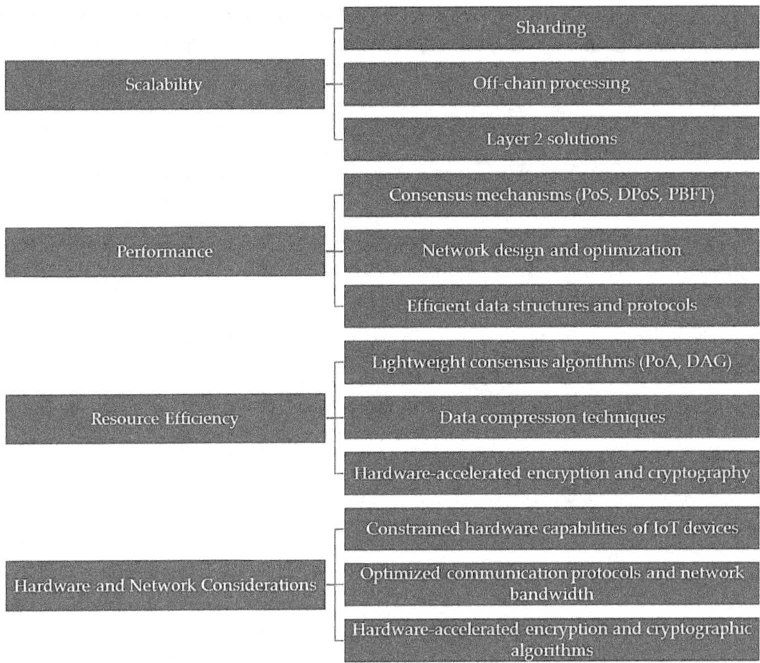

Figure 7.2 Considerations for blockchain integration with IoT security.

Transparency and trust in IoT transactions are two additional benefits of smart contracts [44]. All transactional operations are recorded on the shared, unchangeable ledger offered by blockchain, allowing participants to check the accuracy of the transactions and guarantee that contractual commitments are fulfilled. This openness fosters confidence among parties and creates a solid framework for IoT interactions. In addition, smart contracts reduce waiting times by automating contractual procedures and enabling real-time settlement, which boosts productivity and lowers costs.

Oracles enable the incorporation of real-world data, enhancing the possibilities of IoT smart contracts [45]. Oracles link external systems and the blockchain, allowing smart contracts to use real-time conditions for transaction execution. IoT device interactions can now be dynamic and context sensitive thanks to this capability. In addition, predetermined dispute resolution processes can be incorporated into smart contracts to resolve conflicts quickly and openly, obviating the need for manual involvement and third-party mediators. Standardized smart contract protocols and frameworks support a scalable and interoperable IoT ecosystem, allowing seamless integration between various IoT platforms and devices.

While there are many benefits to using smart contracts in IoT transactions, there are still some difficulties, such as dealing with data privacy issues in the public blockchain environment and scaling issues with blockchain technology.

Nevertheless, the union of smart contracts with blockchain presents a potent instrument for protecting and automating IoT transactions. It creates new opportunities for IoT system implementation regarding trust, effectiveness, and innovation.

7.4 LIMITATIONS AND CHALLENGES

Limitations and difficulties in integrating blockchain with IoT security need to be resolved for widespread adoption. Given the vast scope of IoT deployments and the possible impact on blockchain's transaction processing capabilities, scalability is a major worry. To increase the scalability of blockchains in IoT contexts, researchers are actively investigating techniques like sharding and layer-two protocols. Additionally, energy efficiency is a significant factor, particularly for IoT devices with limited resources. Consensus methods can be improved, and energy-saving protocols might be investigated to reduce the energy consumption brought on by blockchain integration in IoT contexts.

7.4.1 Scalability and performance considerations

When integrating blockchain with IoT systems, there are substantial hurdles related to scalability and performance. Due to the decentralized nature of blockchain technology and the consensus procedures used, scalability issues may affect the speed at which transactions are processed and the overall system throughput. It is essential to solve these issues as IoT transactions and devices proliferate. Traditional blockchain networks, like bitcoin and ethereum, have scalability problems since each transaction needs to be processed and validated by every node. As a result, there are restrictions on how many transactions can be handled per second, which causes bottlenecks in IoT systems that experience a huge influx of data. Researchers are investigating several solutions to this problem.

Sharding is a method that divides the blockchain network into smaller shards, each of which can handle a subset of transactions. This parallel processing makes higher transaction throughput possible since multiple shards can handle transactions simultaneously. Another method is to off-load transactions from the main blockchain using layer-two protocols, such as payment channels and side chains, to reduce congestion and improve scalability.

To improve scalability, improvements in consensus techniques are also being pursued. Although secure, the PoW consensus method requires many resources and is computationally expensive. PoS and DPoS, two different consensus algorithms, provide a better transaction throughput while using less energy. These technologies enable quicker transaction processing by allocating block validation responsibilities depending on stake ownership or delegation. Research and development aim to improve blockchain scalability for the IoT by integrating sharding, layer-two protocols, and better consensus methods. With the help of these strategies, blockchain will be able to handle the enormous scale and high transaction volumes that characterize IoT installations and meet the growing demands of IoT systems.

7.4.2 Energy efficiency and resource constraints

When combining blockchain technology with IoT systems, there are substantial hurdles related to energy efficiency and resource limitations. IoT devices frequently run on constrained power sources, like batteries, and have limited computational and memory resources. PoW, a traditional blockchain consensus process, might be energy intensive and unsuitable for IoT devices with limited resources.

Researchers are looking toward more energy-efficient alternatives to the current consensus procedures to overcome these issues. Consensus algorithms like PoS and DPoS are two examples that use much less energy than PoW. To obtain agreement, these strategies take advantage of network users' stake, lowering the computing load and energy usage.

Energy efficiency can also be increased by improving data transmission and storage inside blockchain-enabled IoT architectures. Off-chain transactions, selective data storage, and data compression are some methods that can assist in lowering the energy consumption required for blockchain activities. Energy consumption can be reduced while keeping the security and integrity offered by blockchain technology by outsourcing some tasks to dependable intermediaries or using edge computing capabilities.

IoT devices' limited memory and processing capabilities need to be considered when addressing resource constraints. Blockchain protocol optimization can assist in saving money by decreasing redundant data storage and reducing the size of transaction blocks. Additionally, by adapting the technology to the specific needs and limits of IoT devices, lightweight blockchain frameworks created for the IoT, such as distributed ledger technologies or IoT-oriented blockchains, might alleviate resource constraints.

7.4.3 Regulatory and legal implications

Significant regulatory and legal issues relating to IoT security need to be resolved for a blockchain solution to be implemented successfully. Legal frameworks need to develop at the same rate as blockchain technology to ensure compliance and safeguard the rights and privacy of people and organizations involved in blockchain-enabled IoT systems.

Data protection and privacy are major issues. IoT devices produce enormous volumes of data, and because blockchains are dispersed, managing and restricting access to this data takes work. It becomes essential to ensure adherence to data protection laws, such as the General Data Protection Regulation (GDPR) in the European Union. It is important to work on creating mechanisms that provide people with choices over how their data is utilized and shared in blockchain-based IoT systems. To solve these privacy problems, privacy-enhancing technologies can be investigated, such as zero-knowledge proofs and selective data disclosure methods.

Additionally, legal frameworks must address data ownership and liability concerns in IoT ecosystems powered by blockchain. New legal issues are introduced by smart contracts, which automate and enforce contracts on the blockchain. When employing decentralized blockchain systems, determining legal duty and accountability in the event of contract disputes or failures becomes complicated. To ensure legal compliance and safeguard the interests of all parties concerned, it is essential to establish legal processes to handle these circumstances and specify the parameters of contractual responsibilities within the context of blockchain technology.

Additional regulatory considerations include anti-money laundering (AML) and know-your-customer (KYC) requirements, intellectual property rights, consumer protection, and cross-border transactions. To build clear legislative frameworks that address these complicated challenges while supporting innovation and the secure deployment of blockchain-enabled IoT devices, policymakers, legal experts, and industry stakeholders need to work together. Pilot programs and regulatory sandboxes can promote experimentation and learning while ensuring regulatory oversight and compliance.

7.5 FUTURE RESEARCH OPPORTUNITIES AND DIRECTIONS

The combination of blockchain technology and IoT security opens up various fascinating research possibilities and avenues for further investigation. These research directions can help the field advance and handle the changing difficulties in securing IoT using blockchain. Some significant chances for future research include the following areas.

7.5.1 Privacy-preserving techniques

Privacy-preserving approaches are crucial to guarantee the security and confidentiality of data in blockchain-enabled IoT systems. The importance of protecting privacy increases with the enormous volume of sensitive information created and exchanged by IoT devices. Several research trajectories and methods can be investigated to improve privacy in this situation. Differential privacy is one method to keep individual-level information private by adding controlled noise to the data. Homomorphic encryption is another method that permits computations to be made directly on encrypted data, which protects privacy while handling sensitive data. Zero-knowledge proofs offer a way to validate a claim without disclosing extra information, which can be used to protect privacy in blockchain-based IoT systems. Secure multiparty computation provides collaborative analysis while safeguarding sensitive data by allowing shared computations on private inputs. Without exchanging raw data, devices may train machine learning models thanks to federated learning, and privacy-preserving smart contracts make it possible to carry out critical actions securely without disclosing personal information. Secure data aggregation approaches can also summarize IoT data while protecting individual privacy.

It is possible to dramatically improve the security and confidentiality of data in IoT deployments by researching and developing these privacy-preserving solutions. Individual-level information is kept private through the use of differential privacy approaches, and computations on encrypted data are made secure through the use of homomorphic encryption. Data can be verified using secure multiparty computing and zero-knowledge proofs without revealing sensitive information. Federated learning and blockchain technology can be combined to enable collaborative model training while protecting the privacy of individual data. Smart contracts that protect privacy guarantee the safe execution of transactions. Last, but not least, safe data aggregation algorithms provide data summaries while protecting individual privacy, allowing for useful insights without revealing sensitive data.

Collectively, these privacy-preserving methods help address privacy issues in blockchain-enabled IoT systems. Users can have more faith in the security and privacy of their data by adopting and developing these measures. This makes it easier for blockchain-enabled IoT solutions to be adopted more widely, enabling the IoT's advantages while protecting personal information. The development of strong and privacy-enhancing solutions at the convergence of blockchain and IoT security will be aided by ongoing research and innovation in privacy-preserving strategies.

7.5.2 Scalability solutions

Due to the enormous amount of data and numerous linked devices, scalability is a significant hurdle when integrating blockchain with IoT. To overcome this issue, researchers are looking into several scaling methods. Sharding dramatically boosts transaction performance by dividing the blockchain network into smaller chunks known as shards. To increase scalability, side chains, independent blockchains compatible with the main blockchain, off-load transactions. Peer-to-peer off-chain transactions are made possible through state channels, which lighten the load on the central network. Optimizations to consensus protocols like PoW and PoS are being looked into to boost throughput and cut energy use. Scalability is improved through layer-two solutions like payment channels and state channel networks, which off-load transactions from the primary blockchain. Hybrid strategies integrate many tactics to play to each one's strengths successfully. Researchers are also looking into innovative blockchain architectures, including DAG structures, to attain great scalability.

However, it is crucial to consider each solution's trade-offs and constraints. To assess these scalability solutions' efficiency, security, and usability in various IoT use cases, ongoing research, standardization initiatives, and collaboration within the blockchain community are required. These solutions address scaling issues, paving the path for the efficient and safe integration of blockchain with IoT systems, allowing the full promise of blockchain technology to protect the IoT to be realized.

7.5.3 Interoperability and standards

Interoperability and standards are essential to guarantee seamless integration and cooperation between IoT devices and blockchain platforms. Future studies should concentrate on standardizing message structures, data formats, and communication protocols to promote interoperability between various blockchain networks and IoT devices. To enable smooth data sharing and integration, efforts should also be focused on building methods for translating, manipulating, and mapping data. Research should also look at the frameworks and interoperability protocols that allow various blockchain networks to communicate with one another and share information, promoting cross-chain interoperability in IoT contexts. To do this, problems with identity management, authentication, access control, and interoperability of smart contracts need to be addressed. Establishing common interoperability standards and certification procedures is crucial for ensuring the compatibility and compliance of IoT devices and blockchain platforms. Collaborative industry initiatives and testing methodologies are also crucial.

7.5.4 Smart contract security

IoT systems powered by blockchain have smart contracts as essential components that automate and enforce contracts and transactions. However, protecting their security is necessary to avoid monetary losses, privacy violations, and service interruptions. Future studies on smart contract security should concentrate on formal verification, secure development practices, auditing and certification, upgradeability and maintenance, user-friendly interfaces and tools, standardization, security frameworks, and initiatives to raise public awareness. These initiatives will help create robust and trustworthy smart contracts, lowering the risk of vulnerabilities and improving the overall security of IoT systems using smart contracts.

Automated technologies for vulnerability discovery, formal verification methods for thorough validation, and secure development procedures for producing trustworthy smart contracts are all needed in smart contract security. While investigating methods for upgradeability and developing user-friendly interfaces will promote the development and management of safe contracts, auditing and certification processes can ensure adherence to security standards. Standardized security frameworks can be established through collaborative efforts with industry stakeholders, and the adoption of best practices will be encouraged through education and awareness campaigns.

7.5.5 Edge computing and blockchain integration

Edge computing and blockchain integration is a new field of study that combines the advantages of localized data processing and storage at the network edge with the security and immutability provided by blockchain technology. This integration has the potential to provide various benefits for IoT systems, including reduced latency, increased efficiency, enhanced privacy, and increased scalability. Researchers are investigating architectural models for integrating edge computing and blockchain and frameworks for efficient data processing, storage, and consensus processes at the network's edge. Additionally, existing distributed ledger technologies (DLTs), such as blockchain or DAG platforms, are being researched for edge computing. To address the restricted resources and dynamic nature of edge devices, optimization of consensus methods, data synchronization protocols, and storage techniques is required.

Edge computing and blockchain integration requires privacy-preserving approaches. Researchers are investigating solutions such as secure multiparty computation, federated learning, and differential privacy to protect data privacy and confidentiality in edge-based blockchain systems. The goal is to provide systems that allow efficient data sharing and analysis while protecting sensitive IoT data privacy. Researchers focus on approaches such as partitioning, caching, and data compression to solve scalability issues and enhance blockchain data storage and processing at the edge. Lightweight consensus algorithms are being developed to reduce blockchain consensus edge devices' computational and energy needs. The implications for trust and security are also being investigated, focusing on potential vulnerabilities, attack vectors, and remedies specific to edge-enabled blockchain networks. Secure communication protocols, identity management methods, and rigorous authentication schemes are being developed to protect the integrity and trustworthiness of blockchain network edge devices.

Finally, academics look into real-world applications and use cases for various industries' edge computing and blockchain integration. Edge computing with blockchain can provide major benefits in sectors such as supply chain management, energy grids, autonomous vehicles, and smart cities. This research aims to create efficient, scalable, and secure systems that take advantage of the benefits of edge computing and blockchain technologies in the context of IoT deployments.

7.5.6 Governance models

Governance models are of the utmost importance to guarantee the efficient functioning, management, and security of blockchain-enabled IoT systems. It is necessary to create suitable governance models that handle this developing industry's particular difficulties and demands as the integration of blockchain and the IoT progresses. First, studies should investigate decentralized governance structures compatible with the decentralized nature of blockchain and the IoT. With the use of mechanisms for consensus-based decision-making, community-driven policies, and dispute resolution, these models enable network participants to decide on policies and enforce them jointly.

Reward schemes are another important factor in encouraging active engagement, data exchange, and security contributions in blockchain-based IoT networks. IoT device owners and network validators, among other stakeholders, should be encouraged to contribute their resources and skills by designing and implementing appropriate incentive models, according to research that should be done in this area. These approaches can improve network security and dependability by coordinating participant interests with system-wide objectives.

Additionally, governance models for blockchain-enabled IoT should consider privacy and data governance on a fundamental level. Research should examine methods for privacy-preserving governance systems that safeguard sensitive IoT data while facilitating open and transparent governance procedures. To reduce risks and preserve trust inside the network, identity verification, reputation systems, and accountability mechanisms are also considered.

In addition, governance models need to be flexible and adaptable to consider new technical developments, legislative changes, and network requirements. Research should concentrate on creating flexible frameworks to accommodate novel consensus processes, take participant feedback into account, and accommodate shifting network dynamics. Exploring collaborative governance models will help players in the blockchain-enabled IoT ecosystem work together and coordinate. To achieve this, industry, academia, governments, and end users need to work together to create governance frameworks that are inclusive, open, and responsive.

7.5.7 AI and machine learning

When combined with blockchain technology, AI and machine learning offer a huge potential to improve the security of the IoT. To develop blockchain-enabled IoT security, research in this field might concentrate on several important directions. Anomaly detection is a key area of study, which uses AI and machine learning algorithms to examine data trends and spot IoT device and network activity that deviates from expected patterns. This proactive strategy makes real-time detection of potential security issues or malicious activity possible. Additionally, by analyzing enormous volumes of data gathered from IoT devices, blockchain transactions, and outside sources, AI and machine learning approaches can identify emerging threats and predict possible weaknesses.

Intelligent decision-making is another critical area that can benefit from AI and machine learning in the context of blockchain-enabled IoT security. Access control, anomaly response, and threat mitigation decisions can be made using AI models by analyzing data from various sources, such as blockchain transactions, IoT device data, and security logs. This method of intelligent decision-making enables more effective and efficient security measures. AI that protects privacy is another important research area. AI and machine learning computations on sensitive IoT data can be made possible while maintaining privacy thanks to federated learning, secure multiparty computation, and differential privacy. This guarantees that IoT data privacy is upheld even when performing AI-driven security assessments.

Combining blockchain and AI can let IoT devices share threat intelligence. IoT devices may safely exchange threat intelligence in a trusted and rewarding way by using the decentralized nature of blockchain and utilizing smart contracts, adding to a collective defense against emerging dangers. Research can also examine how AI and machine learning algorithms are vulnerable to adversarial assaults in IoT security and develop strong defenses to guarantee the dependability and robustness of AI models used in blockchain-enabled IoT systems.

7.6 CONCLUSION

A thorough study of the dangers, current security precautions, and the possibilities of blockchain technology in IoT device security is provided in the chapter on IoT security challenges and vulnerabilities. It underscores the significance of addressing the risks of infected IoT devices and the requirement for strong security measures. The chapter investigates how blockchain technology can improve IoT security by utilizing its security properties like decentralization, immutability, and transparency. It highlights the principal uses of blockchain in IoT security, such as safe data exchange, device identity management, and secure firmware updates. The pros and cons of various blockchain implementations for IoT security are shown through a comparative study. The chapter also covers the application of smart contracts in the IoT environment for safe and automated transactions. It illustrates how smart contracts may uphold confidence, do away with intermediaries, and guarantee the accuracy of exchanges between IoT devices.

However, it is critical to recognize the constraints and difficulties associated with applying blockchain technology in the IoT space. Significant barriers to wider adoption include the necessity for consensus methods, high processing requirements, and scalability difficulties. The chapter emphasizes the importance of additional research and development to solve these issues. Future research possibilities and directions for improving IoT security are highlighted in this chapter. To improve the security of IoT devices, it recommends investigating unique consensus methods, creating lightweight blockchain solutions, and incorporating other cutting-edge technologies like AI and machine learning.

REFERENCES

[1] J. Gubbi, R. Buyya, S. Marusic, and M. Palaniswami, "Internet of Things (IoT): A vision, architectural elements, and future directions," *Future Generation Computer Systems*, vol. 29, no. 7, pp. 1645–1660, 2013.

[2] A. Al-Fuqaha, M. Guizani, M. Mohammadi, M. Aledhari, and M. Ayyash, "Internet of things: A survey on enabling technologies, protocols, and applications," *IEEE Communications Surveys & Tutorials*, vol. 17, no. 4, pp. 2347–2376, 2015.

[3] P. Cui, U. Guin, A. Skjellum, and D. Umphress, "Blockchain in IoT: Current trends, challenges, and future roadmap," *Journal of Hardware and Systems Security*, vol. 3, pp. 338–364, 2019.

[4] A. Zanella, N. Bui, A. Castellani, L. Vangelista, and M. Zorzi, "Internet of things for smart cities," *IEEE Internet of Things Journal*, vol. 1, no. 1, pp. 22–32, 2014.

[5] S. R. Islam, D. Kwak, M. H. Kabir, M. Hossain, and K.-S. Kwak, "The internet of things for health care: A comprehensive survey," *IEEE Access*, vol. 3, pp. 678–708, 2015.

[6] H. Taherdoost, "Blockchain-based internet of medical things," *Applied Sciences*, vol. 13, no. 3, p. 1287, 2023.

[7] P. J. Taylor, T. Dargahi, A. Dehghantanha, R. M. Parizi, and K.-K. R. Choo, "A systematic literature review of blockchain cyber security," *Digital Communications and Networks*, vol. 6, no. 2, pp. 147–156, 2020.

[8] D. Minoli and B. Occhiogrosso, "Blockchain mechanisms for IoT security," *Internet of Things*, vol. 1, pp. 1–13, 2018.

[9] P. Sunhare, R. R. Chowdhary, and M. K. Chattopadhyay, "Internet of things and data mining: An application oriented survey," *Journal of King Saud University-Computer and Information Sciences*, vol. 34, no. 6, pp. 3569–3590, 2022.

[10] H. Hellani, L. Sliman, A. E. Samhat, and E. Exposito, "On blockchain integration with supply chain: Overview on data transparency," *Logistics*, vol. 5, no. 3, p. 46, 2021.

[11] H. F. Atlam, A. Alenezi, M. O. Alassafi, and G. Wills, "Blockchain with internet of things: Benefits, challenges, and future directions," *International Journal of Intelligent Systems and Applications*, vol. 10, no. 6, pp. 40–48, 2018.

[12] O. I. Abiodun, E. O. Abiodun, M. Alawida, R. S. Alkhawaldeh, and H. Arshad, "A review on the security of the internet of things: Challenges and solutions," *Wireless Personal Communications*, vol. 119, pp. 2603–2637, 2021.

[13] M. Litoussi, N. Kannouf, K. El Makkaoui, A. Ezzati, and M. Fartitchou, "IoT security: Challenges and countermeasures," *Procedia Computer Science*, vol. 177, pp. 503–508, 2020.

[14] L. Tawalbeh, F. Muheidat, M. Tawalbeh, and M. Quwaider, "IoT Privacy and Security: Challenges and Solutions," *Appl. Sci.* vol. 10, no. 12, p. 4102, 2020, https://doi.org/10.3390/app10124102.

[15] M. B. Barcena and C. Wueest, "Insecurity in the Internet of Things," *Security Response, Symantec*, vol. 20, 2015.

[16] C. Maple, "Security and privacy in the Internet of Things," *Journal of Cyber Policy*, vol. 2, no. 2, pp. 155–184, 2017.

[17] S. Sokolov, V. Gaskarov, T. Knysh, and A. Sagitova, "IoT security: threats, risks, attacks," in *Proceedings of the XIII International Scientific Conference on Architecture and Construction 2020: Commemorating the 90th anniversary of Novosibirsk State University of Architecture and Civil Engineering*, 2020, pp. 47–56: Springer.

[18] P. Kumari and A. K. Jain, "A comprehensive study of DDoS attacks over IoT network and their countermeasures," *Computers & Security*, p. 103096, 2023.

[19] Y. Al-Hadhrami and F. K. Hussain, "DDoS attacks in IoT networks: A comprehensive systematic literature review," *World Wide Web*, vol. 24, no. 3, pp. 971–1001, 2021.

[20] I. Stellios, P. Kotzanikolaou, M. Psarakis, C. Alcaraz, and J. Lopez, "A survey of IoT-enabled cyberattacks: Assessing attack paths to critical infrastructures and services," *IEEE Communications Surveys & Tutorials*, vol. 20, no. 4, pp. 3453–3495, 2018.

[21] M. Asam *et al.*, "IoT malware detection architecture using a novel channel boosted and squeezed CNN," *Scientific Reports*, vol. 12, no. 1, pp. 1–12, 2022.

[22] F. Thabit, O. Can, A. O. Aljahdali, G. H. Al-Gaphari, and H. A. Alkhzaimi, "A comprehensive literature survey of cryptography algorithms for improving the IoT security," *Internet of Things*, p. 100759, 2023.

[23] A. Kumar, K. Abhishek, M. Ghalib, A. Shankar, and X. Cheng, "Intrusion detection and prevention system for an IoT environment," *Digital Communications and Networks*, vol. 8, no. 4, pp. 540–551, 2022.

[24] Q.-D. Ngo, H.-T. Nguyen, V.-H. Le, and D.-H. Nguyen, "A survey of IoT malware and detection methods based on static features," *ICT Express*, vol. 6, no. 4, pp. 280–286, 2020.

[25] A. R. Kairaldeen, N. F. Abdullah, A. Abu-Samah, and R. Nordin, "Peer-to-peer user identity verification time optimization in IoT blockchain network," *Sensors*, vol. 23, no. 4, p. 2106, 2023.

[26] A. Rashid, A. Masood, and A. U. R. Khan, "Zone of trust: Blockchain assisted IoT authentication to support cross-communication between bubbles of trusted IoTs," *Cluster Computing*, vol. 26, no. 1, pp. 237–254, 2023.

[27] S. P. Sankar, T. Subash, N. Vishwanath, and D. E. Geroge, "Security improvement in block chain technique enabled peer to peer network for beyond 5G and internet of things," *Peer-to-Peer Networking and Applications*, vol. 14, no. 1, pp. 392–402, 2021.

[28] F. Casino, T. K. Dasaklis, and C. Patsakis, "A systematic literature review of blockchain-based applications: Current status, classification and open issues," *Telematics and Informatics*, vol. 36, pp. 55–81, 2019.

[29] A. Khatoon, "A blockchain-based smart contract system for healthcare management," *Electronics*, vol. 9, no. 1, p. 94, 2020.

[30] H. Taherdoost and M. Madanchian, "Blockchain-based new business models: A systematic review," *Electronics*, vol. 12, no. 6, p. 1479, 2023.

[31] O. M. Latvala *et al.*, "Proof-of-Concept for a Granular Incident Management Information Sharing Scheme," in *2022 IEEE World AI IoT Congress (AIIoT)*, 2022, pp. 515–520: IEEE.

[32] J. Seppälä, "The role of trust in understanding the effects of blockchain on business models," Master's thesis, 2016.

[33] R. Zhang, R. Xue, and L. Liu, "Security and privacy on blockchain," *ACM Computing Surveys (CSUR)*, vol. 52, no. 3, pp. 1–34, 2019.

[34] S.-C. Cha, T.-Y. Hsu, Y. Xiang, and K.-H. Yeh, "Privacy enhancing technologies in the Internet of Things: Perspectives and challenges," *IEEE Internet of Things Journal*, vol. 6, no. 2, pp. 2159–2187, 2018.

[35] Y. Wen, F. Lu, Y. Liu, P. Cong, and X. Huang, "Blockchain consensus mechanisms and their applications in IoT: A literature survey," in *Algorithms and Architectures for Parallel Processing: 20th International Conference, ICA3PP 2020, New York City, NY, USA, October 2–4, 2020, Proceedings, Part III 20*, 2020, pp. 564–579: Springer.

[36] S. Aggarwal and N. Kumar, "Cryptographic consensus mechanisms," in *Advances in Computers*, vol. 121: Elsevier, 2021, pp. 211–226.

[37] S. M. H. Bamakan, A. Motavali, and A. B. Bondarti, "A survey of blockchain consensus algorithms performance evaluation criteria," *Expert Systems with Applications*, vol. 154, p. 113385, 2020.

[38] N. K. Akrasi-Mensah *et al.*, "An overview of technologies for improving storage efficiency in blockchain-based IIoT applications," *Electronics*, vol. 11, no. 16, p. 2513, 2022.

[39] B. Zhou, M. Egele, and A. Joshi, "High-performance low-energy implementation of cryptographic algorithms on a programmable SoC for IoT devices," in *2017 IEEE High Performance Extreme Computing Conference (HPEC)*, 2017, pp. 1–6: IEEE.

[40] T. Dasaklis and F. Casino, "Improving vendor-managed inventory strategy based on Internet of Things (IoT) applications and blockchain technology," in *2019 IEEE International Conference on Blockchain and Cryptocurrency (ICBC)*, 2019, pp. 50–55: IEEE.

[41] Y. Hu, M. Liyanage, A. Mansoor, K. Thilakarathna, G. Jourjon, and A. Seneviratne, "Blockchain-based smart contracts-applications and challenges," arXiv preprint arXiv:1810.04699, 2018.

[42] N. Teslya and I. Ryabchikov, "Blockchain-based platform architecture for industrial IoT," in *2017 21st Conference of Open Innovations Association (FRUCT)*, 2017, pp. 321–329: IEEE.

[43] A. Cohn, T. West, and C. Parker, "Smart after all: Blockchain, smart contracts, parametric insurance, and smart energy grids," *Geo. L. Tech. Rev.*, vol. 1, p. 273, 2016.

[44] E. S. Negara, A. N. Hidayanto, R. Andryani, and R. Syaputra, "Survey of smart contract framework and its application," *Information*, vol. 12, no. 7, p. 257, 2021.

[45] G. Caldarelli and C. Rossignoli, "Beyond Oracles–A critical look at real-world blockchains," Doctoral Thesis, 2022.

Chapter 8

Data traffic management in AI-IoT network to reduce congestion

Muhammad Usama, Ubaid Ullah, Zaid Muhammad, Muhammad Bux, Dr. Inam Ullah, Muhammad Rouf, and Taminul Islam

8.1 INTRODUCTION

The internet of things (IoT), a revolutionary technology that links diverse objects and allows them to interact and share data easily, has arisen in recent years [1, 2]. The rapid proliferation of artificial intelligence–internet of things (AI-IoT) devices, such as sensors, actuators, and intelligent appliances, has exponentially increased data traffic within AI-IoT networks [3]. However, this surge in data transmission poses significant challenges, including congestion and network inefficiencies [4]. Effective data traffic management ensures AI-IoT networks' smooth functioning and optimal performance [5]. Congestion, caused by the overwhelming amount of data being transmitted simultaneously, can lead to latency issues, packet loss, reduced throughput, and compromised reliability [6]. Therefore, developing robust strategies and mechanisms to mitigate congestion is paramount in the AI-IoT landscape [7].

In an AI-IoT network, data traffic management is crucial in reducing congestion and optimizing device communication [8]. The volume of data created and transferred by the growing number of connected devices in AI-IoT networks can soon exceed the network infrastructure, causing congestion and performance problems [9]. To address this challenge, several approaches can be implemented. One common strategy is prioritization and quality of service (QoS), which involves assigning different priority levels to various data types [10]. This ensures that critical data is delivered promptly while less time-sensitive data is given lower priority, effectively managing traffic flow. Another approach is traffic segmentation, where the network is divided into logical segments or VLANs, allowing different types of devices or applications to operate independently [11]. This helps isolate traffic and prevents congestion from affecting the entire network [12]. Bandwidth management techniques are also employed to allocate resources effectively [13]. By setting limits on bandwidth for devices or applications, fair sharing of resources is ensured, preventing any single entity from monopolizing the available bandwidth and causing congestion [14]. The volume of data created and transferred by the growing number of connected devices in AI-IoT networks can soon exceed

DOI: 10.1201/9781003496410-10

the network infrastructure, causing congestion and performance problems. Compressing data before transmission and aggregating smaller data packets reduces the overall traffic volume, minimizing congestion. Overall, efficient data traffic management in an AI-IoT network reduces congestion and ensures smooth and reliable communication between devices [15].

This chapter explores AI-IoT networks' data traffic management concept and its role in reducing congestion [16]. It will delve into the challenges associated with congestion in AI-IoT networks, examining the unique characteristics and requirements that differentiate AI-IoT traffic from traditional network traffic [17]. Furthermore, the chapter will present various approaches and techniques to address congestion issues and enhance data transmission efficiency in AI-IoT networks [18]. The significance of this research lies in its potential to optimize AI-IoT network performance, facilitate real-time data processing, and enable the smooth functioning of AI-IoT applications and services [19]. By reducing congestion and improving data traffic management, the reliability and responsiveness of AI-IoT networks can be enhanced, allowing for seamless communication and interaction among interconnected devices [19, 20].

The subsequent sections of this chapter will delve into the current state of data traffic management in AI-IoT networks, exploring existing congestion control mechanisms, resource allocation strategies, and QoS techniques [20]. Furthermore, emerging trends, challenges, and future directions in this field will be discussed, providing valuable insights for researchers, practitioners, and policymakers working toward advancing AI-IoT network infrastructure [21, 22]. In summary, this chapter highlights the importance of data traffic management in AI-IoT networks to alleviate congestion-related challenges [21]. By investigating existing congestion control techniques and exploring potential solutions, it aims to contribute to the development of more efficient and reliable AI-IoT network architectures [23, 24].

Ultimately, reducing congestion and improving data traffic management will play a crucial role in harnessing the full potential of the IoT and ensuring its seamless integration into our daily lives [25, 26]. Figure 8.1 provides a valuable reference for understanding the complexity and interconnectivity involved in harnessing the potential of the IoT.

8.2 LITERATURE

Data traffic management in AI-IoT networks reduces congestion and ensures efficient communication among connected devices [27]. Congestion has become a significant challenge with the proliferation of AI-IoT devices and the exponential growth of the data they generate. Consequently, researchers and industry experts have focused on developing strategies and techniques to address this issue [28]. Several literature sources have explored the topic of data traffic management in AI-IoT networks. One such study by Zhang et al. proposed a congestion control

Figure 8.1 Elements encompassed within the internet of things (AI-IoT) ecosystem.

mechanism designed explicitly for AI-IoT environments. The authors suggested a distributed approach that employed adaptive rate control and prioritization techniques to regulate the flow of data packets. Their system effectively reduced congestion by dynamically adjusting the data transmission rates based on network conditions and device requirements [29, 30].

Another research article by Wang et al. investigated software-defined networking (SDN) for congestion management in AI-IoT networks. The authors proposed an SDN-based architecture that centralized network control and allowed for dynamic traffic routing and load balancing. Their approach effectively optimized data flow by leveraging real-time traffic monitoring and intelligent decision-making algorithms, reducing congestion and improving network performance [31]. Additionally, a study conducted by Lei Bai et al. focused on traffic prediction and scheduling techniques for congestion management in AI-IoT networks. The authors developed a predictive model that analyzed historical data patterns to forecast traffic loads. Based on these predictions, they proposed an adaptive scheduling algorithm that efficiently allocated network resources and managed data traffic to prevent congestion and ensure smooth communication among AI-IoT devices [32].

Furthermore, research by Cheng Feng et al. explored the application of edge computing in data traffic management for AI-IoT networks. The authors suggested an edge-based architecture that transferred data processing responsibilities from main cloud servers to edge units near the data source. This method minimized network congestion and delay by reducing the data transmitted across the network. By distributing computational tasks and optimizing data traffic, their solution improved overall network performance and reduced congestion-related issues [33]. Hon Fong Chong et al. present the development of an AI-IoT device specifically designed for a traffic management system. The device integrates various sensors, communication modules, and data processing capabilities to collect real-time traffic data, analyze it, and provide intelligent insights for efficient traffic management. It highlights the importance of AI-IoT technology in enhancing traffic management systems and discusses the developed device's design considerations and functionalities. The AI-IoT device enables the system to monitor traffic conditions, detect congestion points, and optimize traffic flow through timely interventions. The study contributes to intelligent transportation systems (ITS) by implementing AI-IoT technology practically for effective traffic management [34, 35]. Table 8.1 provides a concise overview of the standard techniques employed in data traffic management in AI-IoT networks to reduce congestion, along with a description and limitations.

Table 8.1 Standard techniques used in data traffic management in AI-IoT networks to reduce congestion, along with a description and limitations

References	*Technique used*	*Description*	*Limitations*
[36, 37]	Queue management	These papers propose a queue management technique for data traffic in AI-IoT networks to reduce congestion. It employs intelligent algorithms to prioritize and manage incoming data packets based on their criticality and network conditions.	Limited scalability in large-scale AI-IoT networks.
[38, 39]	Traffic off-loading	These studies present a traffic off-loading technique that diverts data traffic from congested AI- IoT nodes to less-burdened nodes within the network. It utilizes dynamic routing algorithms to optimize data distribution and reduce congestion.	Requires additional network overhead for routing and coordination.

(Continued)

Table 7.2 (Continued)

References	*Technique used*	*Description*	*Limitations*
[40–42]	Load balancing	These researches explore a load balancing approach for data traffic in AI-IoT networks. It distributes the data load evenly across multiple AI-IoT devices, ensuring efficient utilization of network resources and minimizing congestion.	May introduce additional latency due to the redistribution of data traffic.
[43–46]	Quality-of-service (QoS) mapping	These papers propose a QoS mapping technique that assigns different service levels to data traffic based on priority and requirements. It ensures that critical data receives higher priority, reducing congestion and improving the overall network performance.	Requires a comprehensive understanding of the AI-IoT application's QoS requirements.
[47–49]	Traffic prediction	These studies focus on traffic prediction techniques for AI-IoT networks. They predict future traffic loads by analyzing historical traffic patterns and employing machine learning algorithms. This information enables proactive traffic management and congestion avoidance strategies.	Relies on accurate historical data for effective prediction, which may not always be available.

Note: AI, artificial intelligence; IoT, internet of things.

8.3 OBJECTIVES OF DATA TRAFFIC MANAGEMENT IN AI-IOT NETWORK

The objectives of data traffic management in AI-IoT networks to reduce congestion are as follows.

The first objective is to understand the factors contributing to congestion in AI-IoT networks. This involves studying various elements such as network architecture, data traffic patterns, device density, and resource limitations. Appropriate solutions can be developed by identifying the root causes of congestion [50]. Effective congestion management requires detecting and monitoring congestion in real-time [51]. The objective is to develop robust algorithms and techniques to detect congestion accurately, measure its severity, and monitor network performance metrics [52]. This information is crucial for implementing appropriate congestion control strategies. The primary objective is to develop and implement congestion control mechanisms that dynamically allocate network resources based on demand and prioritize critical data traffic. This involves designing algorithms and protocols that regulate data flow, manage traffic congestion, and ensure fair resource allocation among AI-IoT devices. The aim is to optimize network utilization and reduce congestion-induced performance degradation [53].

Traffic shaping and prioritization are essential in managing data traffic to minimize congestion. The objective is to investigate techniques that shape and prioritize data flows based on predefined policies and QoS requirements [54]. This includes assigning priorities to different types of traffic, applying traffic management policies, and efficiently utilizing network resources to mitigate congestion. Data compression and aggregation techniques can reduce the amount of data transmitted over the network, thereby alleviating congestion [55]. The objective is to explore compression algorithms and aggregation strategies that reduce data size while preserving the necessary information [56]. This helps in minimizing the data traffic volume and optimizing network performance. Computing brings computational capabilities closer to AI-IoT devices, reducing the need for transmitting massive amounts of data to centralized cloud servers. The objective is to investigate how edge computing can be leveraged to perform data processing and analysis at the network edge, thereby reducing data traffic and congestion in AI-IoT networks [57].

The objective is to evaluate and assess the effectiveness of implemented congestion management strategies. This involves analyzing performance metrics such as throughput, latency, packet loss, and QoS to determine the impact of congestion reduction techniques on overall network performance. This evaluation optimizes congestion management strategies. By achieving these objectives, data traffic management in AI-IoT networks can effectively reduce congestion, enhance network performance, ensure reliable data transmission, and improve the overall AI-IoT user experience [58].

8.4 SIGNIFICANCE OF DATA TRAFFIC MANAGEMENT IN AI-IOT NETWORKS

The significance of data traffic management in AI-IoT networks is crucial for several reasons.

8.4.1 Optimal network performance

Effective data traffic management ensures optimal network performance in AI-IoT environments. By efficiently managing the data flow, congestion can be minimized, resulting in improved network responsiveness, reduced latency, and enhanced overall system performance. This is especially important for time-sensitive applications such as real-time monitoring, autonomous vehicles, and industrial control systems, where delays or disruptions can have severe consequences [59].

8.4.2 Reliable data transmission

In AI-IoT networks, reliable data transmission is essential for accurate decision-making and timely actions. Congestion can lead to packet loss, increased transmission errors, and reduced data integrity. By implementing appropriate traffic management techniques, such as congestion control and prioritization, the reliability of data transmission can be enhanced, ensuring that critical data reaches its destination without compromise [60, 61].

8.4.3 Enhanced quality of service

AI-IoT applications often require specific levels of service quality to meet their operational requirements. Data traffic management is crucial in maintaining and delivering the desired QoS in AI-IoT networks. By allocating network resources based on QoS parameters, such as bandwidth, latency, and reliability, traffic management helps meet application-specific requirements and ensures a consistent and satisfactory user experience [62].

8.4.4 Efficient resource utilization

AI-IoT networks typically involve many connected devices with varying resource requirements. Proper data traffic management allows for efficient resource utilization, ensuring network resources, such as bandwidth and processing power, are allocated optimally based on demand. This leads to improved network capacity utilization, reduced resource wastage, and cost-savings [63].

8.4.5 Scalability and scalable growth

Scalability is a critical aspect of AI-IoT networks, as the number of connected devices is expected to grow exponentially. Data traffic management techniques facilitate the scalability of AI-IoT networks by managing the increasing volume of data traffic. Through congestion reduction and efficient resource allocation, these techniques enable networks to handle the growing number of connected devices without compromising performance [64, 65].

8.4.6 Improved security and privacy

Data traffic management can also enhance the security and privacy of AI-IoT networks. By implementing traffic shaping, access control, and encryption mechanisms, traffic management helps protect sensitive data and prevents unauthorized access or malicious activities. It enables the enforcement of security policies and ensures compliance with privacy regulations, safeguarding the integrity and confidentiality of AI-IoT data [66, 67].

8.4.7 Economic and societal impact

Efficient data traffic management in AI-IoT networks can have significant economic and societal benefits. Businesses can improve operational efficiency, enhance productivity, and deliver better customer service by reducing congestion and optimizing network performance. Additionally, in sectors such as health care, transportation, and smart cities, reliable AI-IoT networks enable innovative applications that improve safety, resource management, and overall quality of life [68, 69].

8.4.8 AI-IoT network architecture

AI-IoT network architecture refers to the structure and arrangement of components within an AI-IoT network, including how devices, gateways, cloud platforms, and other elements are organized and interconnected [70]. Figure 8.2 showcases the AI-IoT network architecture, illustrating the corresponding structure and components that enable seamless communication and data exchange within an AI-IoT ecosystem. It serves as a visual representation of the underlying framework that enables the integration and synchronization of AI-IoT technologies to create more innovative and efficient urban environments and homes.

There are several common architectural approaches in AI-IoT networks.

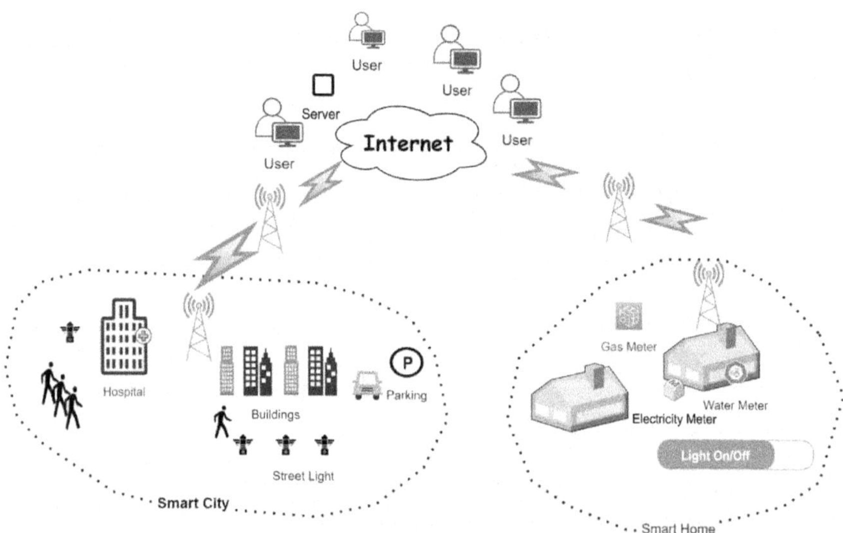

Figure 8.2 AI-IoT network architecture for smart city and smart home integration.

8.4.8.1 Centralized architecture

In a centralized architecture, all the data processing, storage, and control functions are concentrated in a central server or cloud platform. Devices typically send data to the main system, which performs data analysis, storage, and decision-making. This architecture offers simplicity and centralized control but may face scalability and latency challenges when dealing with large volumes of data [71, 72].

8.4.8.2 Edge computing architecture

Edge computing architecture distributes data processing and analytics functions closer to the devices at the network edge. Edge devices or gateways perform local data processing and filtering, reducing the need for transmitting large amounts of raw data to the central system. Edge computing improves latency and bandwidth usage and reduces the dependency on cloud connectivity. It is beneficial in applications requiring real-time or low-latency processing [73].

8.4.8.3 Fog computing architecture

Fog computing architecture extends edge computing, introducing intermediate computing and storage capabilities between the edge and the central cloud. Fog nodes, placed near the devices or gateways, perform local processing and filtering of data. They also provide services like data caching, security, and real-time analytics. This architecture reduces latency, optimizes bandwidth, and provides a more distributed and scalable system than centralized architectures [74].

8.4.8.4 Hybrid architecture

A hybrid architecture combines elements of both centralized and distributed approaches. It leverages cloud platforms for storage, complex analytics, and long-term data retention while utilizing edge or fog computing for real-time processing, local control, and reduced latency. This architecture balances centralized management and distributed processing capabilities [75].

8.4.8.5 Mesh architecture

In a mesh architecture, devices in the AI-IoT network communicate directly, forming a self-configuring and self-healing network. This approach eliminates the need for a central gateway, making the network more resilient and scalable. Mesh architectures are commonly used in AI-IoT applications like smart homes, industrial automation, and smart cities, where devices must communicate efficiently [76].

8.4.8.6 *Hierarchical architecture*

Hierarchical architecture involves organizing devices into multiple tiers or levels based on their proximity to the central system. Each tier may have gateways or aggregation points that collect and transmit data from a subset of devices. This architecture allows for scalable and efficient data aggregation and reduces the burden on the central system. It is commonly used in large-scale deployments, where devices are spread across different geographical areas [77].

The choice of AI-IoT network architecture depends on factors such as the specific use case, data processing requirements, latency constraints, scalability needs, and available resources. Different architectures offer trade-offs in data latency, bandwidth usage, processing capabilities, and system complexity, and the selection should be based on the specific requirements of the AI-IoT application [78, 79].

8.5 OVERVIEW OF AI-IOT NETWORKS

AI-IoT networks are groups of networked gadgets, sensors, and systems that talk to one another and share information online.

These networks allow physical items and digital technology to coexist in various settings, including smart homes, health care, transportation, industrial automation, agriculture, and other areas [80–84].

8.5.1 Overview of AI-IoT network components

8.5.1.1 *AI-IoT devices/things*

AI-IoT devices or things are physical objects embedded with sensors, actuators, and connectivity capabilities. These devices collect data from the surrounding environment or perform actions based on received instructions. Examples include intelligent sensors, wearables, home appliances, industrial machinery, and vehicles [85].

8.5.1.2 *Connectivity technologies*

AI-IoT devices rely on different connectivity technologies to communicate with each other and with other network components. These technologies include Wi-Fi, Bluetooth, cellular networks (3G, 4G, 5G), LPWAN (low power wide area network) technologies like LoRaWAN and NB-AI-IoT, Zigbee, Z-wave, and Ethernet [86].

8.5.1.3 *Network infrastructure*

The network infrastructure consists of routers, switches, gateways, and access points that enable the transmission of data between AI-IoT devices and other network components. It provides the necessary connectivity and supports data routing, traffic management, and network security [87].

8.5.1.4 Cloud services

AI-IoT networks depend heavily on cloud systems. They provide scalable storage, data processing, and computational resources. AI-IoT devices can send data to the cloud for archiving, processing, and integrating with applications. Cloud services facilitate remote device management, software updates, and data sharing among stakeholders [88]. Figure 8.3 showcases the architecture of cloud computing, providing a high-level overview of its components and interactions. It highlights the fundamental elements involved in cloud computing.

8.5.1.5 Edge computing

Edge computing is an essential component of AI-IoT networks, bringing computing capabilities closer to AI-IoT devices. Edge computing nodes, such as edge servers or gateways, perform data processing, analytics, and decision-making tasks locally, near the devices, or at the network edge. This approach reduces latency, improves real-time responsiveness, and reduces the need for transmitting large amounts of data to the cloud [89]. Figure 8.4 visually represents the fundamental differences between edge and cloud computing. It highlights each approach's key characteristics and functionalities, clearly delineating their unique roles in the modern computing landscape, by visually depicting their disparities.

Figure 8.3 Cloud computing architecture overview.

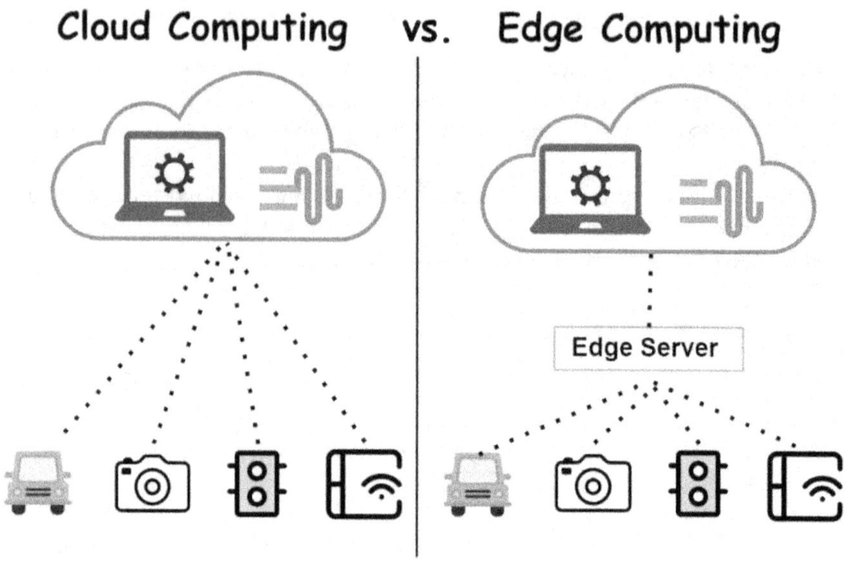

Figure 8.4 Illustrating the distinction between edge and cloud computing.

8.5.1.6 Data analytics and AI-AI-IoT networks

Data analytics and AI-AI-IoT networks generate vast amounts of data. Data analytics and AI techniques extract valuable insights, detect patterns, and make informed decisions based on the collected data. Advanced analytics and AI algorithms can provide predictive maintenance, anomaly detection, optimization, and intelligent automation in AI-IoT applications [89].

8.5.1.7 Security and privacy

Due to interconnected devices' potential vulnerabilities and risks, security is a critical aspect of AI-IoT networks. AI-IoT networks require robust security measures to protect devices, data, and the overall network infrastructure from unauthorized access, data breaches, and cyberattacks. This includes encryption, authentication mechanisms, access control, and secure communication protocols [90].

8.5.1.8 Standards and protocols

AI-IoT networks use standardized protocols to guarantee compatibility and smooth communication between various systems and devices. Message queuing telemetry transport (MQTT), CoAP (constrained application protocol), HTTP (hypertext transfer protocol), and wireless technology–specific protocols like Zigbee and Z-wave are among the most widely used AI-IoT protocols [91].

The overview of AI-IoT networks highlights the interconnectedness of devices; the importance of connectivity technologies, cloud services, edge computing, data analytics, and security; and the standardization efforts required for seamless integration and effective communication among AI-IoT components. These networks have the potential to transform industries, improve efficiency, and enable innovative applications with their ability to gather and analyze real-time data from diverse sources [91, 92].

8.5.2 Importance of congestion reduction in AI-IoT networks

These interconnected systems' effectiveness, dependability, and scalability are crucially dependent on decreasing congestion in AI-IoT networks. To maintain smooth operations and provide seamless user experiences, it is crucial to efficiently manage congestion as the number of connected devices continues to increase quickly [93]. One of the primary reasons why congestion reduction is vital in AI-IoT networks is the sheer volume of data generated and transmitted by interconnected devices. AI-IoT networks encompass many applications, including smart homes, industrial automation, health care monitoring, and smart cities. Each application creates vast amounts of data, from sensor readings to real-time monitoring data, which must be efficiently transmitted and processed. Congestion occurs when the network infrastructure becomes overwhelmed with this massive influx of data, leading to delays, packet loss, and degraded performance [94].

Congestion reduction in AI-IoT networks is of utmost importance due to the significant volume of data generated and the need for low-latency communication, resource efficiency, data integrity, security, and scalability. By implementing congestion reduction techniques and employing robust network management strategies, AI-IoT networks can deliver reliable, efficient, and secure communication, facilitating the seamless integration of interconnected devices and unlocking the full potential of the IoT [95].

8.5.3 Causes of congestion in AI-IoT networks

Congestion in AI-IoT networks can arise due to various factors and causes. Here are some of the common reasons that contribute to congestion in AI-IoT networks.

8.5.3.1 High device density

The increasing number of connected devices within an AI-IoT network can overload the network infrastructure and lead to congestion. As more devices join the network, the available bandwidth gets divided, potentially causing delays and performance degradation [96].

8.5.3.2 Limited bandwidth

AI-IoT networks often operate within constrained bandwidth environments, especially in wireless communication scenarios. The limited bandwidth becomes a bottleneck when numerous devices attempt to transmit data simultaneously, resulting in congestion [97].

8.5.3.3 Bursty traffic patterns

AI-IoT applications generate data sporadically, with bursts of activity during specific periods. For example, multiple devices, such as sensor readings or control commands, in a smart home environment, may simultaneously send data updates. These bursty traffic patterns can overwhelm the network and lead to congestion [97].

8.5.3.4 Inefficient routing

Inefficient routing protocols or network configurations can contribute to congestion in AI-IoT networks. Suboptimal routing decisions may cause data packets to take longer routes or encounter bottlenecks, leading to increased latency and congestion [98].

8.5.3.5 Lack of quality-of-service mechanisms

AI-IoT networks often encompass diverse applications with varying QoS requirements. Without appropriate QoS mechanisms, critical data packets may not receive the necessary prioritization, resulting in congestion and degraded performance for time-sensitive applications [99].

8.5.3.6 Network interference

In wireless AI-IoT networks, interference from other devices or networks operating on the same frequency bands can lead to congestion. Interference can cause packet loss, retransmissions, and increased collisions, reducing the overall network capacity and causing congestion [100].

8.5.3.7 Security attacks

Malicious activities such as distributed denial of service (DDoS) attacks or botnets targeting AI-IoT devices can overwhelm the network infrastructure and cause congestion. These attacks flood the network with malicious traffic, disrupting regular communication and degrading performance [101].

8.5.3.8 Lack of scalability

AI-IoT networks need to accommodate the ever-increasing number of devices and scale efficiently. If the network infrastructure is not designed to handle the growing demands, congestion can occur as the network struggles to cope with the increased traffic and device density [102].

8.5.3.9 Inadequate network management

Poor network management practices, such as insufficient monitoring, inefficient resource allocation, or failure to adapt to changing network conditions, can contribute to congestion in AI-IoT networks. Without proactive management and optimization, congestion-related issues may arise [103].

Addressing congestion in AI-IoT networks requires effective network design, intelligent traffic management, QoS mechanisms, efficient routing protocols, and robust security measures. By understanding the causes of congestion and implementing appropriate mitigation strategies, network operators can ensure smooth operations, reliable communication, and optimal performance in AI-IoT environments [104].

8.5.4 Impact of congestion on AI-IoT network performance

Congestion in AI-IoT networks can significantly impact network performance, leading to various challenges and disruptions. Here are some of the critical effects of congestion on AI-IoT network performance.

8.5.4.1 Increased latency

Congestion can introduce delays in data transmission, resulting in increased latency. The time taken for data packets to travel from the source to the destination can significantly increase, affecting real-time applications that require prompt responses. High latency can hinder critical AI-IoT applications such as remote monitoring, control systems, or autonomous vehicles, where timely actions are essential [105].

8.5.4.2 Packet loss

Congestion can cause packet loss in AI-IoT networks. When the network infrastructure becomes overloaded, it may be unable to handle the incoming data packets effectively. This can result in packets being dropped or discarded, leading to data loss and reduced reliability. Packet loss can adversely affect applications that require accurate and complete data, such as health care monitoring or industrial automation systems [105].

8.5.4.3 Reduced throughput

Congestion limits the available network bandwidth, which can significantly impact the throughput of AI- IoT networks. As more devices contend for the limited resources, the overall data transfer rate decreases, reducing throughput. Slower data transmission can impede the performance of AI-IoT applications that rely on high data rates, such as video streaming or large-scale data analytics [105].

8.5.4.4 Degraded quality of service

AI-IoT networks often host diverse applications with varying QoS requirements. Congestion can disrupt the expected QoS levels, causing degradation in service performance. Time-sensitive applications that require low latency and high reliability may experience compromised QoS, affecting their effectiveness and usability [105, 106].

8.5.4.5 Inefficient resource utilization

Congestion leads to inefficient utilization of network resources in AI-IoT networks. Devices may need to retransmit data due to packet loss or increased latency, consuming additional energy and network capacity. This inefficient resource utilization can increase power consumption, reduce battery life for connected devices, and decrease overall network efficiency [105].

8.5.4.6 Security vulnerabilities

Congestion can create security vulnerabilities in AI-IoT networks. When the network becomes congested, it may become more accessible for attackers to exploit the situation and launch malicious activities. For example, congestion can cover DDoS attacks, where the network's performance is deliberately degraded by flooding it with malicious traffic [97].

8.5.4.7 Scalability challenges

Congestion can hinder the scalability of AI-IoT networks. As the number of connected devices grows, congestion-related issues become more pronounced, limiting the network's ability to handle the increasing demands. Congestion can impede the expansion of the network and the seamless integration of new devices and services. Addressing congestion and its impact on AI-IoT network performance requires implementing effective congestion management techniques, such as traffic shaping, prioritization, and load balancing. Additionally, employing scalable network architectures, optimizing resource allocation, and ensuring efficient routing can help mitigate the adverse effects of congestion, ensuring optimal performance and reliable communication in AI-IoT environments [18, 107].

8.5.5 Challenges in managing congestion in AI-IoT networks

Managing congestion in AI-IoT networks presents several challenges due to these interconnected systems' unique characteristics and requirements. Here are some of the critical challenges in effectively managing congestion in AI-IoT networks.

8.5.5.1 Heterogeneity of devices and applications

AI-IoT networks comprise diverse devices and applications with varying communication requirements, QoS needs, and data patterns. Managing congestion becomes challenging when dealing with different types of devices, generating different amounts of data and data types. Balancing the network resources to accommodate the varying demands of these heterogeneous devices is a complex task [108].

8.5.5.2 Scalability

AI-IoT networks are expected to accommodate many devices, posing scalability challenges in congestion management. As the number of devices increases, the network must efficiently handle the growing traffic without causing congestion. Scaling congestion management mechanisms to address the increasing demands while maintaining performance becomes a significant challenge [109].

8.5.5.3 Limited resources

Many AI-IoT devices operate with limited resources such as power, memory, and processing capabilities. Implementing congestion management techniques in resource-constrained devices can be challenging, as they may have limitations in handling complex algorithms or additional overheads introduced by congestion management mechanisms [109].

8.5.5.4 Real-time requirements

Some AI-IoT applications, such as industrial automation or autonomous vehicles, have strict real-time requirements. Managing congestion while ensuring low-latency communication becomes crucial in these scenarios. Prioritizing time-sensitive data and providing guarantees for timely delivery pose challenges in congested AI-IoT networks [109].

8.5.5.5 Dynamic network conditions

AI-IoT networks often experience dynamic changes in network conditions, such as device mobility, intermittent connectivity, and varying signal strengths.

These conditions can exacerbate congestion management challenges, as the network needs to adapt to changing circumstances and mitigate congestion on the fly [110].

8.5.5.6 Security concerns

Congestion management mechanisms need to coexist with security measures in AI-IoT networks. They ensure secure and authenticated communication while managing congestion, which adds complexity to network management. Balancing congestion management and security can be challenging, as some congestion management techniques may be vulnerable to security threats or attacks [110].

8.5.5.7 Quality-of-service differentiation

AI-IoT networks typically support a wide range of applications with different QoS requirements. Ensuring proper QoS differentiation and resource allocation to prioritize critical data flows over less time-sensitive ones is challenging. Allocating network resources efficiently to meet the diverse QoS requirements while mitigating congestion requires intelligent traffic management mechanisms [110].

8.5.5.8 Interoperability and standards

AI-IoT networks involve devices from various manufacturers and vendors, which may use different communication protocols and standards. Achieving interoperability and standardized congestion management techniques across heterogeneous AI-IoT devices can be challenging. Ensuring seamless communication and congestion management in a multi-vendor AI-IoT environment requires harmonization of protocols and standards. Addressing these challenges requires a holistic approach to congestion management in AI-IoT networks. It involves intelligent traffic management, dynamic resource allocation, scalable architectures, adaptive routing algorithms, and standardized protocols. Moreover, ongoing research and collaboration among industry stakeholders are crucial for developing robust and efficient congestion management solutions tailored to the unique characteristics of AI-IoT networks [110, 111].

8.6 DATA TRAFFIC MANAGEMENT TECHNIQUES

Data traffic management techniques are crucial in optimizing network performance, ensuring efficient resource utilization, and mitigating congestion in various networks, including AI-IoT networks. Here are some commonly used data traffic management techniques.

8.6.1 Traffic shaping

Traffic shaping limits the pace at which packets are sent to manage the data flow. Congestion is less likely since it helps level out erratic traffic patterns and avoids unexpected increases in data transfer. Traffic prioritization, queuing algorithms, and rate-limiting devices are examples of traffic-shaping methods [112].

8.6.2 Quality-of-service differentiation

QoS differentiation allows for prioritizing different types of traffic based on their importance or requirements. Critical or time-sensitive applications can be guaranteed sufficient network resources and minimized latency by assigning different priority levels to data flows. At the same time, non-critical traffic can be allocated lower priority [112].

8.6.3 Load balancing

Load balancing distributes network traffic across multiple paths or resources to prevent congestion and optimize resource utilization. It ensures the workload is evenly distributed, avoiding bottlenecks and maximizing network efficiency. Load balancing techniques can be implemented at various levels, including at the network, transport, or application layer [62].

8.6.4 Traffic engineering

Traffic engineering involves optimizing network resources and paths to enhance performance and minimize congestion. It uses routing protocols, traffic analysis, and network topology information to dynamically adjust traffic flows, reroute traffic, or allocate resources efficiently. Traffic engineering techniques can include route optimization, diversion, and flow control.

8.6.5 Content delivery networks

Content delivery networks (CDNs) are widely used to improve content delivery, especially for web-based applications. CDNs store and distribute content across geographically distributed servers, reducing the distance between users and content servers. CDNs reduce latency, improve responsiveness, and alleviate network congestion by delivering content from servers closer to the users.

8.6.6 Caching

Keeping frequently used data or content close to users or in convenient places throughout the network is known as caching. Through caching, lesser data needs to be retrieved from the source, which lowers network traffic and speeds up responses. Techniques for caching can be used at many levels, such as web caching, content caching, and edge caching [113].

8.6.7 Packet prioritization

Prioritizing packets based on their importance or urgency is crucial for time-sensitive applications. By prioritizing critical data packets, such as real-time sensor data or control signals, congestion-related delays can be minimized, ensuring timely delivery and responsiveness [114].

8.6.8 Deep packet inspection

Deep packet inspection (DPI) is a technique that inspects network packet content and structure at a granular level. DPI allows for intelligent traffic analysis and management based on specific application requirements, protocols, or traffic patterns. It enables traffic shaping, policy enforcement, and identification of potential congestion triggers [115].

8.6.9 Adaptive and machine learning–based techniques

Adaptive techniques, such as adaptive queuing algorithms or adaptive routing, dynamically adjust parameters based on real-time network conditions. Machine learning–based approaches analyze historical traffic patterns and network data to predict congestion or optimize traffic management decisions [114].

8.6.10 Policy-based traffic management

Policy-based traffic management enables the application of specific regulations and guidelines to control traffic behavior. These guidelines specify the network's classification, prioritization, and traffic handling. A policy may be based on the application type, user identification, or service-level agreements. In AI-IoT networks and other network contexts, using an efficient mix of these data traffic management strategies may significantly increase network performance, optimize resource use, and decrease congestion. The approach will rely on the network's needs, traits, limitations, and applications. [116].

8.7 CONGESTION DETECTION AND MONITORING

Congestion detection and monitoring are critical to managing and maintaining network performance in various environments, including AI-IoT networks. By effectively identifying and monitoring congestion, network operators can take proactive measures to mitigate its impact and ensure smooth operations [117, 118].

Here are some standard methods and techniques used for congestion detection and monitoring.

8.7.1 Network traffic analysis

Network traffic analysis involves monitoring and analyzing the flow of data packets within the network. Congestion can be detected by examining network traffic patterns, packet rates, and utilization levels. Packet sniffing, flow analysis, and statistical analysis can provide insights into congestion indicators, including increased packet loss, high latency, or elevated network utilization [118].

8.7.2 Packet delay measurement

Measuring packet delays, precisely the roundtrip time (RTT) or one-way delay, can help identify real-time congestion. By comparing the expected delay with the actual delay experienced by packets, congestion-related delays can be detected. Techniques such as ping, traceroute, or network performance monitoring tools can be used to measure and monitor packet delays [118].

8.7.3 Queue length monitoring

Monitoring the length of network queues can provide indications of congestion. Queues at routers or network switches tend to accumulate packets when congestion occurs. By tracking the queue lengths and observing trends, network administrators can identify periods of congestion and take appropriate measures to alleviate it [118, 119].

8.7.4 Bandwidth utilization monitoring

Monitoring the network's bandwidth utilization can help identify periods of congestion. By measuring the percentage of available bandwidth used, spikes or sustained high utilization levels can indicate congestion. Bandwidth monitoring tools or network performance monitoring systems can provide insights into bandwidth utilization patterns [118, 120].

8.7.5 Performance metrics analysis

Analyzing various performance metrics, such as throughput, latency, packet loss rate, or retransmission rates, can provide valuable information about congestion. By comparing current performance metrics with baseline measurements or predefined thresholds, congestion-related anomalies can be detected, triggering appropriate congestion management actions [118].

8.7.6 Flow-level monitoring

Flow-level monitoring involves analyzing network flows, representing a series of related packets between specific source and destination end points. By examining flow-level information, such as flow rate, duration, or size, congestion patterns and potential congested flows can be identified. Flow-monitoring tools or network flow analyzers can assist in flow-level congestion detection [118, 121].

8.7.7 Anomaly detection

Anomaly detection techniques involve identifying deviations from normal network behavior, which can indicate congestion or other network issues. By establishing baseline models of normal network behavior and comparing current observations with these models, anomalies related to congestion can be detected. Machine learning algorithms or statistical analysis can be applied for anomaly detection [118].

8.7.8 Real-time monitoring and alarms

Real-time monitoring of crucial congestion indicators, such as packet loss, latency, or queue lengths, can trigger alarms or notifications when congestion thresholds are exceeded. Network management systems can be configured to generate alerts, allowing network administrators to take immediate action to address congestion issues [118, 122].

8.7.9 Probing and active measurement

Probing techniques involve injecting test traffic or specific probe packets into the network to measure performance or detect congestion. Active measurement tools can send probe packets and monitor their behavior to assess network conditions and congestion levels. Combining multiple congestion detection and monitoring techniques can provide a comprehensive view of network congestion in AI-IoT networks and enable timely responses to alleviate congestion-related issues. By leveraging these techniques, network operators can optimize network performance, improve user experiences, and ensure the efficient operation of AI-IoT applications [123].

8.8 CONGESTION CONTROL AND MITIGATION STRATEGIES

Network performance management and reliable data transfer rely heavily on congestion control and mitigation measures. Congestion occurs when network resources are overtaxed, which results in subpar performance, packet loss, and increased delay. There are several approaches used to deal with this problem. Network overflow is avoided thanks to traffic shaping, which controls data flow inside a network [124]. QoS techniques prioritize mission-critical data over other forms of traffic. Congestion may be reduced by using traffic engineering, which optimizes resource allocation and makes real-time changes to traffic patterns. To avoid overload, admission control limits the number of concurrent sessions or connections [125]. Devices may regulate transmission rates thanks to technologies that detect and communicate congestion circumstances. Load balancing spreads traffic among many contacts or servers to avoid overload. The discard policy determines which packets should be lost due to congestion. Multipath routing makes use of several pathways to distribute and safeguard traffic. Finally, upgrading infrastructure and using new technologies to increase available bandwidth helps reduce congestion. These congestion management and mitigation measures help the network function effectively and reliably [126].

8.8.1 Congestion control algorithms for AI-IoT networks

Congestion control algorithms are crucial in managing network congestion in AI-IoT networks. Due to the large number of devices and the diverse traffic patterns in AI-IoT environments, traditional congestion control mechanisms may not be directly applicable [8].

Here are some congestion control algorithms designed explicitly for AI-IoT networks.

8.8.1.1 Random early detection

Random early detection (RED) is a commonly used congestion control algorithm that can also be adapted for AI-IoT networks. It monitors the average queue length at routers and selectively drops packets when the queue exceeds a certain threshold. RED can help regulate traffic and prevent congestion by signaling the end points to reduce their transmission rates [127].

8.8.1.2 Congestion control for small packets

Congestion control for small packets (CCSP) is an algorithm designed to handle the unique characteristics of AI-IoT networks, which often involve small-sized packets with low data rates. It aims to maintain fairness and prevent congestion by regulating the transmission rates of devices based on their packet sizes and delay requirements [128].

8.8.1.3 Adaptive beacon congestion control

Adaptive beacon congestion control (ABCC) is a congestion control algorithm designed explicitly for wireless AI-IoT networks. It utilizes periodic beacon messages exchanged between devices to estimate the congestion level in the network.

Based on this information, devices adjust their transmission rates to avoid congestion and ensure efficient data transfer [129].

8.8.1.4 Priority-based congestion control

Priority-based congestion control (PBCC) assigns different priority levels to different types of AI-IoT traffic based on their criticality or importance. Devices with higher priority have a more significant share of network resources, ensuring that crucial data is transmitted with minimal delay and congestion [130].

8.8.1.5 Distributed congestion control

Distributed congestion control (DCC) is a decentralized congestion control algorithm for AI-IoT networks. It enables devices to adjust their transmission rates autonomously based on local congestion indications. By utilizing local information, DCC helps prevent global congestion by distributing the traffic load across devices in the network [131].

8.8.1.6 CoAP-based congestion control

Constrained application protocol is widely used in AI-IoT environments. CoAP-based congestion control algorithms leverage the specific features of CoAP, such as reliable and unreliable transmission modes, to manage congestion effectively. These algorithms regulate transmission rates and adapt to network conditions to prevent congestion and ensure reliable communication [132].

8.8.1.7 Machine learning–based congestion control

Machine learning techniques can be applied to AI-IoT networks for congestion control. By analyzing network traffic patterns, machine learning models can identify congestion indicators and dynamically adjust transmission rates or reroute traffic to mitigate congestion effectively. It's worth noting that congestion control in AI-IoT networks is an active area of research. New algorithms are continuously being developed to address the unique challenges AI-IoT environments present. The choice of congestion control algorithm depends on the specific characteristics of the AI-IoT network, types of devices, traffic patterns, and requirements of the AI-IoT applications being deployed [133].

8.9 DYNAMIC RESOURCE ALLOCATION

Dynamic resource allocation refers to allocating and reallocating computing resources such as CPU, memory, storage, and network bandwidth dynamically and flexibly based on the changing needs of applications or services. It involves efficiently utilizing and optimizing available resources to meet the demands of workload fluctuations and varying priorities. Dynamic resource allocation is particularly relevant in cloud computing environments, where resources are shared among multiple users or tenants. It enables cloud providers to optimize resource utilization, improve performance, and ensure cost-efficiency [134].

Here are some key aspects and techniques associated with dynamic resource allocation.

8.9.1 Elasticity

Elasticity is a fundamental characteristic of dynamic resource allocation. It allows the automatic scaling of resources up or down based on workload demands. When the demand increases, additional resources are allocated to handle the increased load. Conversely, when the demand decreases, resources are released to avoid unnecessary resource consumption [135].

8.9.2 Monitoring and load balancing

Effective dynamic resource allocation requires continuous system and application metrics monitoring to understand the current resource utilization and workload patterns. Load balancing algorithms are then used to distribute the workload evenly across available resources, preventing overloading on specific resources and optimizing resource utilization [135].

8.9.3 Auto-scaling

Auto-scaling is a mechanism that automatically adjusts the resource allocation based on predefined policies or rules. When certain conditions, such as CPU utilization or response time thresholds, are met, auto-scaling triggers the provisioning or de-provisioning of resources to meet the changing workload demands. This ensures that the system can adapt to variations in workload without manual intervention [136].

8.9.4 Virtualization

Virtualization technologies, such as virtual machines (VMs) or containers, enable dynamic resource allocation by providing a layer of abstraction between the physical resources and the applications or services running on top of them. Virtualization allows for flexible allocation and reallocation of resources to different VMs or containers based on demand [136].

8.9.5 Resource scheduling

Resource scheduling algorithms determine how resources are allocated to tasks or jobs based on their priorities, resource requirements, and availability.

Scheduling decisions consider factors such as fairness, performance goals, and resource constraints to optimize the allocation of resources and ensure efficient utilization [136].

8.9.6 Reservation and preemption

Dynamic resource allocation can involve the concept of reservation and preemption. The reservation allows users to allocate resources in advance for specific periods, ensuring their availability when needed. Preemption consists of interrupting lower-priority tasks or jobs to reclaim resources and assign them to higher-priority workloads when necessary [137].

8.9.7 Predictive analytics

Predictive analytics leverages historical data and machine learning techniques to forecast future resource demands accurately. By analyzing patterns and trends in resource utilization, predictive analytics can help anticipate workload changes and facilitate proactive resource allocation. Dynamic resource allocation is critical in ensuring efficient resource utilization, performance optimization, and scalability in modern computing environments. It enables organizations to respond to changing demands effectively, minimize resource wastage, and deliver reliable and responsive services to users [138].

8.9.8 Intelligent routing and network optimization

Intelligent routing and network optimization encompass techniques and algorithms aimed at efficiently directing network traffic and maximizing the performance of communication networks. By leveraging advanced routing protocols, machine learning, and optimization algorithms, intelligent routing dynamically selects the most optimal paths for data transmission, considering factors such as network congestion, latency, bandwidth availability, and QoS requirements. Additionally, network optimization techniques optimize network resources and configurations, improving overall network efficiency, scalability, and reliability. These intelligent approaches enable networks to adapt to changing conditions, optimize resource allocation, and ensure seamless and efficient communication between devices, services, and users [138].

8.10 CASE STUDIES AND BEST PRACTICES

There are several case studies and best practices related to data traffic management in AI-IoT networks aimed at reducing congestion. Here are a few examples.

8.10.1 Edge computing and local processing

Edge computing capabilities installed inside an AI-IoT network may reduce data traffic by processing and analyzing closer to the data source. Congestion is decreased by not having to transfer large amounts of raw data to a single, centralized cloud. For instance, edge devices in an innovative city application may assess sensor data locally, delivering only relevant insights to the cloud and reducing network congestion [139].

8.10.2 Data filtering and aggregation

Implementing data filtering and aggregation techniques at the AI-IoT device level helps reduce unnecessary data transmission. Devices can apply filters only to send relevant or critical data to the network. Aggregating data from multiple devices into a single transmission reduces the number of packets and reduces congestion. This approach is commonly used in applications like environmental monitoring or industrial AI-IoT, where multiple sensors generate data [140].

8.10.3 Traffic prioritization and QoS

Implementing QoS mechanisms in AI-IoT networks allows for prioritizing critical traffic and ensuring timely delivery. By assigning different priority levels to data streams, congestion-sensitive data can be given higher priority, while less-critical data can be allocated lower priority. This practice protects critical traffic from congestion-related delays [54].

8.10.4 Dynamic resource allocation and load balancing

Dynamic resource allocation and load balancing techniques can optimize resource utilization and prevent congestion in AI-IoT networks. By distributing data traffic across available network resources, load balancing ensures that no single device or network link becomes a bottleneck. This can involve techniques like dynamic routing, path selection, and adaptive load balancing algorithms [141].

8.10.5 Predictive analytics and machine learning

Predictive analytics and machine learning algorithms can help anticipate traffic patterns and proactively manage congestion. By analyzing historical data, network operators can forecast future traffic demands and dynamically adjust network resources to meet those demands. This proactive approach reduces the likelihood of congestion and improves network performance [142].

8.10.6 Protocol optimization

Optimizing AI-IoT communication protocols can help reduce overhead and congestion. For instance, protocols like MQTT are designed to be lightweight and efficient, reducing network traffic and minimizing congestion. Optimizing protocol configurations and parameters can enhance performance and reduce unnecessary data transmission [143]. These case studies and best practices demonstrate various strategies for managing data traffic in AI-IoT networks to reduce congestion. It's important to note that the choice of approach depends on the specific AI-IoT application, network architecture, and traffic patterns. Combining these techniques tailored to the particular requirements of the AI-IoT deployment can help alleviate congestion and improve overall network performance.

8.11 SUCCESSFUL IMPLEMENTATIONS OF DATA TRAFFIC MANAGEMENT IN AI-IOT NETWORKS

Several successful implementations of data traffic management in AI-IoT networks have effectively reduced congestion and improved network performance [144].

Here are a few notable examples.

8.11.1 Smart grid systems

Innovative grid implementations leverage AI-IoT technologies to monitor and control energy distribution networks. These systems employ data traffic management techniques to reduce congestion and optimize energy consumption. Only relevant data is transmitted by implementing data filtering and aggregation at the edge devices, reducing network traffic. Additionally, predictive analytics and machine learning algorithms help forecast energy demand, enabling proactive load balancing and resource allocation [145].

8.11.2 Connected vehicles

ITS utilize AI-IoT networks to connect vehicles, infrastructure, and traffic management systems. Data traffic management is crucial to minimize congestion and ensure reliable communication. Traffic prioritization mechanisms are implemented to give priority to critical safety-related information. Additionally, edge computing capabilities enable local processing of vehicle sensor data, reducing the need for transmitting large volumes of raw data to the cloud [145].

8.11.3 Smart health care systems

AI-IoT-based health care applications involve real-time monitoring and managing of patient health data. Traffic prioritization and QoS mechanisms are implemented to reduce congestion and ensure timely delivery of critical health information. Congestion-sensitive health care information can be prioritized by assigning higher priority to life-critical data, leading to more effective health care delivery [145].

8.11.4 Industrial AI-IoT applications

Industrial AI-IoT (IAI-IoT) networks are used in various industrial sectors to monitor and control equipment, optimize production processes, and improve efficiency. Effective data traffic management in IAI-IoT networks prevents congestion and maintains real-time control. Implementing edge computing, data filtering, and aggregation techniques reduces unnecessary data transmission, while dynamic resource allocation and load balancing strategies optimize resource utilization and prevent congestion [145].

8.11.5 Smart cities

AI-IoT-based innovative city applications involve numerous devices and sensors for monitoring and managing various urban services. Successful implementations of data traffic management in smart cities include edge computing for local data processing, data aggregation to reduce network traffic, and predictive analytics for demand forecasting [146]. These approaches help reduce congestion and efficiently manage smart city infrastructure and services. These examples demonstrate the successful implementation of data traffic management techniques in diverse AI-IoT network deployments. By leveraging strategies such as data filtering, aggregation, prioritization, and dynamic resource allocation, congestion can be mitigated, and network performance can be optimized, leading to more reliable and efficient AI-IoT systems.

8.12 LESSONS LEARNED FROM CONGESTION REDUCTION INITIATIVES

Congestion reduction initiatives have provided valuable lessons for effectively managing and mitigating congestion in various contexts. Here are some key lessons learned.

8.12.1 Comprehensive approach

Successful congestion reduction initiatives often require a comprehensive approach that addresses multiple aspects of congestion. This includes infrastructure improvements, ITS, public transportation enhancements, demand management strategies, and effective traffic management techniques. A holistic approach can yield more significant and sustainable results [147].

8.12.2 Data-driven decision-making

Access to accurate and real-time data is crucial for effective congestion management. Initiatives that leverage data from various sources, such as traffic sensors, GPS tracking, and mobile applications, can provide valuable insights into congestion patterns, traffic flows, and bottlenecks. This data can inform decision-making processes and help identify appropriate congestion reduction measures [147].

8.12.3 Multimodal transportation

Encouraging and promoting multimodal transportation options, such as public transit, cycling, and walking, can help alleviate congestion. Providing well-integrated and convenient transportation alternatives reduces the reliance on private vehicles and overall traffic volume. Investments in public transit infrastructure, bike lanes, and pedestrian-friendly infrastructure have shown positive impacts on congestion reduction [148].

8.12.4 Demand management strategies

Implementing demand management strategies can effectively reduce congestion by influencing travel behavior and reducing peak-hour traffic. These strategies may include flexible work arrangements, congestion pricing, carpooling and ride-sharing programs, and promoting alternative travel modes. Reducing the number of vehicles on the road during peak periods can significantly alleviate congestion [149].

8.12.5 Intelligent traffic management systems

ITS and advanced traffic management systems are critical to reducing congestion. Initiatives incorporating technologies such as adaptive traffic signal control, real-time traffic information dissemination, dynamic route guidance, and incident management systems have shown promising results in improving traffic flow, reducing delays, and minimizing congestion [150].

8.12.6 Public engagement and communication

Effective communication and public engagement are vital for successful congestion reduction initiatives. Engaging the public, stakeholders, and community organizations from the early stages of planning and implementation helps build support, gather input, and increase awareness of the benefits of congestion reduction strategies. Transparent communication about congestion reduction initiatives' goals, progress, and impacts fosters trust and encourages participation [150].

8.12.7 Continuous evaluation and adaptation

Congestion reduction initiatives should be evaluated regularly to assess their effectiveness and make necessary adjustments. Monitoring key performance indicators, analyzing data, and gathering feedback from users and stakeholders help identify areas for improvement and fine-tune congestion management strategies. Flexibility and adaptability are crucial to respond to changing transportation patterns and evolving needs. These lessons highlight the importance of a comprehensive, data-driven, and adaptive approach to reducing congestion. By considering multiple strategies, leveraging data, engaging the public, and continuously evaluating and adapting initiatives, congestion can be effectively managed and reduced, improving mobility and quality of life in urban environments [150].

8.13 BEST PRACTICES FOR EFFICIENT DATA TRAFFIC MANAGEMENT

Here are some unique best practices for efficient data traffic management.

8.13.1 Predictive traffic routing

Utilize predictive analytics and machine learning algorithms to forecast traffic patterns and proactively route data traffic accordingly. Analyzing historical data and real-time inputs allows traffic to be directed along the most optimal routes, avoiding potential congestion points and ensuring efficient data transmission [151].

8.13.2 Dynamic bandwidth allocation

Implement dynamic bandwidth allocation techniques that allocate bandwidth resources based on real-time demand. This approach dynamically adjusts the bandwidth for different data flows, prioritizing high-priority or time-sensitive traffic when congestion occurs. By adapting bandwidth allocation in real time, network resources are efficiently utilized, and congestion is mitigated [152].

8.13.3 Network slicing

Deploy network slicing, a technique that partitions the network into multiple virtual networks, each tailored to specific requirements. By creating dedicated slices for different types of traffic or services, such as video streaming, AI-IoT data, or critical communications, resources can be allocated optimally, reducing congestion and ensuring quality of service for each slice [153].

8.13.4 Peer-to-peer communication

Implement peer-to-peer (P2P) communication models, where devices or nodes in the network can directly exchange data without passing through a centralized server. P2P communication reduces the reliance on a single congestion point and distributes data traffic across multiple nodes, improving overall network efficiency and reducing congestion [154]. Figure 8.5 depicts the concept of peer-to-peer communication. In this communication model, participants, known as peers, interact directly with each other without needing a centralized server or intermediary. Figure 8.5 illustrates a network of peers, represented by nodes, connected through various communication channels.

8.13.5 Dynamic protocol selection

Utilize dynamic protocol selection techniques that adaptively switch between communication protocols based on network conditions and congestion levels. For example, if congestion is detected in a specific protocol, the system can automatically switch to a less-congested protocol, reducing data traffic and minimizing congestion [155].

8.13.6 Traffic off-loading to edge devices

Off-load data processing and analytics tasks to edge devices or gateways, reducing the need for transmitting large volumes of raw data over the network. Traffic congestion can be minimized by performing data processing and analysis closer to the data source and optimizing bandwidth consumption [156].

8.13.7 Application-aware traffic management

Implement application-aware traffic management techniques that prioritize data traffic based on different applications' specific requirements and characteristics; by understanding the needs of each application, such as latency sensitivity or

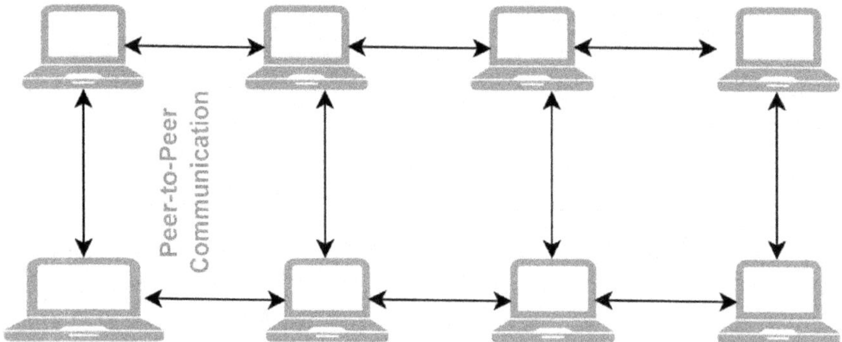

Figure 8.5 Peer-to-peer communication.

bandwidth conditions, traffic management strategies can be tailored to optimize the performance of individual applications and reduce congestion [157].

8.13.8 Adaptive multicast

Utilize adaptive multicast techniques to transmit data to multiple recipients efficiently. Adaptive multicast dynamically adjusts the transmission parameters based on the network conditions, optimizing the data distribution and reducing unnecessary replication and congestion [158].

8.13.9 Dynamic time slot allocation

In time-sensitive AI-IoT networks, implement dynamic time slot allocation mechanisms that allocate transmission time slots based on real-time demand and priority. By efficiently scheduling data transmissions and avoiding collisions, congestion is reduced, and the overall network performance is improved [158].

8.13.10 User-initiated traffic control

Empower end users with the ability to control their data traffic and prioritize their communication needs. By providing user-configurable traffic control options, such as allowing users to define traffic priorities or set data usage limits, congestion can be managed at the individual user level, resulting in a more personalized and efficient data traffic management approach.

These unique best practices can contribute to more efficient data traffic management, reduced congestion, and improved network performance. Organizations should evaluate the applicability and feasibility of these practices based on their specific network environment and requirements [158].

8.14 FUTURE TRENDS AND CHALLENGES

8.14.1 Future trends

Edge Computing for Localized Processing: The proliferation of edge computing capabilities will continue to grow, enabling localized processing and analysis of AI-IoT data at the network edge. This trend will help reduce data traffic by minimizing the need to transmit raw data to centralized cloud servers, thereby alleviating congestion and improving response times.

8.14.2 5G network deployment

The widespread adoption of 5G networks will revolutionize data traffic management in AI-IoT networks. With its high-speed connectivity and low latency, 5G enables faster and more efficient data transmission. The increased capacity and improved network performance will help reduce congestion and enable real-time data processing and decision-making [159].

8.14.3 AI and machine learning–based traffic optimization

Integrating AI and machine learning algorithms into traffic management systems will enable more advanced and dynamic traffic optimization techniques. AI-powered systems can analyze real-time traffic data, predict congestion patterns, and autonomously adjust traffic flows to reduce congestion and improve network efficiency [160].

8.14.4 Security and privacy

AI-IoT networks collect and transmit sensitive data, making security and privacy critical concerns. Implementing robust security measures to protect data in transit and at rest is essential. Balancing the need for data traffic management with user privacy protection poses a challenge that requires careful consideration and implementation of appropriate security measures [161].

8.14.5 Interoperability and standardization

The interoperability of different AI-IoT devices, protocols, and platforms remains challenging. Lack of standardized communication protocols and device interoperability from other manufacturers can hinder efficient data traffic management. Establishing common standards and protocols will enable seamless communication and effective congestion reduction [161].

8.14.6 Real-time decision-making

Managing real-time data traffic to reduce congestion requires timely decision-making and dynamic adjustments. The challenge lies in quickly analyzing large volumes of data, predicting congestion, and implementing traffic management strategies in real time. Developing efficient algorithms and decision-making frameworks will be essential to overcome this challenge [162].

8.15 CONCLUSION

Data traffic management in AI-IoT networks is critical and requires careful attention to reduce congestion and ensure efficient communication among devices. As the number of connected devices continues to skyrocket, it is imperative to implement effective strategies to handle the increasing data load and prevent network congestion. Throughout this chapter, we have explored various techniques for managing data traffic in AI-IoT networks. These techniques include data filtering and aggregation, prioritization schemes, load balancing, and dynamic resource allocation. Each approach plays a crucial role in optimizing data transmission and alleviating congestion.

Data filtering and aggregation techniques help minimize congestion by allowing AI-IoT devices to process and transmit only relevant data, reducing unnecessary traffic on the network. By filtering out redundant or nonessential information and aggregating data at the edge or within the network, bandwidth is conserved, and the volume of data sent to central infrastructure is reduced. Prioritization schemes ensure that critical data or devices receive preferential treatment, guaranteeing timely delivery and minimizing congestion-related delays. The network can allocate resources accordingly and ensure efficient data flow by assigning priority levels to other data types, such as emergency alerts or real-time monitoring information. Load balancing techniques evenly distribute the data traffic across multiple network nodes, preventing any single device or component from becoming overwhelmed. This approach optimizes resource utilization and improves overall network performance, reducing the risk of congestion. Dynamic resource allocation techniques adaptively allocate network resources based on real-time traffic conditions and demands. By continuously monitoring the network and adjusting resource allocation accordingly, these techniques ensure efficient utilization of available resources and effectively manage congestion.

In conclusion, effective data traffic management in AI-IoT networks is crucial for maintaining optimal network performance and reducing congestion. Network administrators can successfully mitigate congestion-related issues and optimize data transmission by combining filtering, aggregation, prioritization, load balancing, and dynamic resource allocation techniques. As AI-IoT continues to expand and revolutionize various industries, ongoing research and development in data traffic management strategies are essential. Advancements in edge computing and fog computing technologies further enhance data traffic management capabilities by bringing computation and storage closer to the network's edge, reducing latency, and minimizing overall data traffic. By prioritizing data traffic management in AI-IoT networks, we can ensure seamless operation, improved efficiency, and reliable device communication. Through these efforts, the full potential of AI-IoT can be realized in transforming our world into a more connected and intelligent ecosystem.

REFERENCES

[1] S. Li, L. Da Xu, and S. Zhao, "The internet of things: A survey," *Inf. Syst. Front.*, vol. 17, pp. 243–259, 2015.

[2] J. Gubbi, R. Buyya, S. Marusic, and M. Palaniswami, "Internet of Things (IoT): A vision, architectural elements, and future directions," *Futur. Gener. Comput. Syst.*, vol. 29, no. 7, pp. 1645–1660, 2013.

[3] S. Hiremath, G. Yang, and K. Mankodiya, "Wearable Internet of Things: Concept, architectural components and promises for person-centered healthcare," in *2014 4th International Conference on Wireless Mobile Communication and Healthcare-Transforming Healthcare Through Innovations in Mobile and Wireless Technologies (MOBIHEALTH)*, IEEE, 2014, pp. 304–307.

[4] O. Vermesan *et al.*, "Internet of things strategic research roadmap," *Internet Things-Glob. Technol. Soc. trends*, vol. 1, no. 2011, pp. 9–52, 2011.

[5] O. Vermesan and P. Friess, *Internet of things: Converging technologies for smart environments and integrated ecosystems*. River Publishers, 2013.

[6] S. Munirathinam, "Industry 4.0: Industrial internet of things (IIOT)," in *Advances in computers*, Elsevier, 2020, pp. 129–164.

[7] E. Manavalan and K. Jayakrishna, "A review of Internet of Things (IoT) embedded sustainable supply chain for industry 4.0 requirements," *Comput. Ind. Eng.*, vol. 127, pp. 925–953, 2019.

[8] L. P. Verma and M. Kumar, "An IoT based congestion control algorithm," *Internet Things*, vol. 9, p. 100157, 2020.

[9] D. Dash, R. Farooq, J. S. Panda, and K. V Sandhyavani, "Internet of Things (IoT): The new paradigm of HRM and skill development in the fourth industrial revolution (Industry 4.0).," *IUP J. Inf. Technol.*, vol. 15, no. 4, 2019.

[10] Z. Bi, L. Da Xu, and C. Wang, "Internet of things for enterprise systems of modern manufacturing," *IEEE Trans. Ind. informatics*, vol. 10, no. 2, pp. 1537–1546, 2014.

[11] J. Jin, J. Gubbi, S. Marusic, and M. Palaniswami, "An information framework for creating a smart city through internet of things," *IEEE Internet Things J.*, vol. 1, no. 2, pp. 112–121, 2014.

[12] P. Raj and A. C. Raman, *The Internet of Things: Enabling technologies, platforms, and use cases*. CRC Press, 2017.

[13] H. Wang, F. Luo, M. Ibrahim, O. Kayiran, and A. Jog, "Efficient and fair multiprogramming in GPUs via effective bandwidth management," in *2018 IEEE International Symposium on High Performance Computer Architecture (HPCA)*, IEEE, 2018, pp. 247–258.

[14] U. I. Ikechukwu, "A survey on bandwidth management techniques via the OSI model network and application layers," *Glob. J. Comput. Sci. Technol.*, vol. 17, no. 4, pp. 1–10, 2017.

[15] N. Mishra, L. P. Verma, P. K. Srivastava, and A. Gupta, "An analysis of IoT congestion control policies," *Procedia Comput. Sci.*, vol. 132, pp. 444–450, 2018.

[16] I. A. Najm, A. K. Hamoud, J. Lloret, and I. Bosch, "Machine learning prediction approach to enhance congestion control in 5G IoT environment," *Electronics*, vol. 8, no. 6, p. 607, 2019.

[17] V. K. Jain, A. P. Mazumdar, P. Faruki, and M. C. Govil, "Congestion control in Internet of Things: Classification, challenges, and future directions," *Sustain. Comput. Informatics Syst.*, vol. 35, p. 100678, 2022.

[18] D. Alghazzawi, O. Bamasaq, S. Bhatia, A. Kumar, P. Dadheech, and A. Albeshri, "Congestion control in cognitive IoT-based WSN network for smart agriculture," *IEEE Access*, vol. 9, pp. 151401–151420, 2021.

[19] C. Lim, "Improving congestion control of TCP for constrained IoT networks," *Sensors*, vol. 20, no. 17, p. 4774, 2020.

[20] A. Betzler, C. Gomez, I. Demirkol, and J. Paradells, "CoAP congestion control for the internet of things," *IEEE Commun. Mag.*, vol. 54, no. 7, pp. 154–160, 2016.

[21] N. Makarem, W. B. Diab, I. Mougharbel, and N. Malouch, “On the design of efficient congestion control for the constrained application protocol in IoT,” *Comput. Networks*, vol. 207, p. 108824, 2022.

[22] K. Upreti, N. Kumar, M. S. Alam, A. Verma, M. Nandan, and A. K. Gupta, “Machine learning- based Congestion Control Routing strategy for healthcare IoT enabled wireless sensor networks,” in *2021 Fourth International Conference on Electrical, Computer and Communication Technologies (ICECCT)*, IEEE, 2021, pp. 1–6.

[23] P. Schulz *et al.*, “Latency critical IoT applications in 5G: Perspective on the design of radio interface and network architecture,” *IEEE Commun. Mag.*, vol. 55, no. 2, pp. 70–78, 2017.

[24] E. D. Ayele, N. Meratnia, and P. J. M. Havinga, “Towards a new opportunistic IoT network architecture for wildlife monitoring system,” in *2018 9th IFIP International Conference on New Technologies, Mobility and Security (NTMS)*, IEEE, 2018, pp. 1–5.

[25] Q. Zhang, M. Jiang, Z. Feng, W. Li, W. Zhang, and M. Pan, “IoT enabled UAV: Network architecture and routing algorithm,” *IEEE Internet Things J.*, vol. 6, no. 2, pp. 3727–3742, 2019.

[26] I. U. Khan *et al.*, “RSSI-controlled long-range communication in secured IoT-enabled unmanned aerial vehicles,” *Mob. Inf. Syst.*, vol. 2021, pp. 1–11, 2021.

[27] J. J. Godwin, B. V. S. Krishna, R. Rajeshwari, P. Sushmitha, and M. Yamini, “IoT based intelligent ambulance monitoring and traffic control system,” *Furth. Adv. Internet Things Biomed. Cyber Phys. Syst.*, pp. 269–278, 2021.

[28] P. Kuppusamy, R. Kalpana, and P. V Venkateswara Rao, “Optimized traffic control and data processing using IoT,” *Cluster Comput.*, vol. 22, no. Suppl 1, pp. 2169–2178, 2019.

[29] Y. Zhang, S. Kasahara, Y. Shen, X. Jiang, and J. Wan, “Smart contract-based access control for the internet of things,” *IEEE Internet Things J.*, vol. 6, no. 2, pp. 1594–1605, 2018.

[30] M. A. Khan, B. A. Alvi, A. Safi, and I. U. Khan, “Drones for good in smart cities: a review,” in *Proceedings of the International Conference on Electrical, Electronics, Computers, Communication, Mechanical and Computing (EECCMC)*, IEEE, 2018, pp. 1–6.

[31] A. Wang, Z. Zha, Y. Guo, and S. Chen, “Software-defined networking enhanced edge computing: A network-centric survey,” *Proc. IEEE*, vol. 107, no. 8, pp. 1500–1519, 2019.

[32] L. Bai, L. Yao, C. Li, X. Wang, and C. Wang, “Adaptive graph convolutional recurrent network for traffic forecasting,” *Adv. Neural Inf. Process. Syst.*, vol. 33, pp. 17804–17815, 2020.

[33] N. Nurelmadina *et al.*, “A systematic review on cognitive radio in low power wide area network for industrial IoT applications,” *Sustainability*, vol. 13, no. 1, p. 338, 2021.

[34] D. Lee and S. N. Yoon, “Application of artificial intelligence-based technologies in the healthcare industry: Opportunities and challenges,” *Int. J. Environ. Res. Public Health*, vol. 18, no. 1, p. 271, 2021.

[35] I. U. Khan, S. B. H. Shah, L. Wang, M. A. Aziz, T. Stephan, and N. Kumar, “Routing protocols & unmanned aerial vehicles autonomous localization in flying networks,” *Int. J. Commun. Syst.*, p. e4885, 2021.

[36] S. T. A. Jafri, I. Ahmed, and S. Ali, "Queue-buffer optimization based on aggressive random early detection in massive NB-IoT MANET for 5G applications," *Electronics*, vol. 11, no. 18, p. 2955, 2022.

[37] M. Beshley, N. Kryvinska, M. Seliuchenko, H. Beshley, E. M. Shakshuki, and A.-U.-H. Yasar, "End-to-End QoS 'smart queue' management algorithms and traffic prioritization mechanisms for narrow-band internet of things services in 4G/5G networks," *Sensors*, vol. 20, no. 8, p. 2324, 2020.

[38] M. S. Bali, K. Gupta, D. Koundal, A. Zaguia, S. Mahajan, and A. K. Pandit, "Smart architectural framework for symmetrical data offloading in IoT," *Symmetry (Basel).*, vol. 13, no. 10, p. 1889, 2021.

[39] A. Ghosh, O. Khalid, R. N. B. Rais, A. Rehman, S. U. R. Malik, and I. A. Khan, "Data offloading in IoT environments: Modeling, analysis, and verification," *EURASIP J. Wirel. Commun. Netw.*, vol. 2019, no. 1, pp. 1–23, 2019.

[40] H.-Y. Kim and J.-M. Kim, "A load balancing scheme based on deep-learning in IoT," *Cluster Comput.*, vol. 20, pp. 873–878, 2017.

[41] W.-C. Chien, C.-F. Lai, H.-H. Cho, and H.-C. Chao, "A SDN-SFC-based service-oriented load balancing for the IoT applications," *J. Netw. Comput. Appl.*, vol. 114, pp. 88–97, 2018.

[42] T. A. Al-Janabi and H. S. Al-Raweshidy, "Optimised clustering algorithm-based centralised architecture for load balancing in IoT network," in *2017 International Symposium on Wireless Communication Systems (ISWCS)*, IEEE, 2017, pp. 269–274.

[43] G. White, V. Nallur, and S. Clarke, "Quality of service approaches in IoT: A systematic mapping," *J. Syst. Softw.*, vol. 132, pp. 186–203, 2017.

[44] M.-A. Nef, L. Perlepes, S. Karagiorgou, G. I. Stamoulis, and P. K. Kikiras, "Enabling QoS in the Internet of Things," *Proc. CTRQ*, 2012.

[45] G. Tanganelli and E. Mingozzi, "Energy-efficient IoT service brokering with quality of service support," *Sensors*, vol. 19, no. 3, p. 693, 2019.

[46] A. N. Abosaif and H. S. Hamza, "Quality of service-aware service selection algorithms for the internet of things environment: A review paper," *Array*, vol. 8, p. 100041, 2020.

[47] S. Neelakandan, M. A. Berlin, S. Tripathi, V. B. Devi, I. Bhardwaj, and N. Arulkumar, "IoT-based traffic prediction and traffic signal control system for smart city," *Soft Comput.*, vol. 25, no. 18, pp. 12241–12248, 2021.

[48] A. R. Abdellah and A. Koucheryavy, "Deep learning with long short-term memory for IoT traffic prediction," in *Internet of Things, Smart Spaces, and Next Generation Networks and Systems: 20th International Conference, NEW2AN 2020, and 13th Conference, ruSMART 2020*, St. Petersburg, 26–28 August, Proceedings, Part I 20, Springer, 2020, pp. 267–280.

[49] M. Lopez-Martin, B. Carro, and A. Sanchez-Esguevillas, "Neural network architecture based on gradient boosting for IoT traffic prediction," *Futur. Gener. Comput. Syst.*, vol. 100, pp. 656–673, 2019.

[50] A. Das, P. Dash, and B. K. Mishra, "An innovation model for smart traffic management system using internet of things (IoT)," in *Cognitive Computing for Big Data Systems Over IoT: Frameworks, Tools and Applications*, Springer, 2017, pp. 355–370.

[51] D. Elkin and V. Vyatkin, "IoT in traffic management: Review of existing methods of road traffic regulation," in *Applied Informatics and Cybernetics in Intelligent*

Systems: Proceedings of the 9th Computer Science On-line Conference 2020, Volume 3, 9, Springer, 2020, pp. 536–551.

[52] A. Sharif, J. P. Li, and M. I. Sharif, "Internet of Things network cognition and traffic management system," *Cluster Comput.*, vol. 22, pp. 13209–13217, 2019.

[53] S. Banerjee, C. Chakraborty, and S. Chatterjee, "A survey on IoT based traffic control and prediction mechanism," *Internet Things Big Data Anal. Smart Gener.*, pp. 53–75, 2019.

[54] J. Martin and N. Feamster, "User-driven dynamic traffic prioritization for home networks," in *Proceedings of the 2012 ACM SIGCOMM Workshop on Measurements Up the Stack*, ACM, 2012, pp. 19–24.

[55] A. M. Bongale, N. Nithin, and L. S. Jyothi, "Traffic prioritization in MPLS enabled OSPF network," in *2012 World Congress on Information and Communication Technologies*, IEEE, 2012, pp. 132–137.

[56] D. Hong and S. S. Rappaport, "Traffic model and performance analysis for cellular mobile radio telephone systems with prioritized and nonprioritized handoff procedures," *IEEE Trans. Veh. Technol.*, vol. 35, no. 3, pp. 77–92, 1986.

[57] J. Erman, A. Mahanti, and M. Arlitt, "Byte me: A case for byte accuracy in traffic classification," in *Proceedings of the 3rd Annual ACM Workshop on Mining Network Data*, IEEE, 2007, pp. 35–38.

[58] D. G. Balan and D. A. Potorac, "Linux HTB queuing discipline implementations," in *2009 First International Conference on Networked Digital Technologies*, IEEE, 2009, pp. 122–126.

[59] S. Kannan *et al.*, "Ubiquitous vehicular ad-hoc network computing using deep neural network with IoT-based bat agents for traffic management," *Electronics*, vol. 10, no. 7, p. 785, 2021.

[60] P. M. Kumar, G. Manogaran, R. Sundarasekar, N. Chilamkurti, and R. Varatharajan, "Ant colony optimization algorithm with internet of vehicles for intelligent traffic control system," *Comput. Networks*, vol. 144, pp. 154–162, 2018.

[61] I. U. Khan, A. Abdollahi, A. Jamil, B. Baig, M. A. Aziz, and F. Subhan, "A novel design of FANET routing protocol aided 5G communication using IoT," *J. Mob. Multimed.*, pp. 1333–1354, 2022.

[62] B. K. J. Al-Shammari, N. Al-Aboody, and H. S. Al-Raweshidy, "IoT traffic management and integration in the QoS supported network," *IEEE Internet Things J.*, vol. 5, no. 1, pp. 352–370, 2017.

[63] U. K. Lilhore *et al.*, "Design and implementation of an ML and IoT based adaptive traffic- management system for smart cities," *Sensors*, vol. 22, no. 8, p. 2908, 2022.

[64] A. Gupta, R. Christie, and R. Manjula, "Scalability in internet of things: features, techniques and research challenges," *Int. J. Comput. Intell. Res*, vol. 13, no. 7, pp. 1617–1627, 2017.

[65] C. Sarkar, S. N. A. U. Nambi, R. V. Prasad, and A. Rahim, "A scalable distributed architecture towards unifying IoT applications," in *2014 IEEE World Forum on Internet of Things (WF-IoT)*, IEEE, 2014, pp. 508–513.

[66] I. D. Addo, S. I. Ahamed, S. S. Yau, and A. Buduru, "A reference architecture for improving security and privacy in internet of things applications," in *2014 IEEE International Conference on Mobile Services*, IEEE, 2014, pp. 108–115.

[67] F. Dalipi and S. Y. Yayilgan, "Security and privacy considerations for IoT application on smart grids: Survey and research challenges," in *2016 IEEE 4th International*

Conference on Future Internet of Things and Cloud Workshops (FiCloudW), IEEE, 2016, pp. 63–68.

[68] N. Pathak, P. K. Deb, A. Mukherjee, and S. Misra, "IoT-to-the-rescue: A survey of IoT solutions for COVID-19-like pandemics," *IEEE Internet Things J.*, vol. 8, no. 17, pp. 13145–13164, 2021.

[69] I. U. Khan, I. M. Qureshi, M. A. Aziz, T. A. Cheema, and S. B. H. Shah, "Smart IoT control-based nature inspired energy efficient routing protocol for flying ad hoc network (FANET)," *IEEE Access*, vol. 8, pp. 56371–56378, 2020.

[70] A. Pekar, J. Mocnej, W. K. G. Seah, and I. Zolotova, "Application domain-based overview of IoT network traffic characteristics," *ACM Comput. Surv.*, vol. 53, no. 4, pp. 1–33, 2020.

[71] S. Rathore, B. W. Kwon, and J. H. Park, "BlockSecIoTNet: Blockchain-based decentralized security architecture for IoT network," *J. Netw. Comput. Appl.*, vol. 143, pp. 167–177, 2019.

[72] I. Mistry, S. Tanwar, S. Tyagi, and N. Kumar, "Blockchain for 5G-enabled IoT for industrial automation: A systematic review, solutions, and challenges," *Mech. Syst. Signal Process.*, vol. 135, p. 106382, 2020.

[73] A. V. Dastjerdi, H. Gupta, R. N. Calheiros, S. K. Ghosh, and R. Buyya, "Fog computing: Principles, architectures, and applications," in *Internet of Things*, Elsevier, 2016, pp. 61–75.

[74] C. Chakraborty, S. Ben Othman, F. A. Almalki, and H. Sakli, "FC-SEEDA: Fog computing-based secure and energy efficient data aggregation scheme for Internet of healthcare things," *Neural Comput. Appl.*, pp. 1–17, 2023.

[75] H. Tran-Dang and D.-S. Kim, "Fog resource aware framework for task offloading in IoT systems," in *Cooperative and Distributed Intelligent Computation in Fog Computing: Concepts, Architectures, and Frameworks*, Springer, 2023, pp. 47–82.

[76] X. He and F. Deng, "Research on architecture of internet of things platform based on service mesh," in *2020 12th International Conference on Measuring Technology and Mechatronics Automation (ICMTMA)*, IEEE, 2020, pp. 755–759.

[77] H. Tran-Dang and D.-S. Kim, "An information framework for internet of things services in physical internet," *IEEE Access*, vol. 6, pp. 43967–43977, 2018.

[78] A. I. Alsalibi, M. K. Y. Shambour, M. A. Abu-Hashem, M. Shehab, and Q. Shambour, "Internet of Things in health care: A survey," *Hybrid Artif. Intell. IoT Healthc.*, pp. 165–200, 2021.

[79] M. Kowal and K. Staniec, "Investigations of the wireless M-bus system resilience under challenging propagation conditions," *Electronics*, vol. 12, no. 4, p. 907, 2023.

[80] H. Tran-Dang, N. Krommenacker, P. Charpentier, and D.-S. Kim, "Toward the internet of things for physical internet: Perspectives and challenges," *IEEE Internet Things J.*, vol. 7, no. 6, pp. 4711–4736, 2020.

[81] S. Madakam, V. Lake, V. Lake, and V. Lake, "Internet of Things (IoT): A literature review," *J. Comput. Commun.*, vol. 3, no. 5, p. 164, 2015.

[82] P. Gokhale, O. Bhat, and S. Bhat, "Introduction to IoT," *Int. Adv. Res. J. Sci. Eng. Technol.*, vol. 5, no. 1, pp. 41–44, 2018.

[83] M. Usama *et al.*, "Predictive modelling of compression strength of waste GP/FA blended expansive soils using multi-expression programming," *Constr. Build. Mater.*, vol. 392, p. 131956, 2023.

[84] M. U. Farooq, M. Waseem, S. Mazhar, A. Khairi, and T. Kamal, "A review on internet of things (IoT)," *Int. J. Comput. Appl.*, vol. 113, no. 1, pp. 1–7, 2015.
[85] S. Huh, S. Cho, and S. Kim, "Managing IoT devices using blockchain platform," in *2017 19th International Conference on Advanced Communication Technology (ICACT)*, IEEE, 2017, pp. 464–467.
[86] L. Xiao, X. Wan, X. Lu, Y. Zhang, and D. Wu, "IoT security techniques based on machine learning: How do IoT devices use AI to enhance security?," *IEEE Signal Process. Mag.*, vol. 35, no. 5, pp. 41–49, 2018.
[87] W. S. Chin, H. Kim, Y. J. Heo, and J. W. Jang, "A context-based future network infrastructure for IoT services," *Procedia Comput. Sci.*, vol. 56, pp. 266–270, 2015.
[88] S. Mubeen, S. A. Asadollah, A. V. Papadopoulos, M. Ashjaei, H. Pei-Breivold, and M. Behnam, "Management of service level agreements for cloud services in IoT: A systematic mapping study," *IEEE Access*, vol. 6, pp. 30184–30207, 2017.
[89] I. Sittón-Candanedo, R. S. Alonso, Ó. García, L. Muñoz, and S. Rodríguez-González, "Edge computing, IoT and social computing in smart energy scenarios," *Sensors*, vol. 19, no. 15, p. 3353, 2019.
[90] M. M. Abur, O. S. Adewale, and S. B. Junaidu, "Cloud computing challenges: a review on security and privacy issues," in *Proceedings of the ACM International Conference on Computer Science Research and Innovations (CoSRI), Ibadan*, 2015, pp. 89–92.
[91] P. K. Dhillon and S. Kalra, "A lightweight biometrics based remote user authentication scheme for IoT services," *J. Inf. Secur. Appl.*, vol. 34, pp. 255–270, 2017.
[92] I. U. Khan *et al.*, "Intelligent detection system enabled attack probability using Markov chain in aerial networks," *Wirel. Commun. Mob. Comput.*, vol. 2021, pp. 1–9, 2021.
[93] S. Majumdar, M. M. Subhani, B. Roullier, A. Anjum, and R. Zhu, "Congestion prediction for smart sustainable cities using IoT and machine learning approaches," *Sustain. Cities Soc.*, vol. 64, p. 102500, 2021.
[94] S. Paiva, M. A. Ahad, G. Tripathi, N. Feroz, and G. Casalino, "Enabling technologies for urban smart mobility: Recent trends, opportunities and challenges," *Sensors*, vol. 21, no. 6, p. 2143, 2021.
[95] W. Alsafery, B. Alturki, S. Reiff-Marganiec, and K. Jambi, "Smart car parking system solution for the internet of things in smart cities," in *2018 1st International Conference on Computer Applications & Information Security (ICCAIS)*, IEEE, 2018, pp. 1–5.
[96] N. B. Soni and J. Saraswat, "A review of IoT devices for traffic management system," in *2017 International Conference on Intelligent Sustainable Systems (ICISS)*, IEEE, 2017, pp. 1052–1055.
[97] H. A. A. Al-Kashoash, H. Kharrufa, Y. Al-Nidawi, and A. H. Kemp, "Congestion control in wireless sensor and 6LoWPAN networks: Toward the Internet of Things," *Wirel. Networks*, vol. 25, pp. 4493–4522, 2019.
[98] A. Rayes and S. Salam, *Internet of Things from Hype to Reality*, Springer, 2017.
[99] G.-C. Deng and K. Wang, "An application-aware QoS routing algorithm for SDN-based IoT networking," in *2018 IEEE Symposium on Computers and Communications (ISCC)*, IEEE, 2018, pp. 186–191.
[100] S. G. H. Soumyalatha, "Study of IoT: understanding IoT architecture, applications, issues and challenges," in *1st International Conference on Innovations in*

Computing & Net-working (ICICN16), CSE, RRCE. International Journal of Advanced Networking & Applications, IEEE, 2016.

[101] M. R. Islam *et al.*, "Smart parking management system to reduce congestion in urban area," in *2020 2nd International Conference on Electrical, Control and Instrumentation Engineering (ICECIE)*, IEEE, 2020, pp. 1–6.

[102] K. C. Okafor, I. E. Achumba, G. A. Chukwudebe, and G. C. Ononiwu, "Leveraging fog computing for scalable IoT datacenter using spine-leaf network topology," *J. Electr. Comput. Eng.*, vol. 2017, 2017.

[103] Y.-C. Chang and Y.-H. Lai, "Campus edge computing network based on IoT street lighting nodes," *IEEE Syst. J.*, vol. 14, no. 1, pp. 164–171, 2018.

[104] S. Rekha, L. Thirupathi, S. Renikunta, and R. Gangula, "Study of security issues and solutions in Internet of Things (IoT)," *Mater. Today Proc.*, vol. 80, pp. 3554–3559, 2023.

[105] K. S. Bhandari, A. S. M. S. Hosen, and G. H. Cho, "CoAR: Congestion-aware routing protocol for low power and lossy networks for IoT applications," *Sensors*, vol. 18, no. 11, p. 3838, 2018.

[106] I. U. Khan *et al.*, "Reinforce based optimization in wireless communication technologies and routing techniques using internet of flying vehicles," in *The 4th International Conference on Future Networks and Distributed Systems (ICFNDS)*, 2020, pp. 1–6.

[107] S. Gheisari and E. Tahavori, "CCCLA: A cognitive approach for congestion control in Internet of Things using a game of learning automata," *Comput. Commun.*, vol. 147, pp. 40–49, 2019.

[108] A. H. Hussein, "Internet of things (IOT): Research challenges and future applications," *Int. J. Adv. Comput. Sci. Appl.*, vol. 10, no. 6, 2019.

[109] Y. Mehmood, F. Ahmad, I. Yaqoob, A. Adnane, M. Imran, and S. Guizani, "Internet-of-things- based smart cities: Recent advances and challenges," *IEEE Commun. Mag.*, vol. 55, no. 9, pp. 16–24, 2017.

[110] S. K. Sharma and X. Wang, "Toward massive machine type communications in ultra-dense cellular IoT networks: Current issues and machine learning-assisted solutions," *IEEE Commun. Surv. Tutorials*, vol. 22, no. 1, pp. 426–471, 2019.

[111] M. Adil, "Congestion free opportunistic multipath routing load balancing scheme for Internet of Things (IoT)," *Comput. Networks*, vol. 184, p. 107707, 2021.

[112] A. K. M. Al-Qurabat, Z. A. Mohammed, and Z. J. Hussein, "Data traffic management based on compression and MDL techniques for smart agriculture in IoT," *Wirel. Pers. Commun.*, vol. 120, no. 3, pp. 2227–2258, 2021.

[113] T. Q. Duong, K. J. Kim, Z. Kaleem, M.-P. Bui, and N.-S. Vo, "UAV caching in 6G networks: A Survey on models, techniques, and applications," *Phys. Commun.*, vol. 51, p. 101532, 2022.

[114] K. Ni'amah, I. N. A. Ramatryana, and K. Anwar, "Coded random access prioritizing human over machines for future IoT networks," in *2018 2nd International Conference on Telematics and Future Generation Networks (TAFGEN)*, IEEE, 2018, pp. 19–24.

[115] A. Daly, "The legality of deep packet inspection," *Int. J. Commun. Law Policy*, no. 14, 2011.

[116] N. C. Damianou, *A Policy Framework for Management of Distributed Systems*, Citeseer, 2002.

[117] C.-Y. Wan, S. B. Eisenman, and A. T. Campbell, "Energy-efficient congestion detection and avoidance in sensor networks," *ACM Trans. Sens. Networks*, vol. 7, no. 4, pp. 1–31, 2011.
[118] C.-Y. Wan, S. B. Eisenman, and A. T. Campbell, "CODA: Congestion detection and avoidance in sensor networks," in *Proceedings of the 1st International Conference on Embedded Networked Sensor Systems*, ACM, 2003, pp. 266–279.
[119] N. Chen, Y. Yuan, and S. Zhou, "Performance analysis of queue length monitoring of M/G/1 systems," *Nav. Res. Logist.*, vol. 58, no. 8, pp. 782–794, 2011.
[120] K. Košťál, P. Helebrandt, M. Belluš, M. Ries, and I. Kotuliak, "Management and monitoring of IoT devices using blockchain," *Sensors*, vol. 19, no. 4, p. 856, 2019.
[121] A. Sivanathan, H. H. Gharakheili, and V. Sivaraman, "Inferring IoT device types from network behavior using unsupervised clustering," in *2019 IEEE 44th Conference on Local Computer Networks (LCN)*, IEEE, 2019, pp. 230–233.
[122] P.-T. Chen, F. Chen, and Z. Qian, "Road traffic congestion monitoring in social media with hinge-loss Markov random fields," in *2014 IEEE International Conference on Data Mining*, IEEE, 2014, pp. 80–89.
[123] Y. Wiseman, "Computerized traffic congestion detection system," *Int. J. Transp. Logist. Manag.*, vol. 1, no. 1, pp. 1–8, 2017.
[124] A. Ghaffari, "Congestion control mechanisms in wireless sensor networks: A survey," *J. Netw. Comput. Appl.*, vol. 52, pp. 101–115, 2015.
[125] W. Yue, C. Li, G. Mao, N. Cheng, and D. Zhou, "Evolution of road traffic congestion control: A survey from perspective of sensing, communication, and computation," *China Commun.*, vol. 18, no. 12, pp. 151–177, 2021.
[126] K. Kharb, B. Sharma, and C. A. Trilok, "Reliable and congestion control protocols for wireless sensor networks," *Int. J. Eng. Technol. Innov.*, vol. 6, no. 1, p. 68, 2016.
[127] S. Floyd and V. Jacobson, "Random early detection gateways for congestion avoidance," *IEEE/ACM Trans. Netw.*, vol. 1, no. 4, pp. 397–413, 1993.
[128] S. Akhtar, *Congestion Control, in a Fast Packet*, Washington University, 1987.
[129] X. Liu, *Ad Hoc Network Congestion Management Based on Entropy and Third-Party Nodes*, ResearchSpace@Auckland, 2020.
[130] C. Wang, K. Sohraby, V. Lawrence, B. Li, and Y. Hu, "Priority-based congestion control in wireless sensor networks," in *IEEE International Conference on Sensor Networks, Ubiquitous, and Trustworthy Computing (SUTC'06)*, IEEE, 2006, pp. 8-pp.
[131] B. Toghi, M. Saifuddin, Y. P. Fallah, and M. O. Mughal, "Analysis of distributed congestion control in cellular vehicle-to-everything networks," in *2019 IEEE 90th Vehicular Technology Conference (VTC2019-Fall)*, IEEE, 2019, pp. 1–7.
[132] M. Swarna and T. Godhavari, "Enhancement of CoAP based congestion control in IoT network-a novel approach," *Mater. Today Proc.*, vol. 37, pp. 775–784, 2021.
[133] H. Jiang *et al.*, "When machine learning meets congestion control: A survey and comparison," *Comput. Networks*, vol. 192, p. 108033, 2021.
[134] Z. Xiao, W. Song, and Q. Chen, "Dynamic resource allocation using virtual machines for cloud computing environment," *IEEE Trans. Parallel Distrib. Syst.*, vol. 24, no. 6, pp. 1107–1117, 2012.
[135] R. Zhang, Y.-C. Liang, and S. Cui, "Dynamic resource allocation in cognitive radio networks," *IEEE Signal Process. Mag.*, vol. 27, no. 3, pp. 102–114, 2010.

[136] A. Chandra, W. Gong, and P. Shenoy, "Dynamic resource allocation for shared data centers using online measurements," in *Proceedings of the 2003 ACM SIGMETRICS International Conference on Measurement and Modeling of Computer Systems*, ACM, 2003, pp. 300–301.

[137] T. Zhang, J. Zhou, Z. Chen, Z. Tian, W. Wen, and Y. Jia, "Information freshness optimization of multiple status update streams in Internet of things: Generation rate control and service rate reservation," *Digit. Commun. Networks*, vol. 9, no. 4, pp. 971–980, 2022.

[138] P. Galetsi, K. Katsaliaki, and S. Kumar, "Big data analytics in health sector: Theoretical framework, techniques and prospects," *Int. J. Inf. Manage.*, vol. 50, pp. 206–216, 2020.

[139] R. Buyya and S. N. Srirama, *Fog and Edge Computing: Principles and Paradigms*, John Wiley & Sons, 2019.

[140] M. Cruz, S. Mafra, E. Teixeira, and F. Figueiredo, "Smart strawberry farming using edge computing and IoT," *Sensors*, vol. 22, no. 15, p. 5866, 2022.

[141] E. Yanmaz and O. K. Tonguz, "Dynamic load balancing and sharing performance of integrated wireless networks," *IEEE J. Sel. Areas Commun.*, vol. 22, no. 5, pp. 862–872, 2004.

[142] J. D. Kelleher, B. Mac Namee, and A. D'Arcy, *Fundamentals of Machine Learning for Predictive Data Analytics: Algorithms, Worked Examples, and Case Studies*, MIT Press, 2020.

[143] D. Maraver, J. Royo, V. Lemort, and S. Quoilin, "Systematic optimization of subcritical and transcritical organic Rankine cycles (ORCs) constrained by technical parameters in multiple applications," *Appl. Energy*, vol. 117, pp. 11–29, 2014.

[144] Z. M. Fadlullah *et al.*, "State-of-the-art deep learning: Evolving machine intelligence toward tomorrow's intelligent network traffic control systems," *IEEE Commun. Surv. Tutorials*, vol. 19, no. 4, pp. 2432–2455, 2017.

[145] A. Rego, L. Garcia, S. Sendra, and J. Lloret, "Software defined network-based control system for an efficient traffic management for emergency situations in smart cities," *Futur. Gener. Comput. Syst.*, vol. 88, pp. 243–253, 2018.

[146] P. Chinnasamy, C. Vinothini, S. Arun Kumar, A. Allwyn Sundarraj, S. V Annlin Jeba, and V. Praveena, "Blockchain technology in smart-cities," in *Blockchain Technology: Applications and Challenges*, Springer, 2021, pp. 179–200.

[147] A. P. Ambrosy *et al.*, "The global health and economic burden of hospitalizations for heart failure: lessons learned from hospitalized heart failure registries," *J. Am. Coll. Cardiol.*, vol. 63, no. 12, pp. 1123–1133, 2014.

[148] A. Sumalee, K. Uchida, and W. H. K. Lam, "Stochastic multi-modal transport network under demand uncertainties and adverse weather condition," *Transp. Res. Part C Emerg. Technol.*, vol. 19, no. 2, pp. 338–350, 2011.

[149] Y. Kim and S.-C. Kang, "Innovative traffic demand management strategy: expressway reservation system," *Transp. Res. Rec.*, vol. 2245, no. 1, pp. 27–35, 2011.

[150] P. Manikonda, A. K. Yerrapragada, and S. S. Annasamudram, "Intelligent traffic management system," in *2011 IEEE Conference on Sustainable Utilization and Development in Engineering and Technology (STUDENT)*, IEEE, 2011, pp. 119–122.

[151] Y. Kim, P. Wang, Y. Zhu, and L. Mihaylova, "A capsule network for traffic speed prediction in complex road networks," in *2018 Sensor Data Fusion: Trends, Solutions, Applications (SDF)*, IEEE, 2018, pp. 1–6.

[152] A. Paris, I. Del Portillo, B. Cameron, and E. Crawley, "A genetic algorithm for joint power and bandwidth allocation in multibeam satellite systems," in *2019 IEEE Aerospace Conference*, IEEE, 2019, pp. 1–15.

[153] C. Campolo, A. Molinaro, and V. Sciancalepore, "5G network slicing for V2X communications: Technologies and enablers," *Radio Access Netw. Slicing Virtualization 5G Vert. Ind.*, pp. 239–257, 2021.

[154] R. Steinmetz and K. Wehrle, *Peer-to-Peer Systems and Applications*, vol. 3485, Springer, 2005.

[155] P. R. L. Almeida, L. S. Oliveira, A. S. Britto Jr, and R. Sabourin, "Adapting dynamic classifier selection for concept drift," *Expert Syst. Appl.*, vol. 104, pp. 67–85, 2018.

[156] Y. Wang, X. Han, and S. Jin, "MAP based modeling method and performance study of a task offloading scheme with time-correlated traffic and VM repair in MEC systems," *Wirel. Networks*, vol. 29, no. 1, pp. 47–68, 2023.

[157] A. Mahgoub *et al.*, "{SONIC}: Application-aware data passing for chained serverless applications," in *2021 USENIX Annual Technical Conference (USENIX ATC 21)*, USENIX, 2021, pp. 285–301.

[158] S.-Y. Chang and H.-T. Chiao, "Adaptive streaming schemes for MPEG-DASH overWiFi multicast," in *2013 15th IEEE International Conference on Communication Technology*, IEEE, 2013, pp. 168–173.

[159] Y. O. Imam-Fulani *et al.*, "5G frequency standardization, technologies, channel models, and network deployment: Advances, challenges, and future directions," *Sustainability*, vol. 15, no. 6, p. 5173, 2023.

[160] Y. Huang and K.-W. Chin, "A three-tier deep learning based channel access method for WiFi networks," *IEEE Trans. Mach. Learn. Commun. Netw.*, vol. 15, no. 6, p. 5173, 2023.

[161] M. Sajwan and S. Singh, "14 challenges with Industry 4.0 security," *Intell. Anal. Ind. 4.0 Appl.*, p. 195, 2023.

[162] J. Shim *et al.*, "Real-time optimal route planning by deep reinforcement learning and validation with flight test," in *AIAA AVIATION 2023 Forum*, ARC, 2023, p. 3623.

Chapter 9

Artificial intelligence–enabled anomaly IDS for IoT network

Trends, Solutions, and Challenges

Farhood Nishat

9.1 INTRODUCTION

Nowadays, the internet of things (IoT) is a trending technology that can be used to transfer information from one place to another. Wireless communication technologies are considered the backbone for IoT networks [1, 2]. IoT networks can be deployed in many fields of studies, including smart grid, smart cities [3], the transportation industry [4, 5], health care [6], and farming [7], and for military [8, 9] applications as well. Integration of artificial intelligence (AI) with IoT will revolutionize the concept of smart cities. Therefore, intelligent, cutting-edge smart solutions will make use of AI, machine learning (ML) [10], deep learning (DL) [11], and federated learning [12]. Also, AI-enabled data traffic monitoring applications need to be introduced in the near future. The IoT network consists of sensors, networks, and applications. Figure 9.1 presents real-time applications of the IoT network. Interconnectivity, bandwidth, and end-to-end delay are prime issues in an IoT network. Due to these problems in an IoT, an intruder can easily influence the overall network. Also, IoT nodes are quite vulnerable to cyberattacks. Denial of service (DoS), distributed denial of service (DDoS), domain name system (DNS), ping of death, sinkhole and Sybil attacks are dangerous threats for an IoT network. Intruder nodes send fake data packets continuously to disrupt the overall communication channels within the network. Weak wireless signals also facilitate the intruder to jam or hijack the IoT network [13]. Traditional systems cannot guarantee or secure the IoT network. Cyberattack detection can be made possible with the help of an intrusion detection system (IDS). There exist at least seven major types of IDS: anomaly, signature, hybrid, network IDS, host IDS, protocol-based IDS, and application-based IDS. This chapter's prime focus is anomaly-based IDS for IoT networks. Anomaly-based IDS is based on a specific threshold through which cyberattacks can be easily detected. The signature-based IDS has a database of attack patterns. The hybrid IDS is a combination of anomaly and signature. The main contribution points of this chapter are as follows:

- AI- and ML-based anomaly-based IDS techniques and solutions.
- AI-enabled anomaly-based IDS applications and use cases.

DOI: 10.1201/9781003496410-11

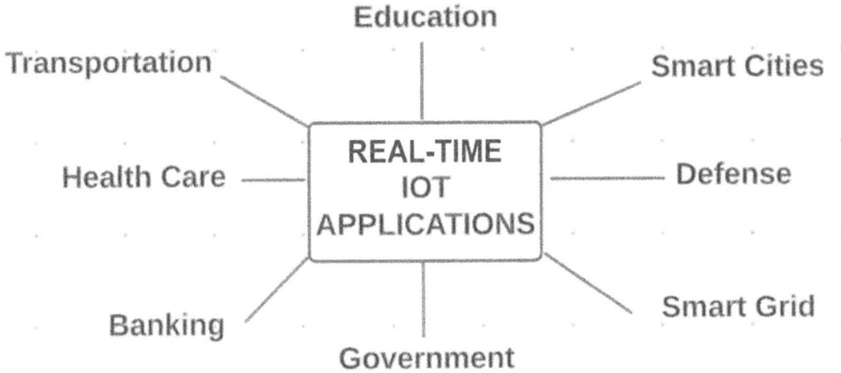

Figure 9.1 Real-time IoT applications.

- Balancing high accuracy using AI is incorporated.
- Future trends, challenges, and solutions are properly discussed.

9.2 LITERATURE STUDY

This section incorporates the limitations/drawbacks of traditional techniques for anomaly-based IDS for the IoT-enabled network:

> Anomaly based IDS is considered best possible practice for almost every field of study. This technique is quite helpful to detect possible cyber-attacks. Georgios Zachos et al., proposed anomaly based IDS for internet of medical thing's (IoMT's). There exist heterogeneous network resources in IoMT's which make it quite vulnerable. Although, proposed methodology easily use to collect history log files from IoMT sensors. However, machine learning techniques are utilized in simulation to reduce overheads. Apart from that IoMT's edge computing scenario is presented which consist of blood pressure sensor, body temperature sensor, motion sensor and insulin pump actuator. Therefore, data traffic monitoring need to be ensured in overall IoT-network. During experimentation binary classification is used. While, heavy weighted devices, high computation overhead and delay are the problems which need to be addressed properly [14].

Muaadh A. Alsoufi et al.'s IoT-enabled DL-based IDS is designed and compared with unsupervised and semi-supervised learning techniques. Anomaly-based IDS can be applied to many fields of studies like transportation, smart grid, smart cities, agriculture, and others to identify possible attacks. Zero-day attacks are the main focus for detection using anomaly IDS. A DL-based IDS is formulated to remove redundant records, and quality assessment is performed. Time complexity and training time need to be further optimized in the near future [15].

Saeid Sheikhi et al. proposed a novel IDS using particle swarm optimization, gray wolf optimization, and genetic algorithm to enhance cyberattack identification modeling. This method is simulated for the security of next-generation communication networks. The network security laboratory (NSL-KDD) dataset is used for simulation. Also, the backpropagation neural network model is simulated during experimentation. Still there is loophole regarding high accuracy and lower error rate, which must be improved. However, the proposed method is basically a hybrid combination of algorithms to detect cyber threats. Quality-of-service metrics like F1-score, recall, precision, and accuracy are used for performance evaluation [16].

Sujeong Kim et al. simulated the IoT-enabled environment to design special anomaly-based IDS. IoT networks have many security issues, especially log-based detection, which are mostly caused by malware. Local outlier factor and autoencoder can easily detect illegal and normal data packets in IoT networks. With the help of this research study, around eight new attacks are detected. The proposed model presents better simulation results and reduces false alarm. However, minimizing the false positive rate still needs to be balanced properly [17].

Ketan Kotecha et al. introduced a network intrusion detection model to detect anomalies. The University of New South-Wale Network-Based 2015 (UNSW-NB15) dataset is used for simulation experimentation. The proposed model performed with a high level of accuracy to detect unknown data packets. For a huge amount of data in a real-time environment, the threat of a cyberattack like zero-day is considered serious. Data packet flow must be properly observed during experimentation to for check normal or illegal information. Also, the idea of network-IPS should be introduced in the near future [18].

Illegal data packets can easily unbalance the entire IoT network. Sending fake data packets cause disruption in the network. DoS/DDoS increase the overhead. Malicious data packets unbalance the false positive and negative rates, through which IoT networks can be easily hijacked [19].

Wang et al. proposed a novel game tree model that is based on players, strategies, and a tree sort of structure. The game theory model utilizes attackers with different strategies using a tree-based model to detect cyber threats [20].

These traditional techniques are used to detect cyberattacks. Therefore, a novel threshold in an anomaly-based IDS can easily solve many related problems. Also, many dangerous cyber threats can be easily identified with the help of anomaly-based IDS.

9.3 AI-ENABLED ANOMALY INTRUSION DETECTION SYSTEM FOR IOT NETWORK

IoT network topology is quite versatile but is exposed to cyberattacks. Therefore, the IoT network needs proper security to secure communication channels.

Utilizing IDS monitoring of data packets is easily possible. AI for IDS is considered a new concept to detect cyber threats. Traditional IDS is specialized to specific cyberattacks. Due to this, ML techniques can be used for anomaly-based IDS to identify cyber threats. AI-/ML-based IDS is an effective approach that uses a huge amount of data. IoT sensor nodes collect information easily, which can be utilized in decision-making processes. AI-/ML-based IDS has the learning capacity to identify illegal data packets. A training and testing process is involved in AI/ML techniques. Traditional methods in comparison with AI/ML techniques are less popular. A high level of accuracy can be achieved using AI-based IDS. Also, DL is a category of machine learning that can also be deployed to detect cyber threats. In addition, AI-enabled IDS relies on feature selection, ML pipelining, data preprocessing, data labeling, data cleaning, data loading, and unbalancing datasets [21–24].

9.4 MACHINE LEARNING–BASED ANOMALY IDS

Machine learning is basically a data analytics approach. Also, ML is a subcategory of AI. There are two main types of ML, including supervised and unsupervised learning. Instances with labeled data come in the category of supervised learning. However, unsupervised learning has unlabeled data. ML also has approaches like semi-supervised and reinforcement learning. More interestingly, reinforcement techniques learn directly from the environment. ML techniques can be either used for classification or regression. Figure 9.2 illustrates the types of ML like supervised, unsupervised, semi-supervised, and reinforcement learning. During experimentation, binary or multilevel classification can be performed. Overfitting problem has been observed in ML techniques. Supervised learning–based IDS can easily predict or identify network traffic flow [25–27]. More interestingly, there are parametric and nonparametric models like logistic regression, naive Bayes, *k*-nearest neighbor, decision tree, and support vector machine (SVM) that can be used for cyberattack detection. Unsupervised clustering algorithms and soft computing models like artificial neural networks (ANNs), fuzzy logic, genetic model, ant colony optimization, and E-AntHocNet can also help to identify cyber threats [28–35].

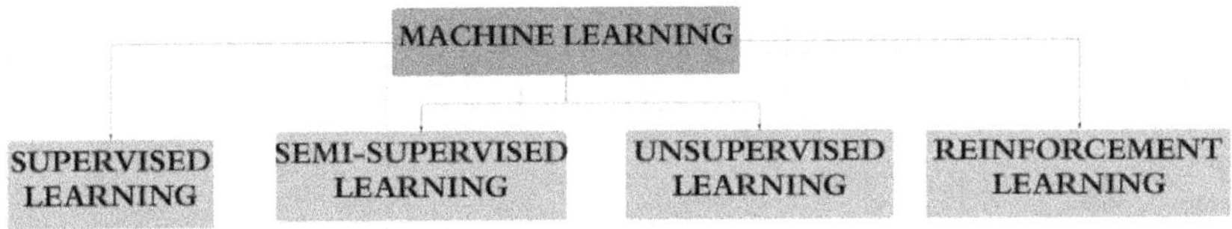

Figure 9.2 Types of machine learning (ML).

9.4.1 SVM-enabled cyberattack detection for IoT network

SVM is a widely used technique for attack detection. Feature selection process can be made possible with the help of SVM. SVM has two stages: training and testing/evaluation. Suad Mohammed Othman et al. proposed a novel technique called spark chi SVM model. Data preprocessing using spark is performed. The feature selection and extraction processes are simulated. The SVM model is the optimal technique to detect cyber threats in IoT environments [36–38].

9.4.2 Decision tree–enabled cyberattack detection for IoT network

Decision tree is a machine learning classifier to extract information. Decision tree can be used for data mining and cyber threat detection as well. IoT network is at high risk these days. However, decision tree is considered fast/reliable to identify cyberattacks from IoT network traces. Also, deployment of decision tree classifier can be made possible on either binary or multiclass classification. Hybridization of decision tree with genetic algorithm will be quite helpful for the feature selection process [39–43].

9.4.3 Random forest–enabled cyberattack detection for IoT network

Random forest (RF) is a machine learning technique that can be utilized for attack detection. Dangerous cyber threats like DoS/DDoS can be easily identified using RF classifier. Liguo Chen et al. proposed a novel model using RF for data traffic management. Also, other flooding attacks like user datagram protocol (UDP) and transmission control protocol (TCP) attacks can be detected with the help of RF. But RF classifier is time consuming. RF is based on the decision tree, which can do classification and regression [44–48].

9.4.4 Logistic regression–based cyberattack detection for IoT network

Logistic regression (LR) classifier is a machine learning approach that can be used for binominal or multiclass classification. However, difficult tasks like attack detection can be possible with the help of LR technique. LR is considered a lightweight classifier that can be applied in many applications like smart cities, drone communication, internet of medical things (IoMTs), internet of things (IoTs), smart grid, smart homes, and others [49–54].

9.5 AI-BASED ANOMALY IDS APPLICATIONS

There exist many AI-based anomaly IDS applications.

9.5.1 AI-based anomaly IDS for smart cities

Recently, the concept of smart cities is trending, and many researchers, practitioners, and engineers are attracted to the idea. Therefore, the infrastructure of smart cities needs to be secured. As IoT-enabled smart cities mainly rely on wireless communication networks, AI/ML techniques can be utilized to safeguard smart cities. With anomaly-based IDS, a novel threshold can be either based on AI/ML/DL. With the help of these techniques, overall smart city communication will be properly secured. Also, through AI techniques, by using IoT-enabled IDS, cyberattacks can be easily detected [55].

9.5.2 AI-based anomaly IDS for future transportation

Nowadays, the world is moving toward future connected transportation systems like drone taxi, autonomous vehicles, electric vehicles, and solar-based vehicles. Accordingly, security is much need for these systems. AI-/ML-based anomaly IDS will help to identify cyber threats and balance quality-of-service metrics [56]. Also, vehicular ad hoc networks can be utilized to improve communication within future connected transportation networks [57].

9.5.3 AI-based anomaly IDS for smart grid

IoT-enabled smart grid system has efficient electric connections. The smart grid mainly improves energy conservation, demand response, and smart metering. Due to complex IoT-enabled connectivity, wireless technology is used. Therefore, the smart grid has many vulnerabilities, through which an intruder can send fake data packets. Moreover, AI-/ML-based IDS can be used to identify cyber threats on the smart grid [58].

Apart from that, AI-based anomaly IDS will detect cyber threats in smart agriculture [59], smart homes [60], the health care industry [61], and educational organizations, and the military [62] as well [63].

9.6 AI-ENABLED ANOMALY DETECTION USE CASES

In this section, AI-enabled anomaly detection use cases are presented.

9.6.1 AI-anomaly detection for health care industry

AI tools can be deployed in the health care industry to detect cyber threats. The use of AI-enabled technologies will secure communication channels in wireless body area sensor networks. Also, AI techniques will be used for diagnoses, detection, prediction, and analyzing. Moreover, clinical trials using AI tools will help medical doctors and patients. AI/ML with natural language processing will be utilized for medical image processing. ML techniques can be used for genetic applications. Classical AI/ML methods can be directly applied to find anomalies. AI tools and techniques are quite helpful in the early diagnosis of cancerous cells, heart attacks, pregnancies, and others [64].

In addition, AI-enabled robots will do remote surgeries/operations. Also, in the near future, AI-enabled robots will change entire dynamics of hospitals. Due to this, the cybersecurity of these robots will be much needed. So AI/ML/DL techniques will help humans in a better way.

9.6.2 AI-anomaly detection for banking

AI techniques can be used in the banking sector to secure the entire network. AI/ML tools are helpful to detect illegal information, which disrupts communication. Moreover, AI software's algorithms can detect fraud and other illegal activities within the bank's network. ML models can detect and prevent attacks to the overall banking system. AI-based real-time detection mechanisms have been developed with the help of researchers/engineers to accurately identify credit card theft and forged documents [65].

9.6.3 AI-anomaly detection for defense and government

AI-enabled robust detection models will be designed especially for defense and government communication. Intruders can send fake information to obtain legitimate data from government agencies. Cyberattacks using malware/viruses and unknown flooding of data can be utilized to obtain direct access to secret defense networks. So AI/ML classifiers can be simulated to identify unknown anomalies [66].

9.7 REDUCING FALSE POSITIVE RATE USING AI

Due to the increasing demand of IoT devices, they must be secured. AI-enabled IDS model can easily observe/monitor data traffic to detect unknown activity. ML and DL techniques are widely used to balance high-accuracy metrics. However, still there is the main problem of balancing false positive/negative rates. AI-based IDS always has the problem of high false positive rate. Due to this, fake messages will be generated to hijack/take control of the entire network. False negative rate shows a cyber threat like a normal data pattern that affects the whole network. Regular network traffic monitoring can be made possible through AI-based IDS. More interestingly, AI/ML/DL techniques can reduce the false positive rate [67, 68]. Figure 9.3 presents the balancing of high-accuracy metrics like false negative/positive and true positive/negative.

HIGH-ACCURACY METRICS	POSITIVE	NEGATIVE
POSITIVE	TRUE POSITIVE PATTERN MATCHED & ATTACK PRESENT	FALSE POSITIVE PATTERN MATCHED & NO ATTACK PRESENT
NEGATIVE	FALSE NEGATIVE NO PATTERN MATCHED & NO ATTACK PRESENT	TRUE NEGATIVE NO PATTERN MATCHED & ATTACK PRESENT

Figure 9.3 High-accuracy metrics.

9.8 FUTURE TRENDS AND CHALLENGES

AI-based IDS plays an important role in the security of IoT networks. Anomaly-based IDS using AI techniques are much better than signature and hybrid IDS. AI-/ML-based IDS in the near future must be less costly and achieve a high level of accuracy to detect dangerous cyberattacks [69]. Cyber threats like DNS [70], DoS, DDoS [71], Sybil [72], TCP/UDP and man-in-the-middle attacks can be easily identified using optimization techniques, game theory, fuzzy logic, ANNs, and ML/DL.

9.9 CONCLUSION

This chapter provides detailed information about AI-/ML-based IDS for IoT networks. Many IoT- and IDS-related issues are highlighted. ML techniques like SVM, decision tree, LR, and RF are discussed in detail. These techniques improve level accuracy in the detection of cyberattacks. AI/ML techniques can be only applied on a large number of datasets. Various real-time applications of AI-based IDS are incorporated. Also, AI-/ML-based IDS can be helpful in both civil and military domains. Real-time case studies about health care, defense, government, and banking using AI-based IDS are discussed. False positive/negative rate needs to be balanced with the help of AI/ML techniques. Apart from this, cybersecurity solutions and future trends are presented properly in this chapter.

REFERENCES

[1] Abdollahi, Asrin, and Mohammad Fathi. "An intrusion detection system on ping of death attacks in IoT networks." Wireless Personal Communications 112 (2020): 2057–2070.

[2] Khan, Inam Ullah, Asrin Abdollahi, Ryan Alturki, Mohammad Dahman Alshehri, Mohammed Abdulaziz Ikram, Hasan J. Alyamani, and Shahzad Khan. "Intelligent detection system enabled attack probability using Markov chain in aerial networks." Wireless Communications and Mobile Computing 2021 (2021): 1–9.

[3] Kyriazis, Dimosthenis, Theodora Varvarigou, Daniel White, Andrea Rossi, and Joshua Cooper. "Sustainable smart city IoT applications: Heat and electricity management & Eco-conscious cruise control for public transportation." In 2013 IEEE 14th International Symposium on "A World of Wireless, Mobile and Multimedia Networks" (WoWMoM), pp. 1–5. IEEE, 2013.

[4] Malche, Timothy, and Priti Maheshwary. "Internet of Things (IoT) for building smart home system." In 2017 International Conference on I-SMAC (IoT in Social, Mobile, Analytics and Cloud) (I-SMAC), pp. 65–70. IEEE, 2017.

[5] Zantalis, Fotios, Grigorios Koulouras, Sotiris Karabetsos, and Dionisis Kandris. "A review of machine learning and IoT in smart transportation." Future Internet 11, no. 4 (2019): 94.

[6] Mezghani, Emna, Ernesto Exposito, and Khalil Drira. "A model-driven methodology for the design of autonomic and cognitive IoT-based systems: Application to healthcare." IEEE Transactions on Emerging Topics in Computational Intelligence 1, no. 3 (2017): 224–234.

[7] Zhao, Ji-Chun, Jun-Feng Zhang, Yu Feng, and Jian-Xin Guo. "The study and application of the IOT technology in agriculture." In 2010 3rd International Conference on Computer Science and Information Technology, vol. 2, pp. 462–465. IEEE, 2010.

[8] Ou, Qinghai, Yan Zhen, Xiangzhen Li, Yiying Zhang, and Lingkang Zeng. "Application of internet of things in smart grid power transmission." In 2012 Third FTRA International Conference on Mobile, Ubiquitous, and Intelligent Computing, pp. 96–100. IEEE, 2012.

[9] Bichi, Bashir Yusuf, Saif ul Islam, Anas Maazu Kademi, and Ishfaq Ahmad. "An energy-aware application module for the fog-based internet of military things." Discover Internet of Things 2, no. 1 (2022): 4.

[10] Da Costa, Kelton A.P., João P. Papa, Celso O. Lisboa, Roberto Munoz, and Victor Hugo C. de Albuquerque. "Internet of Things: A survey on machine learning-based intrusion detection approaches." Computer Networks 151 (2019): 147–157.

[11] Wang, Zheng. "Deep learning-based intrusion detection with adversaries." IEEE Access 6 (2018): 38367–38384.

[12] Agrawal, Shaashwat, Sagnik Sarkar, Ons Aouedi, Gokul Yenduri, Kandaraj Piamrat, Mamoun Alazab, Sweta Bhattacharya, Praveen Kumar Reddy Maddikunta, and Thippa Reddy Gadekallu. "Federated learning for intrusion detection system: Concepts, challenges and future directions." Computer Communications 195 (2022): 346–361.

[13] Y. Shah and S. Sengupta, "A survey on classification of cyber-attacks on IoT and IIoT devices." In 2020 11th IEEE Annual Ubiquitous Computing, Electronics & Mobile Communication Conference (UEMCON), New York, NY, pp. 0406–0413, 2020, doi: 10.1109/UEMCON51285.2020.9298138.

[14] Zachos, Georgios, Ismael Essop, Georgios Mantas, Kyriakos Porfyrakis, José C. Ribeiro, and Jonathan Rodriguez. "An anomaly-based intrusion detection system for internet of medical things networks." Electronics 10, no. 21 (2021): 2562.

[15] Alsoufi, Muaadh A., Shukor Razak, Maheyzah Md Siraj, Ibtehal Nafea, Fuad A. Ghaleb, Faisal Saeed, and Maged Nasser. "Anomaly-based intrusion detection

systems in IoT using deep learning: A systematic literature review." Applied Sciences 11, no. 18 (2021): 8383.
[16] Sheikhi, Saeid, and Panos Kostakos. "A novel anomaly-based intrusion detection model using PSOGWO-optimized BP neural network and GA-based feature selection." Sensors 22, no. 23 (2022): 9318.
[17] Kim, Sujeong, Chanwoong Hwang, and Taejin Lee. "Anomaly based unknown intrusion detection in endpoint environments." Electronics 9, no. 6 (2020): 1022.
[18] Kotecha, Ketan, Raghav Verma, Prahalad V. Rao, Priyanshu Prasad, Vipul Kumar Mishra, Tapas Badal, Divyansh Jain, Deepak Garg, and Shakti Sharma. "Enhanced network intrusion detection system." Sensors 21, no. 23 (2021): 7835.
[19] Sedjelmaci, Hichem, and Sidi Mohammed Senouci. "An accurate and efficient collaborative intrusion detection framework to secure vehicular networks." Computers & Electrical Engineering 43 (2015): 33–47.
[20] Wang, Kun, Miao Du, Dejun Yang, Chunsheng Zhu, Jian Shen, and Yan Zhang. "Game-theory-based active defense for intrusion detection in cyber-physical embedded systems." ACM Transactions on Embedded Computing System (2016): 1–21.
[21] Kanimozhi, V., and T. Prem Jacob. "Artificial intelligence based network intrusion detection with hyper-parameter optimization tuning on the realistic cyber dataset CSE-CIC-IDS2018 using cloud computing." In 2019 International Conference on Communication and Signal Processing (ICCSP), pp. 0033–0036. IEEE, 2019.
[22] Morel, Benoît. "Anomaly based intrusion detection and artificial intelligence." Intrusion Detection Systems 10 (2011): 14103.
[23] Patil, Shruti, Vijayakumar Varadarajan, Siddiqui Mohd Mazhar, Abdulwodood Sahibzada, Nihal Ahmed, Onkar Sinha, Satish Kumar, Kailash Shaw, and Ketan Kotecha. "Explainable artificial intelligence for intrusion detection system." Electronics 11, no. 19 (2022): 3079.
[24] Alrajeh, Nabil Ali, and Jaime Lloret. "Intrusion detection systems based on artificial intelligence techniques in wireless sensor networks." International Journal of Distributed Sensor Networks 9, no. 10 (2013): 351047.
[25] Saheed, Yakub Kayode, Aremu Idris Abiodun, Sanjay Misra, Monica Kristiansen Holone, and Ricardo Colomo-Palacios. "A machine learning-based intrusion detection for detecting internet of things network attacks." Alexandria Engineering Journal 61, no. 12 (2022): 9395–9409.
[26] Dang, Quang-Vinh. "Studying machine learning techniques for intrusion detection systems." In Future Data and Security Engineering: 6th International Conference, FDSE 2019, Nha Trang City, 27–29 November, Proceedings 6, pp. 411–426. Springer International Publishing, 2019.
[27] Halimaa, Anish, and K. Sundarakantham. "Machine learning based intrusion detection system." In 2019 3rd International Conference on Trends in Electronics and Informatics (ICOEI), pp. 916–920. IEEE, 2019.
[28] Chitrakar, Roshan, and Huang Chuanhe. "Anomaly detection using Support Vector Machine classification with k-Medoids clustering." In 2012 Third Asian Himalayas International Conference on Internet, pp. 1–5. IEEE, 2012.
[29] Khan, Inam Ullah, Ijaz Mansoor Qureshi, Muhammad Adnan Aziz, Tanweer Ahmad Cheema, and Syed Bilal Hussain Shah. "Smart IoT control-based nature inspired energy efficient routing protocol for flying ad hoc network (FANET)." IEEE Access 8 (2020): 56371–56378.

[30] Khan, Inam Ullah, Syeda Zillay Nain Zukhraf, Asrin Abdollahi, Shahbaz Ali Imran, Ijaz Mansoor Qureshi, Muhammad Adnan Aziz, and Syed Bilal Hussian Shah. "Reinforce based optimization in wireless communication technologies and routing techniques using internet of flying vehicles." In Proceedings of the 4th International Conference on Future Networks and Distributed Systems, ACM, pp. 1–6, 2020.

[31] Khan, I. Ullah, M. Abul Hassan, Muhammad Fayaz, Jeonghwan Gwak, and M. Adnan Aziz. "Improved sequencing heuristic DSDV protocol using nomadic mobility model for FANETS." Computers, Materials & Continua 70, no. 2 (2022): 3653–3666.

[32] Syarif, Iwan, Adam Prugel-Bennett, and Gary Wills. "Unsupervised clustering approach for network anomaly detection." In Networked Digital Technologies: 4th International Conference, NDT 2012, Dubai, 24–26 April. Proceedings, Part I 4, pp. 135–145. Springer, 2012.

[33] Aung, Khin Moh, and Nyein Oo. "Association rule pattern mining approaches network anomaly detection." PhD diss., MERAL Portal, 2015.

[34] Hamamoto, Anderson Hiroshi, Luiz Fernando Carvalho, Lucas Dias Hiera Sampaio, Taufik Abrão, and Mario Lemes Proença Jr. "Network anomaly detection system using genetic algorithm and fuzzy logic." Expert Systems with Applications 92 (2018): 390–402.

[35] Tufan, Emrah, Cihangir Tezcan, and Cengiz Acartürk. "Anomaly-based intrusion detection by machine learning: A case study on probing attacks to an institutional network." IEEE Access 9 (2021): 50078–50092.

[36] Ioannou, Christiana, and Vasos Vassiliou. "Network attack classification in IoT using support vector machines." Journal of Sensor and Actuator Networks 10, no. 3 (2021): 58.

[37] Anwer, Maryam, S. Mahmood Khan, and M. Umer Farooq. "Attack detection in IoT using machine learning." Engineering, Technology & Applied Science Research 11, no. 3 (2021): 7273–7278.

[38] Othman, Suad Mohammed, Fadl Mutaher Ba-Alwi, Nabeel T. Alsohybe, and Amal Y. Al-Hashida. "Intrusion detection model using machine learning algorithm on Big Data environment." Journal of Big Data 5, no. 1 (2018): 1–12.

[39] Poddar, Rahul, and Hari Babu. "Decision tree based IoT attack detection in programmable data plane using P4 language." In International Conference on Advanced Information Networking and Applications, pp. 671–683. Springer International Publishing, 2022.

[40] Panigrahi, Ranjit, Samarjeet Borah, Akash Kumar Bhoi, Muhammad Fazal Ijaz, Moumita Pramanik, Yogesh Kumar, and Rutvij H. Jhaveri. "A consolidated decision tree-based intrusion detection system for binary and multiclass imbalanced datasets." Mathematics 9, no. 7 (2021): 751.

[41] Wilborne, Deanna. "Application of decision tree classifier in detection of specific denial of service attacks with genetic algorithm based feature selection on NSL-KDD." arXiv preprint arXiv:2210.10232 (2022).

[42] Su, Jingyi, Shan He, and Yan Wu. "Features selection and prediction for IoT attacks." High-Confidence Computing 2, no. 2 (2022): 100047.

[43] Lakshminarasimman, S., S. Ruswin, and K. Sundarakantham. "Detecting DDoS attacks using decision tree algorithm." In 2017 Fourth International Conference on Signal Processing, Communication and Networking (ICSCN), pp. 1–6. IEEE, 2017.

[44] Najar, Ashfaq Ahmad, and S. Manohar Naik. "DDoS attack detection using MLP and Random Forest Algorithms." International Journal of Information Technology 14, no. 5 (2022): 2317–2327.
[45] Chen, Liguo, Yuedong Zhang, Qi Zhao, Guanggang Geng, and ZhiWei Yan. "Detection of DNS DDoS attacks with random forest algorithm on spark." Procedia Computer Science 134 (2018): 310–315.
[46] Chen, Yini, Jun Hou, Qianmu Li, and Huaqiu Long. "DDoS attack detection based on random forest." In 2020 IEEE International Conference on Progress in Informatics and Computing (PIC), pp. 328–334. IEEE, 2020.
[47] Shi, Zhan, Yutu Liang, and Bo Li. "Research on attack detection based on random forest improved algorithm for power line carrier communication." In 2022 7th International Conference on Computer and Communication Systems (ICCCS), pp. 784–788. IEEE, 2022.
[48] Alduailij, Mona, Qazi Waqas Khan, Muhammad Tahir, Muhammad Sardaraz, Mai Alduailij, and Fazila Malik. "Machine-learning-based DDoS attack detection using mutual information and random forest feature importance method." Symmetry 14, no. 6 (2022): 1095.
[49] De Caigny, Arno, Kristof Coussement, and Koen W. De Bock. "A new hybrid classification algorithm for customer churn prediction based on logistic regression and decision trees." European Journal of Operational Research 269, no. 2 (2018): 760–772.
[50] Arabameri, Alireza, Biswajeet Pradhan, Khalil Rezaei, Mojtaba Yamani, Hamid Reza Pourghasemi, and Luigi Lombardo. "Spatial modelling of gully erosion using evidential belief function, logistic regression, and a new ensemble of evidential belief function–logistic regression algorithm." Land Degradation & Development 29, no. 11 (2018): 4035–4049.
[51] Böhning, Dankmar. "Multinomial logistic regression algorithm." Annals of the Institute of Statistical Mathematics 44, no. 1 (1992): 197–200.
[52] Chaudhuri, Kamalika, and Claire Monteleoni. "Privacy-preserving logistic regression." Advances in Neural Information Processing Systems 21 (2008).
[53] Srimaneekarn, Natchalee, Anthony Hayter, Wei Liu, and Chanita Tantipoj. "Binary response analysis using logistic regression in dentistry." International Journal of Dentistry 2022 (2022).
[54] Jullian, Olivia, Beatriz Otero, Eva Rodriguez, Norma Gutierrez, Héctor Antona, and Ramon Canal. "Deep-learning based detection for cyber-attacks in IoT networks: A distributed attack detection framework." Journal of Network and Systems Management 31, no. 2 (2023): 33.
[55] Elsaeidy, Asmaa, Kumudu S. Munasinghe, Dharmendra Sharma, and Abbas Jamalipour. "Intrusion detection in smart cities using restricted Boltzmann machines." Journal of Network and Computer Applications 135 (2019): 76–83.
[56] Aloqaily, Moayad, Safa Otoum, Ismaeel Al Ridhawi, and Yaser Jararweh. "An intrusion detection system for connected vehicles in smart cities." Ad Hoc Networks 90 (2019): 101842.
[57] Bhatia, Tarandeep Kaur, Ramkumar Ketti Ramachandran, Robin Doss, and Lei Pan. "Data congestion in VANETs: Research directions and new trends through a bibliometric analysis." The Journal of Supercomputing 77 (2021): 6586–6628.
[58] Radoglou Grammatikis, Panagiotis, Panagiotis Sarigiannidis, Georgios Efstathopoulos, and Emmanouil Panaousis. "ARIES: A novel multivariate intrusion detection system for smart grid." Sensors 20, no. 18 (2020): 5305.

[59] Zhao, Ji-Chun, Jun-Feng Zhang, Yu Feng, and Jian-Xin Guo. "The study and application of the IOT technology in agriculture." In 2010 3rd International Conference on Computer Science and Information Technology, vol. 2, pp. 462–465. IEEE, 2010.

[60] Malche, Timothy, and Priti Maheshwary. "Internet of Things (IoT) for building smart home system." In 2017 International Conference on I-SMAC (IoT in Social, Mobile, Analytics and Cloud) (I-SMAC), pp. 65–70. IEEE, 2017.

[61] Mezghani, Emna, Ernesto Exposito, and Khalil Drira. "A model-driven methodology for the design of autonomic and cognitive IoT-based systems: Application to healthcare." IEEE Transactions on Emerging Topics in Computational Intelligence 1, no. 3 (2017): 224–234.

[62] Bichi, Bashir Yusuf, Saif ul Islam, Anas Maazu Kademi, and Ishfaq Ahmad. "An energy-aware application module for the fog-based internet of military things." Discover Internet of Things 2, no. 1 (2022): 4.

[63] Bhavsar, Mansi, Kaushik Roy, John Kelly, and Odeyomi Olusola. "Anomaly-based intrusion detection system for IoT application." Discover Internet of Things 3, no. 1 (2023): 5.

[64] Samariya, Durgesh, Jiangang Ma, Sunil Aryal, and Xiaohui Zhao. "Detection and explanation of anomalies in healthcare data." Health Information Science and Systems 11, no. 1 (2023): 20.

[65] Dumitrescu, Bogdan, Andra Băltoiu, and Ştefania Budulan. "Anomaly detection in graphs of bank transactions for anti money laundering applications." IEEE Access 10 (2022): 47699–47714.

[66] Morel, Benoît. "Anomaly based intrusion detection and artificial intelligence." Intrusion Detection Systems 10 (2011): 14103.

[67] Mijalkovic, Jovana, and Angelo Spognardi. "Reducing the false negative rate in deep learning based network intrusion detection systems." Algorithms 15, no. 8 (2022): 258.

[68] Liu, Hongyu, and Bo Lang. "Machine learning and deep learning methods for intrusion detection systems: A survey." Applied Sciences 9, no. 20 (2019): 4396.

[69] Martins, Ines, Joao S. Resende, Patricia R. Sousa, Simao Silva, Luis Antunes, and Joao Gama. "Host-based IDS: A review and open issues of an anomaly detection system in IoT." Future Generation Computer Systems 133 (2022): 95–113.

[70] Rosenthal, Gilad, Ofir Erets Kdosha, Kobi Cohen, Alon Freund, Avishay Bartik, and Aviv Ron. "ARBA: Anomaly and reputation based approach for detecting infected IoT devices." IEEE Access 8 (2020): 145751–145767.

[71] Hussain, Faisal, Syed Ghazanfar Abbas, Muhammad Husnain, Ubaid U. Fayyaz, Farrukh Shahzad, and Ghalib A. Shah. "IoT DoS and DDoS attack detection using ResNet." In 2020 IEEE 23rd International Multitopic Conference (INMIC), pp. 1–6. IEEE, 2020.

[72] Rajan, Anjana, J. Jithish, and Sriram Sankaran. "Sybil attack in IOT: Modelling and defenses." In 2017 International Conference on Advances in Computing, Communications and Informatics (ICACCI), pp. 2323–2327. IEEE, 2017.

Part III

Big data analytics applying in current applications

Chapter 10

Big data intelligence in health care

Abinaya Keerthana S. and Sabitha Banu A

10.1 INTRODUCTION

The research into illness diagnostics is crucial in the realm of health care. A disease is defined as any factor or set of events that produces suffering, ailment, malfunction, or even death in humans. Both the physical and mental well-being of a person can be damaged by diseases, which also significantly alter the affected individual's way of life. The medical field has a lengthy history of embracing new technology before their time. As of right now, machine learning (ML) and deep learning (DL), two subsets of artificial intelligence (AI), are on their approach to replacing humans as a dominant force in the health care industry. They are already utilized to create new health check activities and manage patient accounts and information.

10.2 ALZHEIMER'S DISEASE

Alzheimer's disease (AD) is a neurodegenerative condition that is irreversible and gradually reduces intellectual capacity until dementia emerges. Objective cognitive abnormalities (which are often severe memory impairments) are used to make the clinical diagnosis of AD. In certain circumstances, AD may manifest with unusual symptoms, notably deficits in non-amnesic areas. The DL algorithm correctly identified Alzheimer's patients 90.2% of the time.

10.2.1 AI for AD diagnosis

Thanks in part to the use of ML algorithms, complex, extremely dimensional systems may be managed by AI technology that are beyond the capabilities of personal information processing. By combining data from recently invented tools with data from health records, net promoter score (NPS) examination, brain scanning, and biologic markers, ML has been applied in the diagnosis of a number of illnesses, including Alzheimer's, which is given in Figure 10.1. Through the use of a magnetic resonance imaging PET, 18F-fluorodeoxyglucose-Positron, the

DOI: 10.1201/9781003496410-13

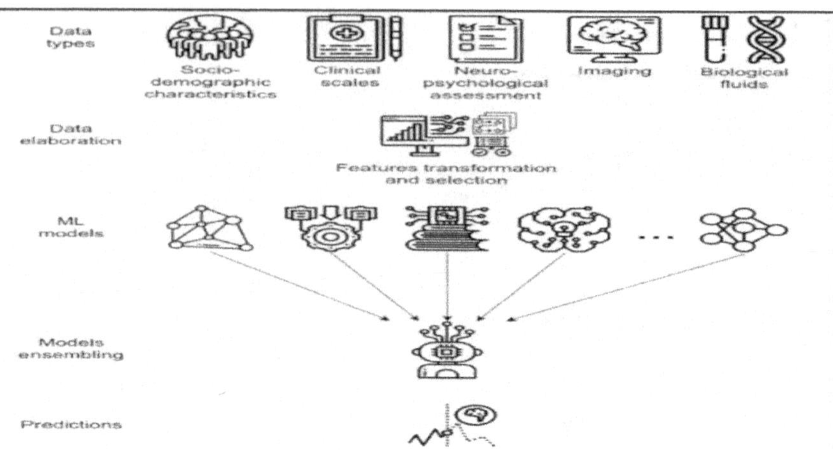

Figure 10.1 AI in alzheimer's disease.

release tomography (FDG-PET), and distribution tensor visualizing (DTI), comprehensive information about the structure and function of the brain is obtained [1]. This information enables the identification of characteristics that support the determination of the condition, such as wasting, amyloid testimony, or microstructural destruction. Brain tumors or cerebrovascular illness are two examples of pathogenic processes that can impair cognitive function but are not caused by AD that can be distinguished using neuro images.

10.3 CARDIOVASCULAR DISEASE

Human health is seriously threatened by cardiovascular diseases (CVD), which are the world's leading killers. The number of CVD-related deaths rose by 14.5% from 2006 to 2016, to 17.6 million. Tragically, especially in emerging nations, the death and morbidity rates of CVD are rising every year. Research has shown that nations with low to middle incomes account for around 80% of fatalities caused by CVD. These mortalities also occur earlier than in nations with high incomes as a result of environmental modifications and unhealthy lifestyles brought on by these nations' rapid economic transition. In emerging nations, elderly populations might raise CVD incidence and cardiovascular risk factors.

10.3.1 AI for CVD diagnosis

Cardiovascular medicine currently uses accurate medicine, therapeutic prediction, cardiac imaging research, and smart robotics. The potential for AI in cardiovascular treatment is promising [2]. With AI's help, it is likely to be possible to develop a precise medical plan that specifically tailors treatment for each patient. Clinicians won't be replaced by AI, according to the mainstream agreement [3]. On the other hand, practitioners should be informed about the use of AI. Through the study of large data, AI technology can help physicians better diagnose and treat patients with cardiovascular disease. This will hasten the arrival of the precision medicine era [3]. AI will probably help with precision medicine, which will individualize therapy for each patient. It is anticipated that even after AI, clinicians would still be required.

10.3.2 Cardiac imaging analysis

The study of cardiac imaging provides a number of intriguing future applications because deep learning has advanced recently. It is feasible to use DL to analyze coronary angiography, echocardiograms, and electrocardiograms (ECG) [4]. The utilization of cardiac intervention, involving CVD and ACS (acute coronary syndrome) treatment, has significantly increased in recent years. Thanks to deep learning, AI will soon be ready to identify cardiac plaques that are atherosclerotic with greater accuracy than physicians [2]. Moreover, AI may be used to the interpretation of echocardiographic imagery, including the automatic estimation of the size of each heart chamber and the evaluation of left ventricular function [4]. It may also be utilized to evaluate structural sickness such as valve disease to aid in disease staging. Various areas of cardiac imaging are given in Figure 10.2.

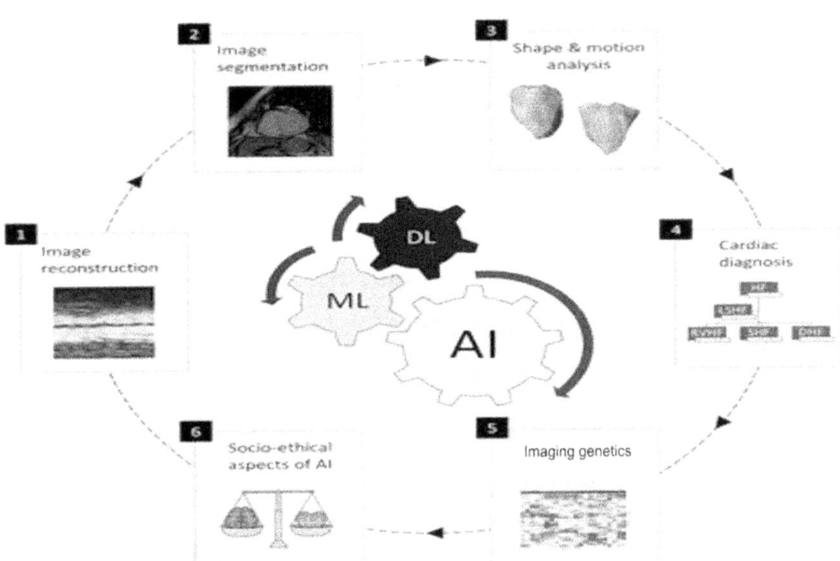

Figure 10.2 Different areas of cardiac imaging.

10.4 ARTIFICIAL INTELLIGENCE IN CANCER DISEASE

The biggest medical issue on the planet is cancer. The significant degree of heterogeneity in cancer makes it possible for people with the same tumor to respond differently to the same medications or surgical procedures, requiring the development of more precise tumor-specific therapeutic approaches and individualized patient care [5, 6]. Being able to build patient-specific treatment plans requires developing a thorough knowledge of the changes that tumors go through, including changes to their genes, proteins, and cancer cell morphologies. This is because the exact treatment of tumors is crucial. Big data–based AI can glean relevant knowledge, significant facts, and hidden patterns from the vast volume of data.

10.4.1 Magnetic resonance imaging

Anatomical features may be visualized in three dimensions with great detail using magnetic resonance imaging (MRI). It is frequently used for illness detection, diagnosis, and monitoring. Water, which makes up living tissues, includes protons that may be triggered and detected by cutting-edge technology. Based on relaxation pathways, MRI images may be divided into longitudinal (T1) and transverse (T2) weighted images [5, 6]. Contrast chemicals can enhance the clarity and understanding of MRI images. For T2, magnetic nanoparticles (MNPs) are frequently employed as contrast agents, while T1 uses paramagnetic complexes [7]. According to a standard technique (Figure 10.3), a radio frequency pulse is administered to the body of a living creature in a static magnetic field to excite hydrogen protons within the body and produce magnetic resonance (MR). As soon as the pulse is interrupted, the protons begin to relax, producing an MR signal. To create MR signals, a number of operations must be carried out, including image reconstruction, spatial coding, and reception. Spin properties are what produce the majority of MR signals. Generally speaking, 1H is one of the most appropriate elements for nuclear MRI and nuclear magnetic resonance (NMR) spectroscopy due to its innate sensitivity.

10.4.2 Future research in cancer disease

The current level of research mostly concentrates on the prediction accuracy of a few applied AI systems. Human-computer interaction, often known as computer-human interaction, is a subject that is discussed substantially less frequently. Investigating how the relationship between AI and humans may or should appear in the context of cancer detection is crucial [7]. There are other potential results, such as substitution, amplification, and gathering. While assemblage refers to the situation in which AI and people are dynamically combined to work as a single entity, augmentation describes the scenario in which AI and humans enhance one another. Last, but not least, substitution denotes the AI system's total replacement of the human.

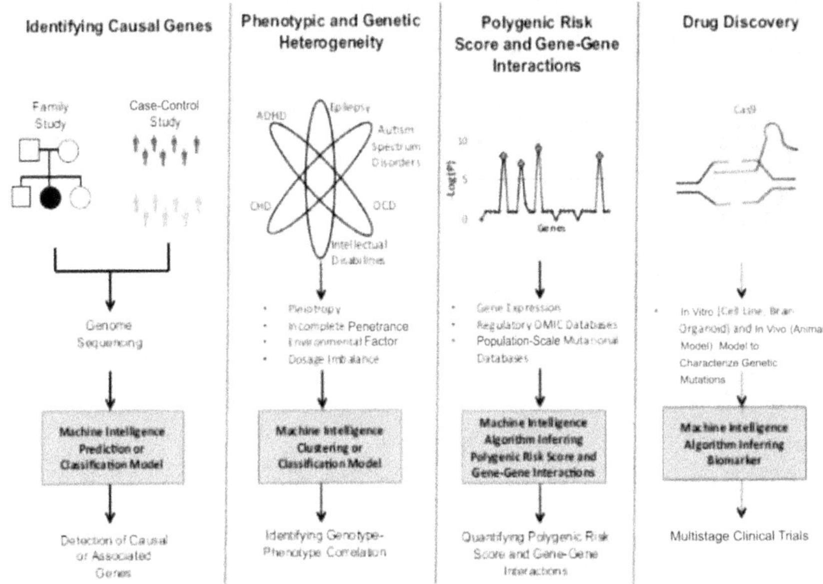

Figure 10.3 Precision medicine for cancer.

10.5 DIABETES DISEASES

The term "diabetes," often known as diabetes mellitus (DM), refers to a number of illnesses that affect how the body uses food for energy. After eating, the body turns the food into the sugar glucose and releases it into circulation. When blood sugar levels climb over normal but not high enough for a doctor to diagnose diabetes, this condition is known as prediabetes. Diabetes type II and heart disease are both more likely in those with prediabetes. These risks can be decreased by increasing exercise and losing weight—often less than 5%–7% of one's body weight, which is excess weight. The following comprise the suggested categories for diabetes management to condense the most recent contributions and outcomes detailed in the papers under review:

- Blood glucose management techniques
- Blood sugar forecast
- Identifying negative glycemic events
- Calculators and advising systems for insulin boluses
- Patient personalization and risk
- Identifying meals, exercise, and errors
- Support for managing diabetes via lifestyle and day-to-day activities

10.5.1 AI in diabetes disease

Whether using ML or DL, there are several ways to create algorithms to recognize the presence of diabetes. ML is a technique in which people train computers to gather certain patterns of characteristics from data to do a particular task. Only the characteristics that the machine has been taught to recognize may be detected using this method. As a result, ML-based systems can predict when trained levels of diabetic retinopathy (DR) will occur [8]. Conversely, DL is a type of machine learning in which an algorithm is developed to instantly recognize the best predictive qualities from sizable datasets that have been labeled with specific details. Artificial neural networks (ANNs) are the foundation of deep learning, which mimics how the human brain operates and discovers aspects directly from data. An example of a prediction model is given in Figure 10.4.

10.5.2 Future AI prediction in diabetes

Prediction and risk stratification make up the fourth category of AI applications in the diagnosis and management of diabetes. In preemptive medicine, this group may be used to precisely identify members of the general public who are extremely likely to develop a certain disease at the pre-illness stage [8]. Thus, by setting up health care for these individuals at a very early stage, this technique might ultimately eradicate the incidence of diabetes. With the development of ML technologies, diabetes onset cannot be predicted [9].

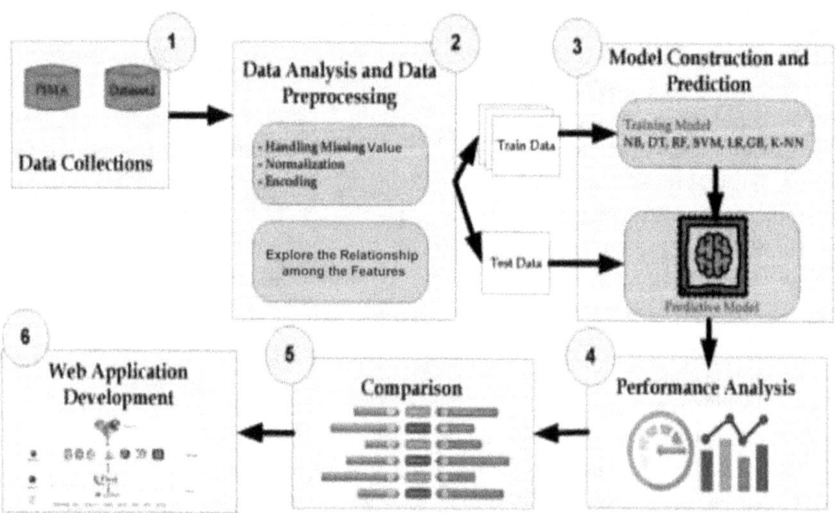

Figure 10.4 Prediction model of diabetes diseases.

10.6 TUBERCULOSIS DISEASES

In addition to killing approximately 1.4 million people each year and the danger that tuberculosis (TB) poses to the public's health, DR-TB cases are on the rise, making it harder to contain the pandemic. Bacterial cultures and acid-fast bacilli are two accurate but time- and condition-consuming bacteria-based detection techniques. The laboratory setting limits and complicates gene testing to detect medication resistance or infection in the pathogen *Mycobacterium tuberculosis*. TB is an infectious illness that transmits from one person to another directly and is a leading cause of morbidity and death around the globe. Anti-TB medications can effectively treat TB patients, and the drug regimen, dosage, and length of the treatment period depend on the exact spot of the infection in the body, whether it is a drug-resistant strain or not, and any coexisting conditions. The majority of anti-TB drugs may be harmful to the liver and cause hepatitis as a side effect.

10.6.1 Pulmonary tuberculosis detection

The World Health Organization (WHO) has recommended using chest radiographs for TB screening in high-risk populations since they are comparatively common, and a large proportion of people diagnosed with lung tuberculosis show unfavorable pulmonary chest x-rays (CXR) with signs of suspicion for pulmonary TB (PTB), such as tooth decay, centrilobular masses, and reorganizations. Similar to this, when PTB happens, anomalies might be seen on CT imaging. A DL model is frequently trained using these typical medical photos to identify PTB sufferers. Using radiographic results, symptoms, and demographic data, in 1999, a machine learning neural network was utilized to forecast active tuberculosis.

10.6.2 Future AI predictions in tuberculosis diseases

The findings of programmed sorting of medical photographs are reported in two categories in this work: with and without TB. Features are extracted for the ResNet-50 ANN with advanced ML for categorization. A pair of grouping scenarios was used: cross-validation and the generation of both testing and training sets. The scenario with the most beneficial results was the one in which the training and test sets were produced with higher than 85% accuracy.

10.7 STROKE DETECTION

In the United States, strokes rank as the sixth leading cause of death. Major vascular occlusion (LVO) and cerebral hemorrhage (ICH) are two stroke subtypes that have documented treatment options. One of the most important factors in optimizing neurological function preservation is prompt treatment. The fact that every minute of treatment delay results in considerable neuronal death and 4.2 days of healthy life lost is important. Tools that speed up and increase the detection and treatment of stroke may improve patient outcomes [10]. AI/ML will be crucial for the development of such technologies. AI/ML use in health care is increasing at a rate of 40% yearly, with the potential to lower costs by USD 150 billion by 2026.

10.7.1 Using AI in stroke disease

Cerebral infarctions are frequently caused by cardio embolic stroke, which is also increasing in frequency in industrialized nations. Stroke and systemic embolism risk are increased by the prevalence of arrhythmias. According to the review's results, particularly when it comes to identifying people who are at a high risk of chaotic atrial fibrillation (AF) or cardiac flutter, ML and AI have the potential to totally revolutionize how stroke danger is detected and managed. Subudhi, Dash, and Sabut highlight the incredible ability of an AI-ECG model, created using more than 500,000 typical sinus rhythm conventional 12-lead ECGs from more than 180,000 people, to detect patients at high risk of chaotic AF or cardiac flutter [10]. The risk of generalization is increased since AI-ECG models are developed for homogenous individuals because it is uncertain how ethnic origin and race may affect the interpretation of ECG data using ML [11].

The findings of this study imply that DL is possibly the technique most often employed for ECG evaluation and meaning. Convolutional neural networks (CNNs) are being employed in increasing numbers in computerized categorization because of their ability to categorize ECG patterns with precision. By applying DL to a straightforward screening test, such as a chest radiograph, cardio embolic infarction classification using supervised learning was competently conducted by making use of a digital medical record collection. A few studies have tried to classify cardio embolic stroke [12]. In separating cardio embolic from non-cardio

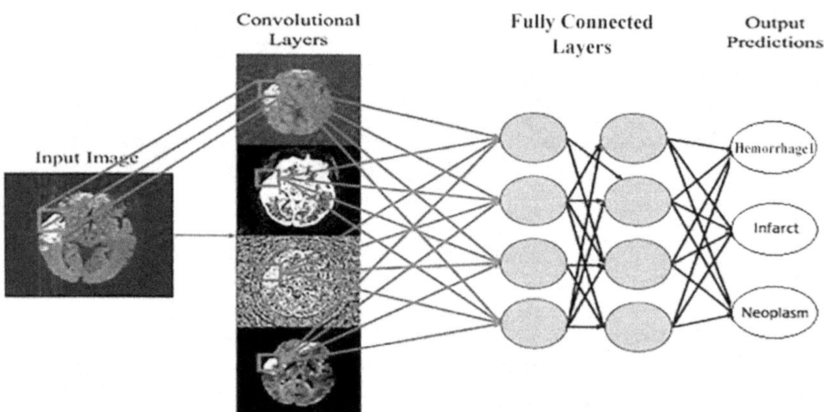

Figure 10.5 Stroke prediction using AI.

embolic stroke, the resultant system performed better in high-risk sources like AF, demonstrating substantial classification capability and biological relevance. Figure 10.5 shows the prediction model of a stroke disease.

10.8 HYPERTENSION DISEASE DETECTION

High blood pressure, sometimes referred to as elevated blood pressure, is a significant public health issue that increases the risk of renal failure, coronary artery disease, and death anywhere on the globe. While the risk of hypertension increases with age, managing it becomes more challenging. A patient's increased blood pressure reading may be related to current stress levels or previous occurrences. To accurately assess a patient's risk of enduring hypertension, clinicians examine a variety of criteria derived from physical tests.

10.8.1 Hypertension disease detection using AI

Hypertension is a widespread circumstance that results in a slew of additional health issues. The diagnosis of hypertension is a complex subject with several risk factors. The risk variables that are inputs into the systems for diagnosing hypertension must be properly chosen with expert guidance. According to the literature study, fuzzy systems and contrary to artificial neural networks, which have all previously employed a broad variety of input variables, integrated neuro-fuzzy algorithms have been effectively used with an aggregate of four incoming indicators of risk [13]. Compared to simple neural network systems or fuzzy systems, a hybrid neuro-fuzzy system can more accurately depict a doctor's competence. Neural networks that are artificial (ANN) have been utilized in several researches to recognize or predict strokes. Data from stroke patients show that ANNs can identify the risk of stroke. In that work, backpropagation learning was used to increase prediction consistency and diagnostic accuracy. We used 300 research results and ANNs to determine the potential incidence of attack. An accurate model for matching the data of stroke patients was created by the experiment, scoring 95.33%.

A system that only focuses on forecast accuracy makes it challenging to understand the mechanics of its operation. In Figure 10.6, the first module manages electromyography (EMG) data, does offline processing, and creates learning models utilizing DL and ML. Using real-time EMG bio signals gathered from daily activities, the second module does online processing in addition to early detection and prediction of strokes. According to the cycle defined by the system, the EMG bio-signals data, acquired in real time during daily walking in the offline module, is updated in the repository. The acquired EMG bio signals are preprocessed using ML and DL techniques to build a learning model. DL approaches are utilized to construct models of learning for promptly identifying and forecasting stroke issues by choosing a property sample from the cleaned EMG.

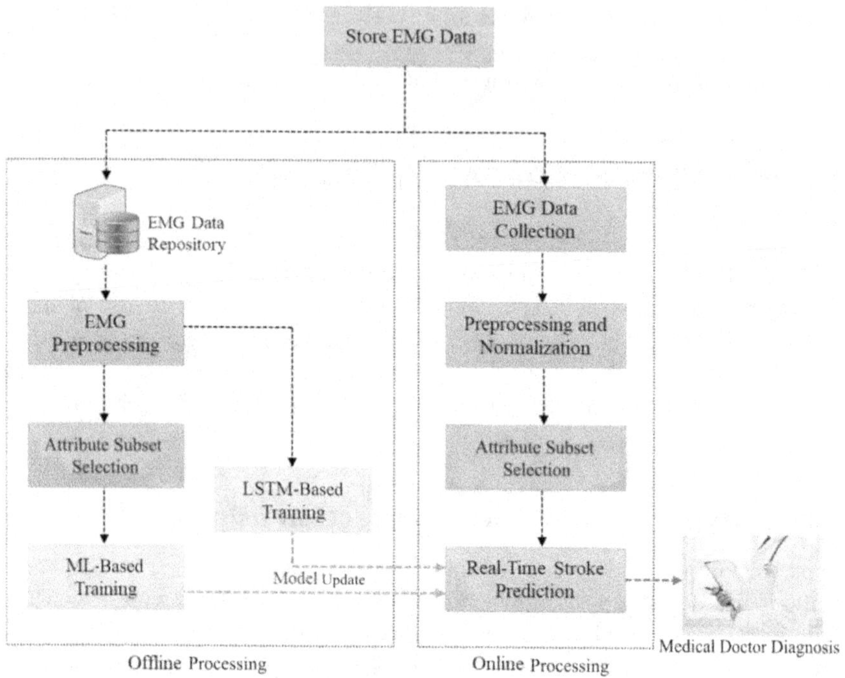

Figure 10.6 EMG bio signals.

10.9 SKIN DISEASE

An adult's skin weighs 3.6 kg and has a surface area of 2 m^2, making it the largest organ in the human body. The body's skin protects it from temperature extremes, harmful UV rays, and dangerous substances. The epidermis, dermis, and hypodermis make up this highly organized system, which serves the roles of protection, sensibility, and thermoregulation. The skin's epidermis, which is its outermost layer, serves as a superb barrier against external assault. Sweat glands, hair follicles, and stiff connective tissue located in the dermis under the epidermis help to differentiate the look of the skin [14]. Skin conditions can have a variety of biological and physical causes, such as insect bites, allergic conditions, and even viral infections, as well as environmental influences like light, temperature, and friction.

10.9.1 AI in skin disease

Once the program is in place, pattern detection of photos using the DL approach may be done automatically. A CNN can accept high-fidelity image input and automatically extract key characteristics. As a result, this method eliminates the need for information extraction from pictures before learning. Shallow layers allow for the learning of basic elements like picture edges. More complicated high-order features are learned at deep levels close to the output layer.

DL techniques have been created for the detection of dermatological illness and have been demonstrated to be useful in a variety of sectors. Various researchers, institutions, and challenges are working on the automatic diagnosis of skin diseases. Figure 10.7 shows the four steps of picture capture, image preprocessing, feature extraction and classification, and criterion evaluation that commonly make up the diagnosis of skin disorders [14].

The process of skin categorization begins with the capture of photographs, and more photos are often a sign of higher accuracy and flexibility. For improved training, lesions are segmented, and photos are cropped, zoomed, and otherwise preprocessed. By using information on color, texture, and border, feature extraction primarily obtains the characteristics of the skin lesions.

10.9.2 Future detection in skin disease

Recently, dermatological AI has advanced quickly, but there are still a number of pressing issues that need to be resolved. These include the following. First, the volume of data now available on skin diseases is still minimal, hospitals don't share much information with one another, and skin imaging standards and quality vary. High-quality picture data is difficult to obtain, which makes study findings suspect. Second, it is quite difficult to find people with complicated AI and medical skills together. Close collaboration with multidisciplinary experts in computer science, biomedicine, and medicine is essential. Third, dermatology deals with a wide variety of illnesses. Only one or a few particular skin illnesses may be identified using dermatological AI. One of the problems with AI skin disease diagnosis

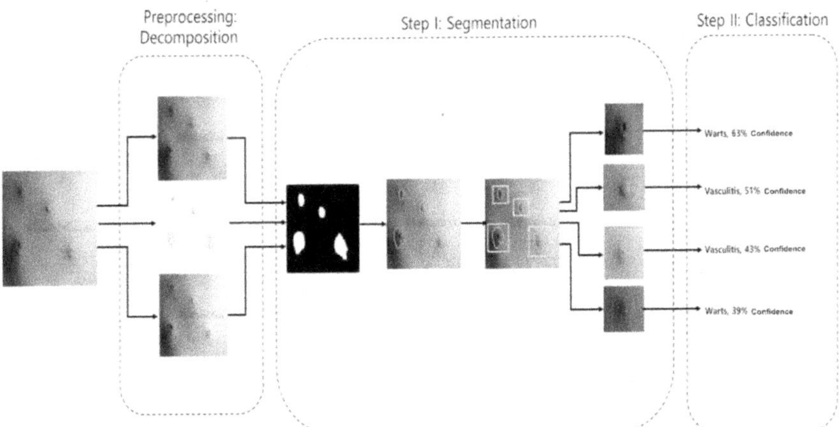

Figure 10.7 AI-based segmentation.

is how to make it recognize more skin illnesses. Fourth, the present AI diagnosis also raises a number of unresolved legal, moral, and data privacy challenges. Fifth, as a result of the uniqueness of AI products, clearance procedures, and database sources, a dermatological product has yet to receive market entry approval, and there are no medical AI products manufactured by a Chinese business that have secured a medical device registration certificate. The commercialization of medical AI is subject to several limitations. Consequently, it is necessary to combine patient data with skin imaging data, and AI is then used to thoroughly analyze this data. As a result, illness diagnosis, therapy selection, and prognosis assessment will all be influenced more by this data in the future. AI cannot replace doctor-patient dialogue or offer patient care or other humanistic services.

Dermatologists and other specialists in related fields are attempting to find solutions to a variety of issues arising from these complex issues. In summary, dermatological AI has gotten unheard-of attention from all governmental levels, including people training, funding for scientific research, technological development, and market capital. In the discipline of dermatology, there are still great prospects for AI. Dermatologists will soon have access to a more sophisticated system. The advancement of AI theory and technology, the growth and quality enhancement of information resources, and multifarious and multidisciplinary teams have all contributed to the creation of precise and customized supplementary diagnosis and therapy. Making AI more useful for physicians and patients is the ultimate objective.

10.10 CHRONIC LIVER DISEASE DETECTION

The 11th biggest cause of mortality worldwide is chronic liver disease and cirrhosis, which account for 1.1 million fatalities per year. Cirrhosis is becoming more common than ever, with 122 million people worldwide having the disease in 2017 compared to 71 million in 1990. Hepatitis B and C illnesses, liver disease brought on by drugs and nonalcoholic steatohepatitis (NASH) are common causes of cirrhosis [15]. There has been a temporal change in the frequency of the causes of cirrhosis over the past ten years, with NASH's prevalence rising sharply while that of other causes steadily declining. Nonalcoholic fatty liver disease (NAFLD) is thought to affect 25% of people globally.

10.10.1 AI for liver disease

The task of evaluating and diagnosing a huge volume of medical pictures presents a challenge for medical professionals; it is prone to human error, especially under pressure of time [16]. There are a number of AI algorithms that use medical images, including ultrasound (US), to diagnose liver disease. With the help of imaging data, several studies have tried to forecast the progression of illnesses [17]. Test results were used to propose a system for dividing severe hepatitis B into phases. Based on periods of time, testing results from severe hepatitis C, and learning algorithms, another study developed an AI model [18]; the outcomes demonstrated a reasonable area under the curve of receiver operating characteristic (AUROC) to anticipate cirrhosis [15]. Using information from the arterial biology of people with chronic hepatitis C, an AI model was built for predicting the outcomes of interferon-ribavirin, a combination treatment; 48,940 B-mode US pictures from chronic patients with hepatitis C were used as training data for four machine learning models, which resulted in the creation of an innovative AI model that predicted the incapacity of direct-acting antiviral medicines to treat patients.

To anticipate issues caused by liver cirrhosis, AI may be applied [17]. It used laboratory results and US as training data for AI models to distinguish between different incidences based on pressure changes between the vein known as the portal and hepatitis vein. It created a neural network prediction model that took into account the level of platelets, lymphoma, and blood vessel width to forecast the emergence of varices [18]. With the use of clinical data from 3,972 individuals, AI was able to predict the development of esophageal varix in patients with liver cirrhosis brought on by the hepatitis C virus (HCV) with a 68.9% accuracy. The endo score for esophageal gastric varices was predicted using ML using blood data and ascites results [16] (Figure 10.8).

10.10.2 Radiology image analysis

Handcrafted feature extraction, sometimes called "radiomics," is a well-known image analysis technique in radiology. To learn features and forecast a target label, supervised DL with neural networks may be utilized as an option. Radiology

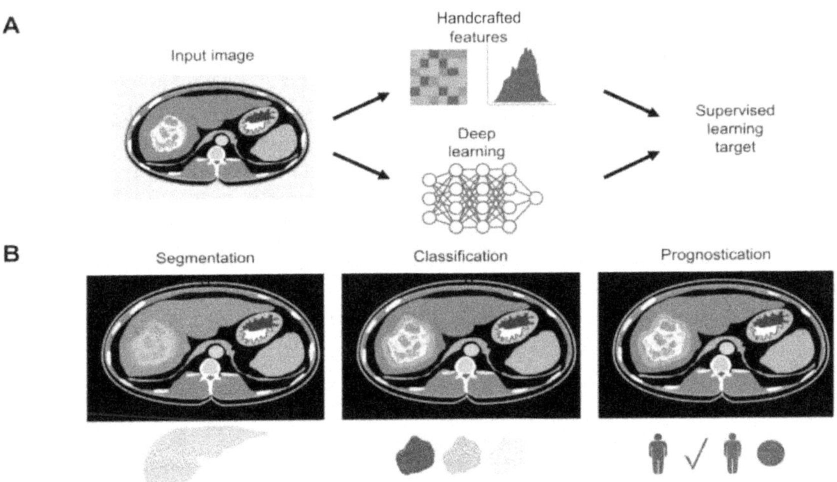

Figure 10.8 Radiology image analysis.

image analysis frequently involves the tasks of prediction, classification, and segmentation [18].

10.10.3 Standardization of image analysis

Although AI holds the potential to revolutionize liver imaging, differences in approach now preclude its integration into clinical decision-making. There is much heterogeneity in the radiomics workflow from data gathering through feature selection, while comparable factors also apply to DL. The imaging modalities of CT, MRI, or US for liver imaging each have unique data gathering requirements.

10.11 CONCLUSION

The capacity of systems constructed using AI to collect, evaluate, and extrapolate information from massive volumes of data is one of its key advantages. In a best-case situation, this may aid in illness prediction, a difficult task for conventional doctors. AI can assist the doctor in determining the chance of getting cancer or the factors that make heart attacks more likely than strokes. In the near future, medical professionals may help their patients avoid these disorders by implementing a long-term plan of patient care. All these developments are anticipated to enhance medical practices in the future. This will happen when health care treatments become more accurate, comprehensive, and potentially more affordable over time by eliminating unnecessary testing, reducing adverse effects, and preventing infections. This is unquestionably fantastic news for everyone involved, including patients, their families, and health care professionals.

REFERENCES

[1] Bahadur T, Verma K, Kumar B, Jain D, Singh S (2020) ***Automatic detection of Alzheimer related abnormalities in chest X-ray images using hierarchical feature extraction scheme***. Expert Syst Appl 158:113514. https://doi.org/10.1016/j.eswa.2020.113514.

[2] Babu BS, Likhitha V, Narendra I, Harika G (2019) ***Prediction and detection of heart attack using machine learning and internet of things***. J Comput Sci 4:105–108.

[3] Wu Y, Benjamin EJ, MacMahon S (2016) ***Prevention and control of cardiovascular disease in the rapidly changing economy of China***. Circulation 133:2545–2560.

[4] Gersh BJ, Sliwa K, Mayosi BM, & Yusuf S. (2010) ***Novel therapeutic concepts: The epidemic of cardiovascular disease in the developing world: Global implications***. Eur Heart J 31:642–648.

[5] Torre LA, Siegel RL, Ward EM, Jemal A (2016) ***Global cancer incidence and mortality rates and trends–an update***. Cancer Epidemiol Biomarkers Prev 25(1):16–27. https://doi.org/10.1158/1055-9965.EPI-15-0578

[6] Bray F, Ren JS, Masuyer E, Ferlay J (2013) ***Global estimates of cancer prevalence for 27 sites in the adult population in 2008***. Int J Cancer 132(5):1133–1145. https://doi.org/10.1002/ijc.27711

[7] Pan C, Schoppe O, Parra-Damas A, Cai R, Todorov MI, Gondi G, et al. (2019) ***Deep learning reveals cancer metastasis and therapeutic antibody targeting in the entire body***. Cell 179:1661–1676.e19. https://doi.org/10.1016/j.cell.2019.11.

[8] Saeedi P, Petersohn I, Salpea P, Malanda B, Karuranga S, Unwin N, Colagiuri S, Guariguata L, Motala AA, Ogurtsova K, Shaw JE, Bright D, & Williams R. (2019) ***Global and regional diabetes prevalence estimates for 2019 and projections for 2030 and 2045: Results from the International Diabetes Federation Diabetes Atlas, 9th edition***. Diabetes Res Clin Pract 157:107843.

[9] Gulshan V, Peng L, Coram M, Stumpe MC, Wu D, Narayanaswamy A, Venugopalan S, Widner K, Madams T, Cuadros J, Kim R, Raman R, Nelson PC, Mega JL, Webster DR. (2016) ***Development and validation of a deep learning algorithm for detection of diabetic retinopathy in retinal fundus photographs***. JAMA 316(22):2402–2410.

[10] Subudhi A, Dash M, Sabut S (2020) ***Automated segmentation and classification of brain stroke using expectation-maximization and random forest classifier***. Biocybern Biomed Eng 40:277–289.

[11] Poorthuis MH, Algra AM, Algra A, Kappelle LJ, Klijn, CJ (2017) ***Female-and male-specific risk factors for stroke: A systematic review and meta-analysis***. JAMA Neurol 29:86–93.

[12] Lee HJ, Lee JS, Choi JC, Cho YJ, Kim BJ, Bae HJ, Kim DE, Ryu WS, Cha JK, Kim DH, Nah HW, Choi KH, Kim JT, Park MS, Hong JH, Sung Il Sohn, Kyusik Kang, Jong-Moo Park, Wook-Joo Kim, Jun Lee, Shin DI, Yeo MJ, Lee KB, Kim JG, Lee SJ, Lee BC, Oh MS, Yu KH, Park TH, Lee J, & Hong KS. (2017) ***Simple estimates of symptomatic intracranial hemorrhage risk and outcome after intravenous thrombolysis using age and stroke severity***. J Stroke 19:229–231.

[13] Chaikijurajai T, Laffin L, Tang W (2020) ***Artificial intelligence and hypertension: Recent advances and future outlook***. Am J Hypertens 33:967–974. https://doi.org/10.1093/ajh/hpaa102

[14] Saba T, Khan MA, Rehman A, Marie-Sainte SL (2019) ***Region extraction and classification of skin cancer: A heterogeneous framework of deep CNN features fusion and reduction***. J Med Syst 43:289.

[15] Battineni G, Sagaro GG, Chinatalapudi N, Amenta F (2020) ***Applications of machine learning predictive models in the chronic disease diagnosis***. J Personal Med. https://doi.org/10.3390/jpm10020021.

[16] Battineni G, Sagaro GG, Chinatalapudi N, Amenta F (2020) ***Applications of machine learning predictive models in the chronic disease diagnosis***. J Personal Med. 10(2):21.

[17] Chuang C (2011) ***Case based reasoning support for liver disease diagnosis***. Artif Intell 53:15–23. https://doi.org/10.1016/j.artmed.2011.06.002

[18] Pesapane F, Volonté C, Codari M, Sardanelli F (2018) ***Artificial intelligence as a medical device in radiology: Ethical and. regulatory issues in Europe and the United States***. Insights Imaging 9:745–753. https://doi.org/10.1007/s13244–018–0645-y

Chapter 11

Trends and challenges in harnessing big data intelligence for health care transformation

H. Arshad, M. Tayyab, M. Bilal, S. Akhtar, and A. M. Abdullahi

11.1 INTRODUCTION

Big data intelligence is a specialized field that harnesses the immense potential of big data to provide not only insightful and illuminating information but also valuable guidance for the decision-making process. The core objective of employing big data intelligence revolves around leveraging its inherent power to generate information that is both enlightening and thought-provoking [1]. The initial step in extracting meaningful insights from vast datasets, such as those from online marketplaces and social media platforms, involves meticulous data cleaning. This pivotal phase, which comes right after data collection, entails the removal of irrelevant features based on data organization. Once the data is refined, it is then transferred to cloud storage to enhance scalability and accessibility [2, 3]. This refined dataset holds the potential to address a spectrum of decision-making challenges, data-driven issues, and various scenarios where outcomes hinge on the analysis of big data. By harnessing the capabilities of big data intelligence, organizations can unlock profound insights that steer their strategies in the right direction.

The digital age has given birth to the revolutionary concept of big data, which has fundamentally altered the landscape of data analytics and driven transformative changes across various industries. This transformation has been accelerated by the rapid evolution of internet technology and the gradual integration of big data analysis into medical enterprise operations. As a result of this integration, extensive repositories of electronic medical records, comprehensive hospital information systems, and vast stores of medical imaging data have been created [4]. These repositories contain a wealth of structured, semi-structured, and unstructured data derived from a variety of sources, ranging from social media interactions to online transactions and sensor-generated data. This diverse set of data sources contributes to a complex tapestry of information with enormous potential for insight and innovation.

The health care sector is currently undergoing a substantial transformation due to the convergence of big data and advancements in medical technologies. This

DOI: 10.1201/9781003496410-14

convergence has brought about a shift in how medical data is collected, processed, and utilized, prominently seen in the growing reliance on big data intelligence. The vast amount of health care data originating from sources like electronic health records (EHRs), medical imaging, wearable devices, and genetic sequencing has triggered a disruptive influence: namely, data analytics. This transformation is not only revolutionizing information management but also yielding significant cost-savings and operational enhancements. As of 2020, medical data had surged to a remarkable 35 zettabytes (ZB), marking a 44-fold increase from the 2009 data volume, as reported by experts [4]. This data explosion presents both challenges and opportunities for health care institutions. By effectively employing data analytics, these establishments can optimize resource allocation, refine processes, and eliminate inefficiencies, necessitating a comprehensive analysis of patient flow, hospital usage, and resource availability. The expansive health care data repository holds diverse information, including medical images, clinical records, prescriptions, physician's notes, computed tomography (CT) scans, magnetic resonance imaging (MRI) images, lab data, pharmacy records, insurance details, electronic patient records (EPRs), and other vital data for administrative workflows [5]. Essentially, the amalgamation of big data and medical technologies is reshaping the health care landscape, extending beyond data management to bolster decision-making, resource optimization, and overall operational excellence in the sector.

The term "big data analytics" denotes the systematic analysis of extensive datasets gathered from diverse sources. To enhance human safety and cost-effective data processing, there's a need to expand the application of big data analytics. This involves classifying abundant data into two main categories [5]. Structured data pertains to systematically organized information with predefined formats and time frames, encompassing numerical values, dates, and text. This study's data originates from sources like mobile devices, computers, sensors, and online logs, including EHRs, home monitoring data, health care prescriptions, and similar cases. On the other hand, unstructured data lacks predefined formats and organization. This type of data, acquired from sources such as social media, mobile devices, videos, and the web, poses challenges for comprehensive analysis within the realm of big data. In the health domain, disorganized data includes health-related content from platforms like Twitter and Facebook, user blogs, physician notes, and prescription records containing instructions.

The field of health care has undergone significant transformation with the integration of big data intelligence. This has unlocked the latent potential within complex health care datasets, providing health care professionals, researchers, and policymakers with valuable insights for informed decision-making. These insights contribute to enhanced patient care, accelerated medical research, and improved efficiency in various health care processes. The ability to effectively analyze extensive data volumes holds the key to creating more efficient health care systems, with the potential to positively impact both individual and population health [6]. However, the realm of health care big data intelligence presents not only opportunities but also distinct challenges. The realization of its benefits relies on addressing crucial issues like workforce preparedness, seamless data interoperability, and upholding high data quality standards. Successfully overcoming these obstacles is of utmost importance for health care practitioners, researchers, and policymakers. This effort not only elevates patient care and drives medical research but also advances public health initiatives. By proactively tackling these challenges, the health care sector can ultimately cultivate improved health outcomes and foster the overall progress of the industry [7].

Figure 11.1 depicts an overview of the major topics covered in this chapter. This chapter is divided into several sections. Section 11.2 delves into big data in health care, including its definition and sources. Section 11.3 discusses the significance of big data intelligence in health care. Section 11.4 examines various big data analytics techniques used in health care. Section 11.5 examines how big data intelligence is being applied in various areas of health care. Section 11.6 delves into trends and opportunities. Section 11.7 discusses challenges and future directions, while Section 11.8 provides a chapter summary.

11.2 UNDERSTANDING BIG DATA IN HEALTH CARE

Cutting-edge analytical techniques, including advanced analytics methods, artificial intelligence, and machine learning, play a pivotal role in revealing intricate patterns, significant trends, and valuable correlations. These insights not only pave the way for groundbreaking medical advancements but also facilitate the delivery of tailored and individualized patient care. Harnessing the full potential of big data in the realm of health care is the ultimate goal [8]. The realm of big data analytics in health care extends far beyond its surface, offering the capacity to

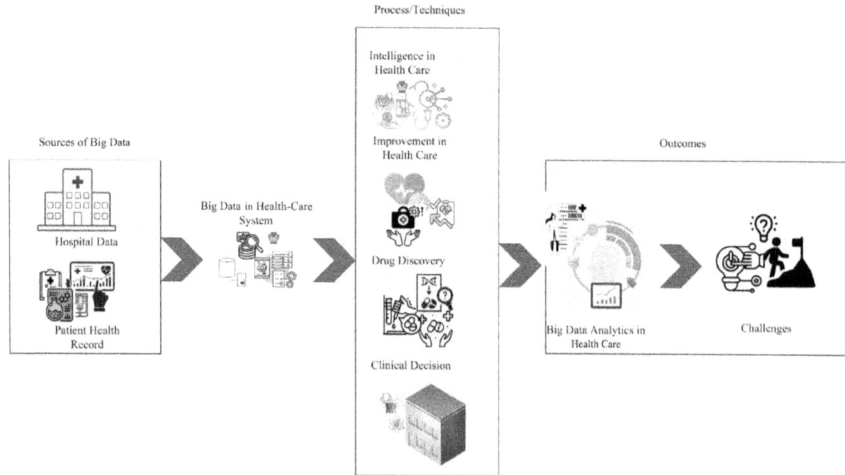

Figure 11.1 Big data intelligence for health care transformation.

pinpoint individuals at high risk, predict and mitigate potential disease outbreaks, and refine existing medical procedures [9]. Moreover, the wealth of information stored within vast patient datasets holds the promise of unveiling remarkable and unexpected revelations that can reshape our understanding of health and medicine.

11.2.1 Definition and characteristics of big data

"Big data" refers to large and complex datasets that are challenging to handle and analyze using traditional data processing approaches. Volume, velocity, variety, and value are the key traits that set it apart. This includes organized, semi-structured, and unstructured data sources [10]. It may reveal significant trends and insights that might help guide data-driven decision-making across a range of businesses. "Big data" also applies to datasets too large for common systems to handle [11]. In May 2011, the concept of big data was widely recognized as a promising domain for competition, advancement, and efficiency. The number of internet users experienced a 7.5% increase from 2016 to 2018, resulting in a total of over 3.7 billion individuals accessing the internet. The global production of data reached approximately 1 ZB at its highest point in 2010 and then increased to 7 ZB by 2014 [12]. The three dimensions, namely, volume, velocity, and variety, were initially proposed in 2001 to delineate the growing characteristics of big data [13]. In 2011, the International Data Corporation (IDC) employed the four dimensions of volume, variety, velocity, and value to establish a comprehensive framework for defining big data. The inclusion of authenticity as a significant characteristic of big data occurred in 2012. The present study emphasizes the five predominant characteristics of big data, as depicted in the accompanying visual representation, notwithstanding the existence of numerous supplementary factors [14].

Volume: Big data is distinguished by its enormous volume, staggering, and unprecedented amount of data. We are currently witnessing the generation of an unprecedented amount of data, ranging from terabytes to exabytes and even exceeding these limits. This deluge of data comes from a variety of sources, including sensor networks that capture real-world phenomena, intricate financial records that document economic activities, the dynamic tapestry of social media interactions, and a variety of other prolific generators. To deal with this data flood, scalable and reliable storage solutions must be implemented alongside cutting-edge processing technologies. These technological advancements are critical in facilitating the storage and comprehensive analysis of these massive data volumes, allowing us to gain valuable insights and make informed decisions [10].

Velocity: It refers to the rate at which massive amounts of data must be generated and processed. In today's interconnected world, data is not only produced but it is frequently produced in real time or near real time. This necessitates the rapid and efficient processing of such data, which has emerged as a requirement. The ability to manage data at such a rapid pace is critical for expeditious decision-making and maintaining relevance in industries characterized by rapid transformations and fluctuations [15]. As a result, the concept of velocity emphasizes the importance of quickly navigating through the data deluge to glean valuable insights and capitalize on emerging opportunities.

Variety: It encompasses the vast realm of big data, which includes a range of data types such as structured, semi-structured, and unstructured data. Semi-structured data, despite the lack of rigid predefined standards, demonstrates a degree of inherent organization, often taking the form of extensible markup language (XML) or JavaScript object notation (JSON). Structured data, on the other hand, adheres to a well-organized framework, allowing for efficient search and retrieval in the same way that traditional databases do. Unstructured data, which includes text, images, and videos, on the other hand, defies preconceived formats. Navigating through this rich data tapestry requires the use of cutting-edge data analytics and machine learning techniques [15]. The use of such methodologies has become critical to extract meaningful insights from these disparate data types and realize their full potential for various applications across industries.

Value: Unlike other characteristics, the idea of value focuses on how important data is for making decisions and how it can be used in real life. Analytics techniques have been used successfully by companies like Facebook, Google, and Amazon to make the best use of big amounts of data in their products. Amazon has several databases that store information about users and what they've bought. This information is used to do research, with the goal of making personalized product suggestions and improving sales and how people interact with Amazon. Google gets information about where Android users are to make location-based services like Google Maps work better. Facebook uses what its users do to suggest friends and show them ads that are relevant to them. These three companies have grown much in recent years. This may be because they have spent much time analyzing large amounts of unprocessed data, which has given them useful insights that help them make better strategic decisions [14].

11.2.2 Sources of big data in health care

The health care industry is among the major producers of big data since it gathers a wide variety of data from many sources. Figure 11.2 depicts the major sources of big data in health care. These sources have a substantial impact on the landscape of health care data analytics, which has an impact on patient care, medical research, and public health initiatives. Big data and associated concepts are frequently mentioned in health care applications in a variety of ways [16]. It is essential to be aware of the extensive range of data sources that are currently accessible to fully comprehend the value of big data applied to health care [17]. The most established sources of big data, which have historically influenced technology adoption and the creation of big data applications, are as follows:

EHRs: These are digital documents that contain important patient information. This information may include a patient's medical history and diagnosis, in addition to specifics regarding medications and laboratory findings. As a direct result of the widespread deployment of EHRs, medical professionals now have access to the data of their patients in a time frame that is closer to real time. In addition, they are in a position to make judgments on individualized care that are based on the data [5, 18]. Although their potential is mostly unexplored, EHRs offer comprehensive and important data about patient treatments and results. Conventional health data centers have the potential to collect and store a sizable amount of structured data related to several aspects of health care, including diagnostics, laboratory testing, drugs, and more clinical data. To analyze and index the semantic contents that underlie each patient report systematically, natural language processing must be used. The use of EHRs can greatly improve clinical knowledge and speed up clinical research projects. The identification of phenotypic data, which aids in the discovery and comprehension of numerous clinical traits, is one noteworthy use [19].

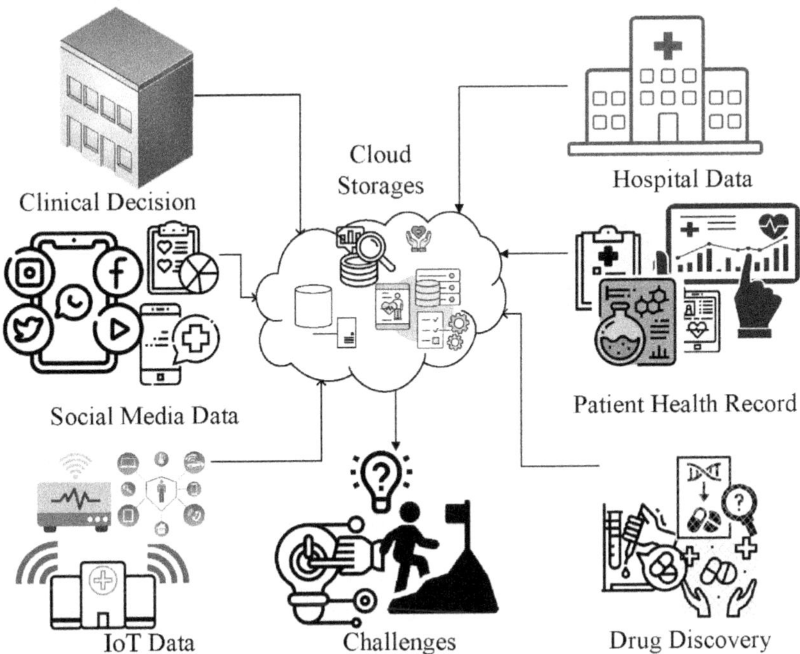

Figure 11.2 Big data in health care intelligence.

Medical Imaging: Extensive amounts of image data are generated by various imaging modalities, such as x-rays, CT scans, and modern medical imaging methods. The availability of such comprehensive imagery greatly simplifies the early detection of a wide range of medical conditions and allows for prompt treatment initiation. This allows for the earliest possible start of treatment for these ailments. Incorporating big data analytics has become an essential part of improving medical image interpretation. As a result, radiologists and medical professionals can now identify recurring irregularities and patterns in these images much more easily [20]. Various imaging techniques are used in health care that involve large-scale capture of high-resolution medical images as well as patient-specific data [21]. Health care professionals such as radiologists and physicians evaluate the medical data contained within these files to identify specific anomalies. However, it is critical to recognize that numerous diseases are hampered by a lack of adequately trained medical personnel. To address this workforce shortage, the development of the picture archiving and communication system (PACS) has emerged as a viable solution. This system makes it easier to store and retrieve medical images and reports [12].

Wearable Devices: In recent years, there has been a rise in the use of wearable technology, like smartwatches and fitness trackers. This has helped to make it easier to collect continuous data from patients, as patients are more likely to be wearing such devices [22]. These electronic gizmos are attracting new followers on a daily basis, which is extremely exciting. By tracking a person's vital signs, such as their heart rate, the amount of movement they get, and the patterns in which they sleep, these devices are able to provide key information regarding a person's health as well as the way that they live. This information may be used to make important decisions regarding a person's health as well as the way that they live [23]. For example, the emergence of wearable sensor devices such as the Apple Watch and sports bracelets has facilitated the continuous monitoring of various physical health indicators, such as blood pressure, height, weight, blood glucose levels, and blood-calcium levels. This constant monitoring enables a comprehensive assessment of the patient's health status. These indicators assist medical practitioners in monitoring patients, so potentially reducing the need for unneeded doctor visits. Simultaneously, patients perceive themselves as more autonomous and have a heightened awareness of their health care state [24].

Genomics Sequencing: The sequencing of deoxyribonucleic acid (DNA) results in the generation of genomics data, which has a major beneficial effect on tailored medicine and precision health care. This effect has been brought about as a result of the sequencing of DNA. It's possible that both these areas will experience this effect. Researchers are now able to establish genetic differences, comprehend the possibility of illness, and come up with treatments that are specific to the individual as a result of the study of genomics data [25]. The human genome encompasses the entirety of genes that constitute a human cell, establishing a comprehensive linkage among them. DNA constitutes the genetic material that comprises the genes. The human genome project (HGP) undertook the sequencing of the complete human genome to identify potential patterns that may contribute to the development of certain illnesses. The comprehensive sequencing of the human genome represents a remarkable effort with the potential to yield life-saving outcomes but with a considerable cost, complexity, and protracted timeline exceeding a decade for completion. The determination of human characteristics, such as eye color, skin tone, illness susceptibility, and others, is governed by a certain sequence of DNA proteins (adenine, thymine, cytosine, and guanine) inside an individual's genome. Hence, the application of analytical techniques such as machine learning on the human genome enables the identification of significant information and facilitates the acquisition of novel insights pertaining to the individual's well-being. The examination of a drug's mechanism of action allows for a comprehensive understanding of its precise impact on an individual. The fundamental basis of customized health care, which facilitates the incorporation of tailored treatments for individual patients, is in the utilization of genomic data [26].

Patient-Generated Data (PGD): The proliferation of PGD represents a significant shift in health care dynamics, facilitated by tools such as mobile apps, patient portals, and online health networks. This trend of patients actively contributing their health-related data is expected to not only continue but also grow in the near future. This surge in PGD includes a wide range of data types, from self-reported symptoms to medication adherence measurements and quality-of-life indicators [27]. This wealth of patient-centric data has enormous potential to transform health care delivery. By leveraging these insights, health care providers will be able to gain a more comprehensive understanding of each patient's unique health trajectory, resulting in personalized and effective treatment strategies. Furthermore, incorporating PGD into clinical workflows has the potential to empower patients, improve communication among health care stakeholders, and ultimately, improve the overall quality of care provided. In essence, PGD emerges as a critical pillar in the modern health care edifice, promising a paradigm shift toward more holistic and patient-centered approaches.

Social Media Interactions: The numerous and diverse social media platforms are currently overflowing with content that is connected in some way to the mental and physical health of users. By analyzing the data obtained from social media platforms, researchers can identify the locations of disease outbreaks, monitor the evolution of public health patterns, and gain insight into patient preferences [28]. Enhancing local health awareness and facilitating crisis communication are just two examples of the many goals that can be achieved by using social network data as a tool for social health aid within a community. Making diagnostic judgments may benefit from the incorporation of social media data into clinical decision support systems [24]. Several studies have examined the effectiveness of social media data in health care communication and wellness profiling [29, 30].

Clinical Trials and Research Studies: The findings from clinical studies and research inquiries are a significant contributor to the development of new knowledge regarding the efficacy of medications, how efficiently they function, and the progression of diseases. This new information has the potential to be put to use in the development of innovative treatments as well as in the prevention of the progression of diseases. It is possible to get study results that are more thorough if they are obtained by merging the data that was gathered from a number of various inquiries. This is a method that has been shown to be effective. Because of this, it is now possible to gain research outcomes that are more precise [31].

11.3 THE ROLE OF BIG DATA INTELLIGENCE IN HEALTH CARE

The emergence of big data intelligence, which is revolutionizing medical research, patient care, and public health initiatives, is causing significant disruption in the health care sector. This transformative impact is due to the use of large and diverse health care databases, which harness the power of big data intelligence to provide valuable insights to medical practitioners, researchers, and policymakers alike. The creation of a "smart health care network"— a platform that improves patient care while lowering costs—exemplifies the convergence of benefits from sensor technologies, the internet of things (IoT), and sophisticated big data analytics. The current challenge in the health care sector is effectively managing and safeguarding the massive amount of data generated to derive valuable insights. To meet this requirement, innovative techniques and strategies for managing the increasing influx of health care data from sources such as IoT devices, EHRs, mobile health apps, and telemedicine evaluations must be developed. A study [32] provided a brief overview of the advancement of significant data analysis in this rapidly evolving field, with a primary focus on its application in remote monitoring for intelligent health care setups.

11.3.1 Improving patient care and outcome

Improving patient care is one of the main goals of big data intelligence in health care. Health care practitioners can fully comprehend patient demographics and medical needs by evaluating substantial patient data. Personalized treatment regimens, improved disease management, and the early identification of potential health concerns are all made possible by data-driven insights. Big data analytics, for instance, can be applied to swiftly identify high-risk patients and take appropriate action, improving patient outcomes and lowering health care costs [33].

11.3.2 Advancing medical research and drug discovery

Big data intelligence advances medicine by allowing for extensive data analysis. Big data can be used by researchers to spot patterns in disease, evaluate the efficacy of already available drugs, and develop new ones. Integrating various data, such as patient outcomes, genomics data, and clinical trial data, promotes more thorough research discoveries and hastens medical improvements [34]. Big data intelligence is also required for projects addressing public health. Public health organizations can track population health trends, spot disease outbreaks early, and implement targeted interventions to stop the spread of diseases by analyzing data from a variety of sources, such as social media interactions, environmental sensors, and disease monitoring systems [35, 36].

11.3.3 Enhancing clinical decision-making

Big data intelligence is enhancing medical research and medication discovery by changing the way researchers analyze health care data. The potential of big data analytics enables researchers to manage and analyze substantial and varied datasets to derive useful knowledge and catalyze significant advancements in the medical field. One of the most significant contributions of big data intelligence to medical research is the identification of disease patterns and trends. By studying extensive patient data, researchers can understand disease progression, identify risk factors related to specific illnesses, and identify disease precursors [37]. These data allow for more precise diagnoses and individualized treatment plans, both of which improve patient outcomes.

11.3.4 Transforming health care operations and management

Big data analytics can be used by health care providers to automate a range of health care processes, enhancing both organizational effectiveness and patient care. Streamlining patient care workflows is one of the key ways big data intelligence impacts health care operations. Health care practitioners may quickly track patient status and treatment outcomes with real-time data processing and analysis, enabling timely interventions and personalized care plans. This proactive approach leads to better patient outcomes and fewer readmissions to hospitals. Resource allocation and resource usage optimization both benefit greatly from big data knowledge. Health care businesses may properly distribute resources and make sure they are used by assessing data on patient flow, hospital use, and resource availability [31].

11.4 BIG DATA ANALYTICS IN HEALTH CARE

11.4.1 Data mining and pattern recognition

"Big data analytics" refers to the complex process of analyzing massive amounts of data derived from various sources and formats to make real-time decisions. To examine the collected data, various analytical techniques such as data mining and artificial intelligence are used [13]. These methodologies for dissecting large amounts of data can efficiently detect anomalies by examining large amounts of data from various sources. Notably, the health care industry deals extensively with large amounts of data, which is often referred to as big data. Medical imagery, clinical records provided by physicians, prescriptions, health care professionals' notes, CT scans, MRI images, laboratory findings, pharmacy records, EPRs for insurance purposes, and other administrative data are all included. Even among international health care conglomerates, this concept is gaining traction. In the realm of big data analytics, large amounts of data from various sources are systematically examined [5]. This data comes in a variety of forms and types, making it amenable to a variety of analytical approaches, such as data mining and artificial intelligence. Using big data analytics tools makes it easier to spot anomalous patterns that emerge from the convergence of disparate data streams [13].

11.4.2 Machine learning and artificial intelligence in health care

Artificial intelligence has made significant contributions to the health care industry. Traditionally, medical experts have guided health care. However, the advancement of artificial intelligence and machine learning, combined with the increased accessibility of cloud computing for extensive data storage, has facilitated the provision of personalized care [38, 39]. The use of machine learning for the comprehensive analysis of large health care datasets has numerous advantages. In terms of flexibility and scalability, this approach outperforms traditional biostatistical methods when dealing with complex health care data. These characteristics allow it to be used for a variety of tasks such as risk assessment, diagnosis, classification, and prognosis [40]. Furthermore, machine learning algorithms are capable of assimilating a wide range of data sources, including laboratory results, demographic information, and imaging data. This assimilation improves the accuracy of predictions about disease risk, diagnosis, prognosis, and appropriate interventions. Artificial intelligence is how machines show intelligence, which is different from how people show intelligence through their natural intelligence. Artificial intelligence can be used for things like machine learning, which gives systems the ability to learn on their own and get better as they are trained without being specifically programmed to do so. Artificial intelligence has helped in many areas of health care, such as cancer research, heart, diabetes, mental health, predicting prognoses, finding Alzheimer's disease, separating clinical groups, finding cardiovascular disease, studying strokes, and so on [5].

11.4.3 Natural language processing for health care data

The use of advanced data analytics has increased the value of intelligent applications in the health and medical fields significantly. When considering patient health records and medical records in digital format, this advancement is clear. Similarly, significant textual information is generated in computer-aided design, engineering, and manufacturing applications. This unstructured data contains a wide range of information categories, including system requirements. The integration of advanced data analytics presents significant opportunities for businesses, particularly those operating in the aerospace industry's aeronautics and astronautics sectors. Given the ongoing expansion, complexity, and rapid pace of change, effectively managing large volumes of real-time numerical and textual data is a constant challenge [41]. Increased efforts in the field of computer-assisted design are required to clarify the concept of large-scale data for individuals. Similarly, numerous notable techniques in the field of language processing are emerging, such as identifying word meanings, disambiguating word senses, marking parts of speech, and employing probabilistic context-free grammar. Textual information analysis has used a variety of approaches from natural language processing, which include extracting information, identifying topics, condensing text, categorizing, grouping, responding to questions, and assessing opinions. Together, these various methods address a wide range of tasks within this domain [42, 43].

11.5 APPLICATIONS OF BIG DATA IN SPECIFIC HEALTH CARE AREAS

11.5.1 Big data in diagnosis and imaging

In recent years, the scientific community has prioritized multimodel medical imaging advancement, recognizing its profound implications for fields such as computer vision, the IoT, and medical diagnostics [44]. The advancement of medical imaging techniques has resulted in a significant increase in the volume of diverse medical image data. X-rays, CT scans, MRI images, optical coherence tomography (OCT) scans, microscope imagery, and positron emission tomography (PET) scans are examples of these. Deep learning has made significant advances in computer vision, and researchers have quickly taken advantage of these advancements to examine medical image data. Such images provide unaltered, direct representations of a patient's medical condition and contain critical information. It is worth noting that certain clinical imaging methodologies outperform others in detecting pathological transformations within various organs. For example, ophthalmic imaging provides a specialized view of ocular conditions, whereas MRI is adept at detecting anomalies in the brain, heart, bones, and joints, with a particular focus on bone and joint issues. While x-rays are excellent for revealing chest and breast structures, CT scans are superior for visualizing abdominal and thoracic organs [20].

11.5.2 Big data in personalized medicine and genomics

The area of precision medicine is growing quickly, just like many other fields. The National Research Council was the first to use the phrase "precision medicine" to describe the work of making a new classification of human diseases based on molecular biology. From a different point of view, the sequencing of the human genome could be seen as the start of a medical care revolution. With the rise of precision medicine, doctors are now able to access and share data that can either confirm or change a medical choice. This change moves away from making medical decisions based only on evidence about the average patient and toward making choices based on what makes each person unique. Doctors find it easy to give each patient care that is tailored to their needs. The development of precision medicine has made it possible for new possibilities to arise that would not have been possible earlier [15].

Recent genomic discoveries, such as single-cell genome and transcriptome sequencing, liquid biopsy for detecting circulating cancer DNA, and metagenomics, are already having an impact on medicine and will soon become standard practice. Particularly when combined with short-read sequencing and visual mapping, long-read sequencing has proven useful for building novel genomes and assessing the makeup of microorganisms [45]. The use of genotype-phenotype data from Genome-Wide Association Studies (GWAS), multi-omics data integration, and literature mining has made a big difference in the development of cutting-edge artificial intelligence methods and solutions in the field of cancer genomics. As a result, this success has given medical professionals the tools they need to use precision medicine to give personalized medical care. The idea of "precision medicine" refers to a modern way to avoid and treat diseases that takes into account each person's unique genetic makeup, as well as environmental factors and dietary preferences. Artificial intelligence systems could help improve personalized medicine by figuring out how different people respond to drugs in different ways. These systems can, among other things, make suggestions based on trends found in large collections of both public and private data sources [38].

11.5.3 Big data in health care IoT and wearables

The term "IoT" was first used in 1999 by Kevin Ashton to describe supply chain management. It suggests a more complex universe with internet access and intelligent creatures [46]. Big data is crucial due to the interconnectedness of IoT devices and the numerous data streams they generate. Numerous IoT devices and apps generate enormous amounts of data. This data is mined using a variety of big data techniques that help decision-making. Since the advent of e-health and m-health, Dlamini, Zita Francies, Hull, and Marima [47] have observed an increase in the usage of technologies in the health care sector. A patient's health is continuously monitored by millions of implanted sensors using a range of physiological, environmental, and behavioral data.

11.6 TRENDS AND OPPORTUNITIES

Nowadays, mobile devices, sensors, patient control devices, the IoT, hospitals, researchers, providers, and organizations are generating numerous datasets, which can be analyzed using different machine learning and artificial intelligence techniques [48]. The biggest challenge is to get datasets, analyze, and find innovations in health care systems to make the user's life healthier and easier, by the induction of new methods which not only cure diseases but also predict outcomes at earlier stages and generate real-time decisions.

11.6.1 The evaluation of big data in health care

The digital medical record, like the digital health record, serves as a storage system for consistent medical and clinical information gathered from individuals seeking health care services. Software for managing medical practices, digital health records, personal health records, and various other components of health care information have the potential to improve health care quality, efficiency, and cost-effectiveness while decreasing medical errors. In the early 2000s, digital health records were gradually adopted. Nonetheless, the use of digital health records has increased significantly since 2009 [49]. Information technology is increasingly being used to manage and utilize health care data. The development and application of technologies for monitoring well-being, as well as the associated software, has progressed in tandem with real-time systems for monitoring biomedical and health aspects. These technologies can be used to generate alerts and share patient health information with the appropriate health care providers. These devices generate large amounts of data that can be analyzed to provide immediate clinical or therapeutic interventions. The widespread use of data in the health care field has the potential to improve health outcomes while lowering costs [50].

11.6.2 Emerging technologies and innovations

Networked medical devices are getting better because they are getting smaller, wireless transmission technology is getting better, processing power is getting better, and computers can do more. These gadgets can gather, make, analyze, and send information. The phrase "internet of medical things" (IoMT) refers to a framework made up of software programs, medical devices, health care services, and computer systems that are all connected to each other. All these gadgets and the data they make are what make up the network in question. Medical emergencies like asthma attacks, heart attacks, diabetes, falls, and other similar events can be monitored in real time by IoMT-connected devices, which could save lives. With the help of networked devices and sensors, health care organizations can handle their clinical workflows and operations well, even when they are working from far away. This improves the quality of patient care [51]. Big data analytics techniques are used in the health care industry to predict epidemics, reduce preventable deaths, and improve the overall quality of life as much as is practical. The ability of big data to make it easier to analyze enormous amounts of health care data is what gives it its worth.

11.7 CHALLENGES AND FUTURE DIRECTIONS

Big data in health care has many advantages along with a number of open challenges that need to be addressed in the near future. Some of the prominent open challenges and future directions are discussed in this section.

11.7.1 Storage

Even though most people think that having a lot of data is difficult, many businesses still prefer to keep their data on-site. The system has many benefits, such as being able to control who can access it, making it easy to use, and making it safer. On the other hand, setting up and running a computer network on-site could be very expensive. Cloud-based storage choices are used by several health care businesses because they think they are reliable and cost effective. Compliance and security are very important in the health care business, so organizations need to be careful when choosing cloud service providers. Also, cloud storage has perks like a lower initial investment, faster recovery from disasters, and easier scaling. A hybrid method to data storage may be the most flexible and useful way for organizations to store and look at data in different ways [52].

11.7.2 Security

Health care organizations place a high priority on data security due to the frequency of data breaches, illegal access, phishing emails, and malware attacks. A thorough list of technical safeguards was created after a number of vulnerabilities in protected health information (PHI) was found [53]. The Health Insurance Portability and Accountability Act (HIPAA) is a series of rules that provide guidance to companies on how to handle data storage, transmission, authentication, access, integrity, and audits, among other things. It can be quite advantageous to use basic security precautions like firewalls, two-factor authentication, and regularly updated antivirus software [52]. Ensuring data security is crucial, encompassing aspects such as privacy, integrity, and authentication.

11.7.3 Opacity of infrastructure

One of the biggest problems with cloud computing has to do with how the technology works. The cloud markets itself as a service-oriented answer with attractive price structures that hide the underlying infrastructure. Doing this gives economic benefits from economies of scale and frees cloud clients from having to keep track of the complicated operations of cloud resources. Even though cloud computing is supposed to have the same features, it can be hard to get the best speed when there isn't enough information about the infrastructure. There are always problems with how well networks work, how well information protocols work, and how well computers work. These problems are caused by things like limited speed and unpredictable wait times when transferring large amounts of data. Also, scientific papers show that worries about the scalability of implemented solutions are widespread, even though the scalability of cloud technology is generally accepted [17].

11.7.4 Data uncertainty

Data uncertainty is a prominent feature of real-world data, typically resulting from insufficient data or a lack of operational context comprehension. The most important aspect of health care is data quality, which is evaluated based on its precision, exhaustiveness, and coherence. Thus, the efficacy of utility decision-making is wholly dependent on the quality of the data. Nonetheless, it is essential to recognize that the collection of real-world data is not error-free, as it is susceptible to errors caused by a variety of factors, such as disturbances and the presence of absent or inconsistent data. Sensor errors and imprecision, communication lags and delays, cyberattacks, physical damage to equipment, time-synchronized data, absent or inconsistent data, and extraneous sounds are among the primary contributors to data uncertainties and compromised data quality [54].

11.7.5 Ethical and legal issues

The use of big data in health care raises many ethical and legal issues because the data being used is so sensitive. Some of the ethical and legal difficulties that arise from using big data include worries about how it might harm people's privacy and freedom as well as how it might affect the public's needs for transparency, trust, and justice. A number of significant technological and infrastructure issues could arise with the implementation of a health care system based on big data. These issues include the fact that data isn't uniform, data security, managing data analytically, and inadequate infrastructure for data storage [55].

11.8 CONCLUSION

The rise of big data intelligence in the field of health care has greatly transformed medical research, patient care, and public health initiatives. It provides essential insights to medical practitioners, academics, and policymakers alike by leveraging huge libraries of health care data. It has a significant impact on patient care, allowing for personalized treatments, early risk detection, and improved patient outcomes. Furthermore, it has an impact on medical research, boosting advances in medication development and public health initiatives through the identification of critical patterns and the integration of diverse data sources. Clinical decision-making is aided by the ability to detect illness trends and patterns, ultimately improving health care operations through the process of automation and optimized resource allocation. The combination of big data analytics, machine learning, and artificial intelligence strengthens the possibility for personalized care, with machine learning serving as a critical component for analysis, risk assessment, and correct diagnosis. It derives useful information from previously unexplored data sources such as unstructured health records using natural language processing. The integration of big data intelligence is poised to transform health care across multiple sectors. The integration of advanced medical imaging techniques, genomics, and IoT wearable devices has catapulted the health care industry into the realm of big data utilization. Medical imaging tools such as MRI and CT improve diagnosis by adopting deep learning, whereas precision medicine led by genomic insights customizes interventions based on individual genetics and health information. In the meantime, IoT devices and wearables add to real-time data streams for continuous monitoring. Despite these great advances, difficulties persist. Data analysis, privacy protection, and the seamless integration of technologies are all areas that require significant attention to optimize health care outcomes. To satisfy storage preferences, a balance between on-site and cloud-based storage solutions becomes critical, taking into account issues such as cost-effectiveness and security. Health care institutions are rigorously establishing technology measures and adhering to compliance rules, including the stringent provisions of HIPAA, to ensure data security. The cloud's inherent opacity, along with the uncertainties associated with real-world data flaws, creates complex issues that need creative solutions. As sensitive data is used, ethical and legal concerns loom large, prompting crucial discussions on the privacy, transparency, and safety of the infrastructure underpinning the big data health care paradigm.

REFERENCES

[1] Sreelakshmi Krishnamoorthy, Amit Dua, and Shashank Gupta. Role of emerging technologies in future IoT-driven healthcare 4.0 technologies: A survey, current challenges and future directions. *Journal of Ambient Intelligence and Humanized Computing*, 14(1):361–407, Jan. 2023.

[2] Yichuan Wang, LeeAnn Kung, and Terry Anthony Byrd. Big data analytics: Understanding its capabilities and potential benefits for healthcare organizations. *Technological Forecasting and Social Change*, 126:3–13, 2018.

[3] Rajeswari Chengoden, Nancy Victor, Thien Huynh-The, Gokul Yenduri, Rutvij H. Jhaveri, Mamoun Alazab, Sweta Bhattacharya, Pawan Hegde, Praveen Kumar

Reddy Maddikunta, and Thippa Reddy Gadekallu. Metaverse for healthcare: A survey on potential applications, challenges and future directions. *IEEE Access*, 11:12765–12795, 2023.

[4] Zhihan Lv and Liang Qiao. Analysis of healthcare big data. *Future Generation Computer Systems*, 109:103–110, 2020.

[5] Z. Faizal Khan and Sultan Refa Alotaibi. Applications of artificial intelligence and big data analytics in m-health: A healthcare system perspective. *Journal of Healthcare Engineering*, 2020:1–15, 2020.

[6] Uzair Iqbal, Teh Ying Wah, Muhammad Habib Ur Rehman, Muhammad Bilal, and Adeel Ahmed. BD-ECG: Identification of myocardial infarction in ECG via behavior coupling. In *Proceedings of the 6th International Conference on Information System and Data Mining*, pp. 76–80, 2022.

[7] Muhammad Bilal, Raja Sher Afgun Usmani, Muhammad Tayyab, Abdullahi Akibu Mahmoud, Reem Mohamed Abdalla, Mohsen Marjani, Thulasyammal Ramiah Pillai, and Ibrahim Abaker Targio Hashem. Smart cities data: Framework, applications, and challenges. *Handbook of Smart Cities*, 1–29, 2020.

[8] Iqra Javed, Uzair Iqbal, Muhammad Bilal, Basit Shahzad, Tae-Sun Chung, Muhammad Attique, et al. Next generation infectious diseases monitoring gages via incremental federated learning: Current trends and future possibilities. *Computational Intelligence and Neuroscience*, 2023:1–12, 2023.

[9] R.S.A. Usmani, T.R. Pillai, I.A.T. Hashem, M. Marjani, R.B. Shaharudin, and M.T. Latif. Artificial intelligence techniques for predicting cardiorespiratory mortality caused by air pollution. *International Journal of Environmental Science and Technology*, 20(3):2623–2634, 2023.

[10] Fernando Martin-Sanchez and Karin Verspoor. Big data in medicine is driving big changes. *Yearbook of Medical Informatics*, 23(1):14–20, 2014.

[11] Mehdi Mohammadi, Ala Al-Fuqaha, Sameh Sorour, and Mohsen Guizani. Deep learning for IoT big data and streaming analytics: A survey. *IEEE Communications Surveys & Tutorials*, 20(4):2923–2960, 2018.

[12] Sabyasachi Dash, Sushil Kumar Shakyawar, Mohit Sharma, and Sandeep Kaushik. Big data in healthcare: Management, analysis and future prospects. *Journal of Big Data*, 6(1):54, Jun 2019.

[13] Jin Yang, Yuanjie Li, Qingqing Liu, Li Li, Aozi Feng, Tianyi Wang, Shuai Zheng, Anding Xu, and Jun Lyu. Brief introduction of medical database and data mining technology in big data era. *Journal of Evidence-Based Medicine*, 13(1):57–69, 2020.

[14] Reihaneh H. Hariri, Erik M. Fredericks, and Kate M. Bowers. Uncertainty in big data analytics: Survey, opportunities, and challenges. *Journal of Big Data*, 6(1):1–16, 2019.

[15] Kevin B. Johnson, Wei-Qi Wei, Dilhan Weeraratne, Mark E. Frisse, Karl Misulis, Kyu Rhee, Juan Zhao, and Jane L. Snowdon. Precision medicine, AI, and the future of personalized health care. *Clinical and Translational Science*, 14(1):86–93, 2021.

[16] Rita Rubin. A precision medicine approach to clinical trials. *JAMA*, 316(19):1953–1955, 2016.

[17] Giuseppe Aceto, Valerio Persico, and Antonio Pescapé. Industry 4.0 and health: Internet of things, big data, and cloud computing for healthcare 4.0. *Journal of Industrial Information Integration*, 18:100129, 2020.

[18] Sherali Zeadally, Farhan Siddiqui, Zubair Baig, and Ahmed Ibrahim. Smart healthcare: Challenges and potential solutions using internet of things (IoT) and big data analytics. *PSU Research Review*, 4(2):149–168, 2020.

[19] Jimeng Sun, Candace D. McNaughton, Ping Zhang, Adam Perer, Aris Gkoulalas-Divanis, Joshua C. Denny, Jacqueline Kirby, Thomas Lasko, Alexander Saip, and Bradley A. Malin. Predicting changes in hypertension control using electronic health records from a chronic disease management program. *Journal of the American Medical Informatics Association*, 21(2):337–344, 2014.

[20] Ying Yu, Min Li, Liangliang Liu, Yaohang Li, and Jianxin Wang. Clinical big data and deep learning: Applications, challenges, and future outlooks. *Big Data Mining and Analytics*, 2(4):288–305, 2019.

[21] Azhar Imran, Arslan Nasir, Muhammad Bilal, Guangmin Sun, Abdulkareem Alzahrani, and Abdullah Almuhaimeed. Skin cancer detection using combined decision of deep learners. *IEEE Access*, 10:118198–118212, 2022.

[22] Delshi Howsalya Devi, Kumutha Duraisamy, Ammar Armghan, Meshari Alsharari, Khaled Aliqab, Vishal Sorathiya, Sudipta Das, and Nasr Rashid. 5G technology in healthcare and wearable devices: A review. *Sensors*, 23(5):2519, 2023.

[23] Victor Chang, Le Minh Thao Doan, Qianwen Ariel Xu, Karl Hall, Yuanyuan Anna Wang, and Muhammad Mustafa Kamal. Digitalization in omnichannel healthcare supply chain businesses: The role of smart wearable devices. *Journal of Business Research*, 156:113369, 2023.

[24] Fan Liang, Wei Yu, Dou An, Qingyu Yang, Xinwen Fu, and Wei Zhao. A survey on big data market: Pricing, trading and protection. *IEEE Access*, 6:15132–15154, 2018.

[25] Euan A. Ashley. Towards precision medicine. *Nature Reviews Genetics*, 17(9):507–522, 2016.

[26] Mark Hoffman, C. Arnoldi, and I. Chuang. The clinical bioinformatics ontology: A curated semantic network utilizing RefSeq information. In *Biocomputing*, pp. 139–150. World Scientific, 2005.

[27] Deepanshu Khanna, Neeru Jindal, Harpreet Singh, and Prashant S. Rana. Applications and challenges in healthcare big data: A strategic review. *Current Medical Imaging*, 19(1):27–36, 2023.

[28] Smail Benzidia, Omar Bentahar, Julien Husson, and Naouel Makaoui. Big data analytics capability in healthcare operations and supply chain management: The role of green process innovation. *Annals of Operations Research*, 1–25, 2023.

[29] Muhammad Bilal, Abdullah Gani, Muhammad Ikram Ullah Lali, Mohsen Marjani, and Nadia Malik. Social profiling: A review, taxonomy, and challenges. *Cyberpsychology, Behavior, and Social Networking*, 22(7):433–450, 2019.

[30] M. Saqib Nawaz, M. Bilal, M. IkramUllah Lali, Raza Ul Mustafa, Waqar Aslam, and Salman Jajja. Effectiveness of social media data in healthcare communication. *Journal of Medical Imaging and Health Informatics*, 7(6):1365–1371, 2017.

[31] Kornelia Batko and Andrzej Slkezak. The use of big data analytics in healthcare. *Journal of Big Data*, 9(1):3, 2022.

[32] Liyakathunisa Syed, Saima Jabeen, S. Manimala, and Hoda A. Elsayed. Data science algorithms and techniques for smart healthcare using IoT and big data analytics. *Smart Techniques for a Smarter Planet: Towards Smarter Algorithms*, pp. 211–241, 2019.

[33] Sri Venkat Gunturi Subrahmanya, Dasharathraj K. Shetty, Vathsala Patil, B.M. Zeeshan Hameed, Rahul Paul, Komal Smriti, Nithesh Naik, and Bhaskar K. Somani. The role of data science in healthcare advancements: Applications, benefits, and future prospects. *Irish Journal of Medical Science (1971–)*, 191(4):1473–1483, 2022.

[34] Óscar Álvarez-Machancoses and Juan Luis Fernández-Martínez. Using artificial intelligence methods to speed up drug discovery. *Expert Opinion on Drug Discovery*, 14(8):769–777, 2019.

[35] Gisbert Schneider. Automating drug discovery. *Nature Reviews Drug Discovery*, 17(2):97–113, 2018.

[36] Mubashir Hassan, Faryal Mehwish Awan, Anam Naz, Enrique J. deAndrés Galiana, Oscar Alvarez, Ana Cernea, Lucas Fernández-Brillet, Juan Luis Fernández-Martínez, and Andrzej Kloczkowski. Innovations in genomics and big data analytics for personalized medicine and health care: A review. *International Journal of Molecular Sciences*, 23(9):4645, 2022.

[37] Khalid Mahmood Aamir, Laiba Sarfraz, Muhammad Ramzan, Muhammad Bilal, Jana Shafi, and Muhammad Attique. A fuzzy rule-based system for classification of diabetes. *Sensors*, 21(23):8095, 2021.

[38] Jia Xu, Pengwei Yang, Shang Xue, Bhuvan Sharma, Marta Sanchez-Martin, Fang Wang, Kirk A. Beaty, Elinor Dehan, and Baiju Parikh. Translating cancer genomics into precision medicine with artificial intelligence: Applications, challenges and future perspectives. *Human Genetics*, 138(2):109–124, 2019.

[39] Pranjal Kumar, Siddhartha Chauhan, and Lalit Kumar Awasthi. Artificial intelligence in healthcare: Review, ethics, trust challenges & future research directions. *Engineering Applications of Artificial Intelligence*, 120:105894, 2023.

[40] Kee Yuan Ngiam and Wei Khor. Big data and machine learning algorithms for healthcare delivery. *The Lancet Oncology*, 20(5):e262–e273, 2019.

[41] Min Chen, Shiwen Mao, and Yunhao Liu. Big data: A survey. *Mobile Networks and Applications*, 19:171–209, 2014.

[42] Jinhyuk Lee, Wonjin Yoon, Sungdong Kim, Donghyeon Kim, Sunkyu Kim, Chan Ho So, and Jaewoo Kang. Biobert: A pre-trained biomedical language representation model for biomedical text mining. *Bioinformatics*, 36(4):1234–1240, 2020.

[43] Laila Rasmy, Yang Xiang, Ziqian Xie, Cui Tao, and Degui Zhi. Med-Bert: Pre-trained contextualized embeddings on large-scale structured electronic health records for disease prediction. *NPJ Digital Medicine*, 4(1):86, 2021.

[44] Wei Tan, Prayag Tiwari, Hari Mohan Pandey, Catarina Moreira, and Amit Kumar Jaiswal. Multimodal medical image fusion algorithm in the era of big data. *Neural Computing and Applications*, pp. 1–21, 2020.

[45] Davide Cirillo and Alfonso Valencia. Big data analytics for personalized medicine. *Current Opinion in Biotechnology*, 58:161–167, 2019.

[46] Wei Li, Yuanbo Chai, Fazlullah Khan, Syed Rooh Ullah Jan, Sahil Verma, Varun G. Menon, F.N.M. Kavita, and Xingwang Li. A comprehensive survey on machine learning-based big data analytics for IoT-enabled smart healthcare system. *Mobile Networks and Applications*, 26:234–252, 2021.

[47] Zodwa Dlamini, Flavia Zita Francies, Rodney Hull, and Rahaba Marima. Artificial intelligence (AI) and big data in cancer and precision oncology. *Computational and Structural Biotechnology Journal*, 18:2300–2311, 2020.

[48] Muhammad Haseeb Arshad, Muhammad Bilal, and Abdullah Gani. Human activity recognition: Review, taxonomy and open challenges. *Sensors*, 22(17):6463, 2022.

[49] Miriam Reisman. EHRS: The challenge of making electronic data usable and interoperable. *Pharmacy and Therapeutics*, 42(9):572, 2017.

[50] Khader Shameer, Marcus A. Badgeley, Riccardo Miotto, Benjamin S. Glicksberg, Joseph W. Morgan, and Joel T. Dudley. Translational bioinformatics in the era of real-time biomedical, health care and wellness data streams. *Briefings in Bioinformatics*, 18(1):105–124, 2017.

[51] Liyakathunisa Syed, Saima Jabeen, S. Manimala, and Abdullah Alsaeedi. Smart health-care framework for ambient assisted living using IOMT and big data analytics techniques. *Future Generation Computer Systems*, 101:136–151, 2019.

[52] Sabyasachi Dash, Sushil Kumar Shakyawar, Mohit Sharma, and Sandeep Kaushik. Big data in healthcare: Management, analysis and future prospects. *Journal of Big Data*, 6(1):1–25, 2019.

[53] Karim Abouelmehdi, Abderrahim Beni-Hessane, and Hayat Khaloufi. Big healthcare data: Preserving security and privacy. *Journal of Big Data*, 5(1):1–18, 2018.

[54] Bishnu P. Bhattarai, Sumit Paudyal, Yusheng Luo, Manish Mohanpurkar, Kwok Cheung, Reinaldo Tonkoski, Rob Hovsapian, Kurt S. Myers, Rui Zhang, Power Zhao, Milos Manic, Song Zhang, and Xiaping Zhang. Big data analytics in smart grids: State-of-the-art, challenges, opportunities, and future directions. *IET Smart Grid*, 2(2):141–154, 2019.

[55] Roberta Pastorino, Corrado De Vito, Giuseppe Migliara, Katrin Glocker, Ilona Binenbaum, Walter Ricciardi, and Stefania Boccia. Benefits and challenges of big data in healthcare: An overview of the European initiatives. *European Journal of Public Health*, 29(Supplement 3):23–27, 2019.

Chapter 12

Improving hepatitis C diagnosis using machine learning techniques

An experimental analysis

Md. Alif Sheakh, Mst. Sazia Tahosin, Taminul Islam, and Rishalatun Jannat Lima

12.1 INTRODUCTION

Hepatitis C is a major global public health issue, affecting a significant number of people on a worldwide level. Hepatitis C is a liver problem connected to the hepatitis C virus (HCV) [1]. The illness can appear as either acute or chronic. Individuals who catch HCV and have been infected for a duration of less than six months suffer from a disease known as acute hepatitis C [2]. Usually, people don't show any symptoms [2]. Around 20% of the people who have been infected will properly rid themselves of the virus within a period of six months, which prevents them from suffering any suffering consequences [3]. It is estimated that the remaining 80% of people infected with HCV will progress to a chronic state of the disease [4]. Individuals infected will continue to have the virus indefinitely, thereby maintaining their infectiousness and posing a potential risk to others. It is estimated that 71 million individuals approximately suffer from chronic HCV infection based on data provided by the World Health Organization [5]. HCV has the potential to induce various liver ailments, such as chronic hepatitis, liver cirrhosis, and hepatocellular carcinoma (HCC) [6]. The precise and prompt identification of hepatitis C is essential for efficiently controlling and mitigating disease advancement. Traditional diagnostic techniques, including antibody tests and polymerase chain reaction (PCR) assays, have garnered extensive utilization [7]. However, these methods possess certain drawbacks about their sensitivity, specificity, and cost-efficiency (Figure 12.1).

The majority of people infected with hepatitis C have no symptoms. The start of symptoms may occur within a period that varies from 14 to 180 days following the initial infection [8]. Yet it is possible that symptoms might begin years after initial exposure, possibly suggesting that the liver damage has reached an advanced stage. Currently, there is a lack of an effective vaccine to prevent hepatitis C [9]. However, various strategies are present to reduce the possibility of developing the infection [10]:

- For drug use, don't share needles, spoons, and so on.
- Sharing personal items like razors and toothbrushes that may come into contact with infectious blood is not recommended.

DOI: 10.1201/9781003496410-15

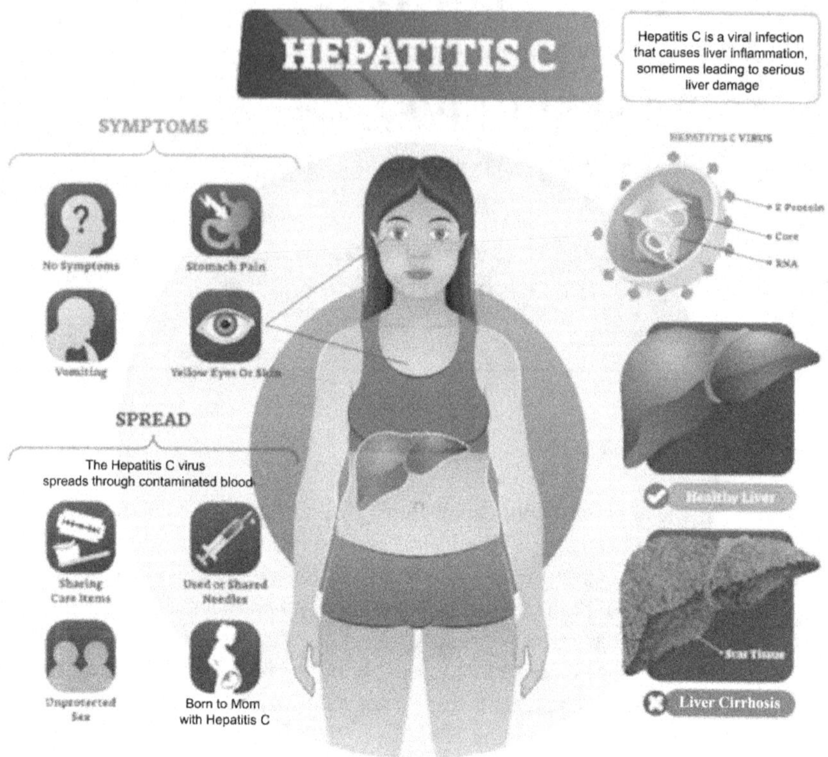

Figure 12.1 Overview of hepatitis C.

- Condoms are required to be used during sexual activity.
- Visit authorized tattoo, acupuncture, and body piercing outlets.

Machine learning has become popular for health care research and has demonstrated promise in a number of medical fields, such as disease diagnosis [11]. Machine learning algorithms can immediately discover relationships and patterns in large datasets [12]. This allows the creation of predictive models to help detect diseases quickly and accurately. Machine learning models may boost hepatitis C diagnostic accuracy and patient care by analyzing complicated data patterns and combining multiple factors.

This study analyzes potential machine learning models that may enhance hepatitis C treatment. Our objective is to use scientific and clinical information to establish accurate models that can help identify hepatitis C infection efficiently. To achieve our research objective, we are going to perform a thorough and methodical analysis of the experiment. The dataset of patients will be utilized in our evaluation. The dataset provided will be used as a starting point for both training and validating our machine learning models. This study aims to utilize a range of machine learning algorithms, such as random forest [13], gradient boosting [14], k-nearest neighbors [15], extreme gradient boosting [16], extra tree [17], AdaBoost [18], LogitBoost [19], CatBoost [20], support vector machine [21], naive Bayes [22], neural networks (multilayer perceptron) [23], and Gaussian process [24] classifier models, to perform an analysis of the dataset and create predictive models for the diagnosis of hepatitis C. Besides all these algorithms, convolutional neural network (CNN) [25] is also applied in various medical problem-solving research. The models will be tested by analyzing accuracy, recall, and precision metrics.

The expected findings of the research are predicted to have major impacts for hepatology and how it's used in hospitals. The possible effect of creating reliable machine learning models for diagnosing hepatitis C is significant, as it can significantly change existing diagnostic methodologies and supply clinicians with valuable guidance resources. Through applying elegant algorithms and an in-depth review of huge amounts of patient data, our objective is to create exact and strong models that can aid in timely recognition and diagnosis of health problems. The findings of this study have the opportunity to make a valuable contribution to the progress of medical practices and better worldwide management of hepatitis C.

12.2 LITERATURE REVIEW

In the research they carried out, Ma et al. [26] evaluated six separate machine learning algorithms to try to predict hepatitis C. In particular, they stated a high-accuracy rate of 95% when using support vector machines (SVMs) and XGBoost algorithms. They employed the NHANES and UCI datasets, which include clinical data from a group of over 5,000 and 1,000 people diagnosed with hepatitis C, respectively. The main finding of the research was the active assessment and contrast of different machine learning methods for predicting the onset of hepatitis C. The result of the research was that SVMs and XGBoost showed the highest degree of accuracy in the present situation.

Wang et al. [27] identified hepatitis B virus (HBV) and HCV status utilizing four machine learning models. Notably, their HCV status analysis displayed a 98.1% accuracy rate. The NHANES dataset, which consists of clinical data from over 5,000 adults in the United States was used by the researchers. This study employed five algorithms based on machine learning to predict hepatitis C status. This study revealed that the k-nearest neighbors (KNN) model was incredibly accurate.

Ahammed et al. [28] employed feature selection and predictive machine learning algorithms to forecast hepatitis C. Their KNN algorithm reached 83% accuracy. They evaluated 1,801 patients and 12 features. This work applied feature selection to enhance hepatitis C predictive machine learning models.

Based on clinical information, Abdelrahman et al. [29] constructed a structure for machine learning for identifying hepatitis C. Their architecture has a 94.88% accuracy rate using the random forest (RF) method. Clinical data from

850 people suffering from hepatitis C have been utilized in the Egyptian HCV dataset. A machine learning construct that accurately foretold real-world hepatitis C cases was its primary contribution. The most significant contribution included a machine learning framework that could predict real-world hepatitis C infections.

Singh et al. [30] put forward a method of ensemble learning to predict hepatitis C disease, reaching a high accuracy of 95.69%. The scientists utilized the UCI dataset, which has clinical data obtained from a group of over 1,000 people suffering from hepatitis C. The main work consisted of the creation of an ensemble learning structure that was able to increase the accuracy of machine learning models in the field of hepatitis C estimation.

Priya et al. [31] evaluated machine learning approaches for identifying hepatitis C. The logistic regression (LR) methodology had 97.9% accuracy. Clinical information of more than 1,000 hepatitis C patients was utilized in the UCI dataset. Evaluating machine learning techniques for hepatitis C prediction was the study's major contribution. The research found that the LR approach was the most accurate.

To predict hepatitis C, Sahoo et al. [32] evaluated multiclass and binary labeling of a dataset. The AdaBoost algorithm provided the researchers with 54.23% accuracy. Clinical information of over 1,000 hepatitis C patients was employed in the PROMISE dataset. This study examined the efficacy of multiclass and binary classifications to predict hepatitis C. The research conducted found that binary class labels appeared more accurate.

In their research, Syafa'ah et al. [33] compared four machine learning classification algorithms, specifically KNN, naive Bayes, neural network, and RF, to predict hepatitis C. A dataset of 1,200 patients identified as having hepatitis C was utilized in the research. The findings showed that the neural network displayed the greatest accuracy, specifically 95.12%.

Alotaibi et al. [34] utilized machine learning methods to identify the presence of liver disease for people suffering from hepatitis C. The researchers employed a dataset comprising 2,038 Egyptian patients from the UCI Machine Learning Repository. Following that, they trained four different machine learning algorithms on this dataset, including RF, gradient boosting machine, extreme gradient boosting, and extra trees model. The extra trees model showed better outcomes compared to the other models, getting an accuracy rate of 96.92%.

In their research, Abd El-Salam et al. [35] used machine learning approaches to project the likelihood of esophageal varices for humans diagnosed with chronic hepatitis C. A dataset of 4,962 patients from Egypt was used, and six different machine learning algorithms were used for learning purposes. The algorithms were neural network, naive Bayes, decision tree, SVM, RF, and Bayesian network. The Bayesian network algorithm displayed excellent results, with an accuracy rate of 74.8%.

In their study, Chen and Ji [36] created a customized machine learning structure to identify hepatitis C in particular patients. The model was trained utilizing patients' individual clinical data, helping it make predictions concerning the possibility of hepatitis C in a given patient. The model's efficiency was examined using a dataset with 615 patients, which gave an accuracy rate of 99% (Table 12.1).

Table 12.1 Comparison with related works

Ref	*Contribution*	*Year*	*Dataset*	*Model*	*Accuracy*
This Work	An experiment with 12 machine learning models to predict hepatitis C	2023	UCI	RF, KNN, GB, XGBoost, Extra Trees, AdaBoost, LogitBoost, CatBoost, SVM, NB, NN, GP	**99% (Extra trees)**
[26]	Compared 6 machine learning algorithms for hepatitis C prediction	2023	NHANES and UCI	SVM, KNN, LR, DT, XGBoost, ANN	95% (SVM, XGBoost)
[27]	Used 4 machine learning models to predict HBV and HCV status	2023	NHANES	KNN, SVM, LR, CART	98.1% (KNN)
[28]	Used a combination of feature selection and machine learning to predict hepatitis C	2023	Dataset of 1,801 patients with 12 features	KNN, LR, DT, RF, NN	83% (KNN)
[29]	Developed a machine learning framework to predict hepatitis C using clinical data	2023	Egyptian HCV dataset	NB, RF, KNN, LR	94.88% (RF)
[30]	Developed an ensemble learning model for predicting hepatitis C	2022	UCI	MLP, Bayesian network, quest, ensemble	95.69% (Ensemble model)
[31]	Compared the performance of different machine learning models for predicting hepatitis C	2022	UCI	KNN, SVM, RF, NN, NB, LR	97.9% (LR)
[32]	Comparisons between multi- and binary class labels of the same dataset	2020	PROMISE dataset	KNN, RF, SVM, GNB, NN, bagging, AdaBoost	54.23% (AdaBoost)

Note: ANN, automated neural net; CART, real-time captioning; DT, diphtheria-tetanus; GB, gradient boost; GP, Gaussian process KNN, k-nearest neighbors; LR, logistic regression; MLP, medical-legal partnership; NB, naive Bayes; NN, neural network; RF, random forest; SVM, support vector machine; XGBoost, extreme gradient boost.

12.3 METHODOLOGY

The proposed methodology for this research uses machine learning algorithms in the early diagnosis of hepatitis C, which involves several phases, as shown in Figure 12.2.

At the very beginning, a dataset that includes people suffering from hepatitis C is collected. The dataset is then subjected to multiple data preprocessing methods. The data preparation procedures involve removing unnecessary columns,

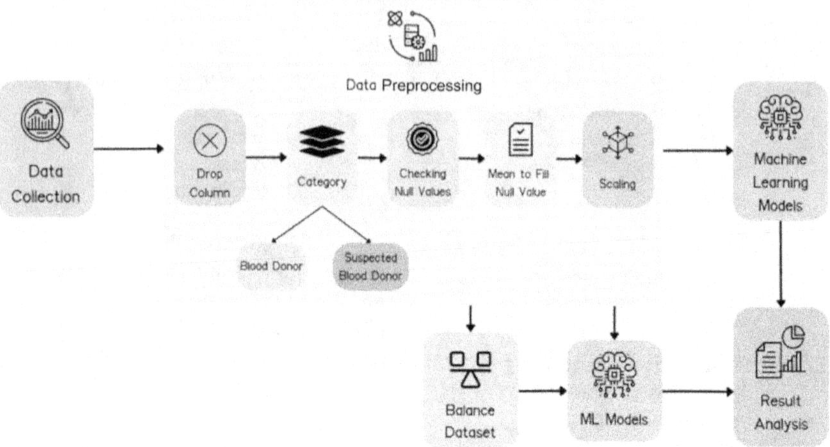

Figure 12.2 Proposed methodology.

changing the "Category" column into a binary class representation (Blood Donor, Suspected Blood Donor), and encoding the "Sex" column as binary values (0 for Male and 1 for Female). The dataset contains rows with mean values that replace null values. The robust scaler is used to connect all attributes to a common scale. The dataset that received preprocessing is split into two subsets: training and testing sets. After that, a comprehensive set of 12 different machine learning models is applied: namely, RF [13], KNN [15], gradient boosting [14], extreme gradient boost [16], extra trees [17], AdaBoost [18], LogitBoost [19], CatBoost [20], SVM [21], naive Bayes [22], neural network (multilayer perceptrons) [23], and Gaussian process [24] classifiers. The comparison of each model involves an evaluation of its classification accuracy, precision, and recall. The dataset is balanced by implementing underestimating techniques to reduce potential class imbalance issues. This entails the undersampling of the majority class to achieve a balanced distribution of samples with the minority class. Then we evaluate the model by calculating multiple statistics, including accuracy, precision, recall, and confusion matrix [37]. The results of the research are subjected to evaluation and translation. This methodology offers an approach for predicting hepatitis C and related liver diseases by implementing machine learning algorithms. This research aims to identify the most efficient model for early detection of hepatitis C by evaluating various models and performance metrics.

12.3.1 Data collection

We take this research's dataset from the UCI Machine Learning Repository [38]. The dataset consists of the records of 615 patients, 75 of whom are healthy and 540 uncertain of hepatitis C. The research sample contains patients across various regions. The dataset has been compiled in Table 12.2. The dataset's target variable, classification class, and sex distribution are shown in Figures 12.3 and 12.4.

12.3.2 Data preprocessing

To simplify the process, we remove unnecessary columns from the dataset that do not assist with the classification task. In the "Category" column, there were multiple categories identified, including "Blood Donor," "Hepatitis," "Fibrosis," and "Cirrhosis." The classes were transformed into binary classes, with "Blood Donor" being categorized as "Healthy Patients" and all other three classes being

Table 12.2 Summary of the dataset

Classes	*Values*
Category	Healthy Patients (0), Suspected Patients (1)
Age	19–77
Sex	Male (0), Female (1)
Albumin (ALB)	14.9–82.2
Alkaline Phosphatase (ALP)	11.3–414.6
Alanine Aminotransferase (ALT)	0.9–325.3
Aspartate Aminotransferase (AST)	10.6–324
Bilirubin (BIL)	0.8–254
Cholinesterase (CHE)	1.42–16.41
Cholesterol (CHOL)	1.43–9.67
Creatinine (CREA)	8–1079.1
Gamma-Glutamyl Transferase (GGT)	4.5–650.9
Protein (PROT)	44.8–90

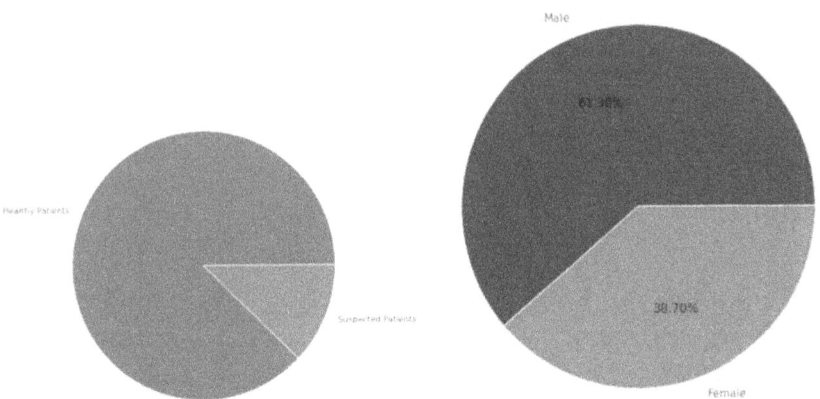

Figure 12.3 Distribution of category class. *Figure 12.4* Distribution of sex class.

classified as "Suspected Patients." Similarly, the "Sex" column, first denoted as "Male" and "Female," was encoded using binary values (0 and 1), where 0 means "Male" and 1 means "Female." After that, we check the dataset to see if any null value exists or not, and if found, they are handled using the mean value of the related feature.

Some of the columns contain outliers, as shown in Figure 12.5. The robust scaler scales data because it is less dependent on dataset outliers. Figure 12.6 shows age data from our dataset. In Figure 12.7, we use a boxplot to explore the variation in age within different groups to identify the number of healthy and insufficient patients.

In Figure 12.8, a scatterplot illustrates the relationship between aspartate aminotransferase (AST) and the variable type within the dataset. After plotting the data, we can observe the range of aspartate aminotransferase (AST) levels across both categories of hepatitis C.

The dataset is processed to identify the correlation between all columns, shown in Figure 12.9. Figure 12.9 analyzing a dataset's variables' connection and interdependence allows us to understand the intricate relationships and patterns within the data, facilitating informed decision-making and deeper insights into underlying trends and correlations.

In the end, in Figure 12.10, our focus moves toward a graph of the distribution of various blood test results through both groups. Histogram plots are used for visualization in this context. This method makes identifying major differences or overlaps in category values easier.

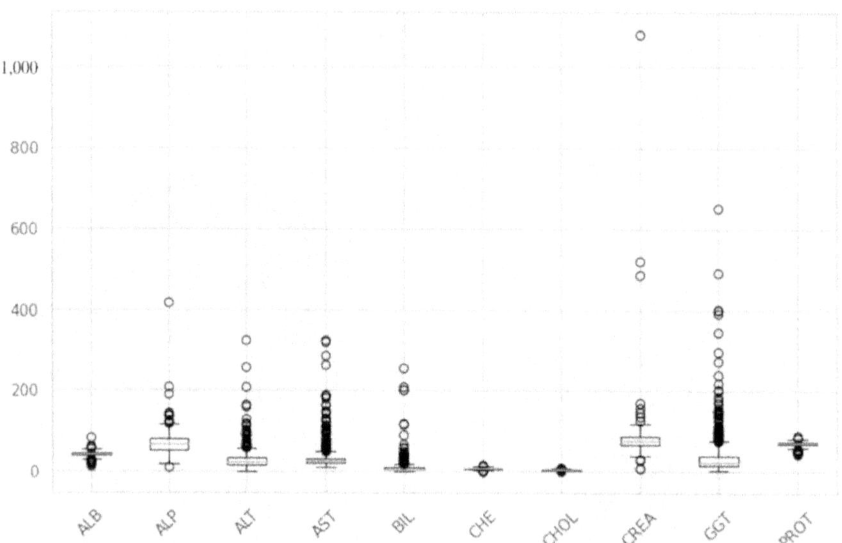

Figure 12.5 Box plot for each column.

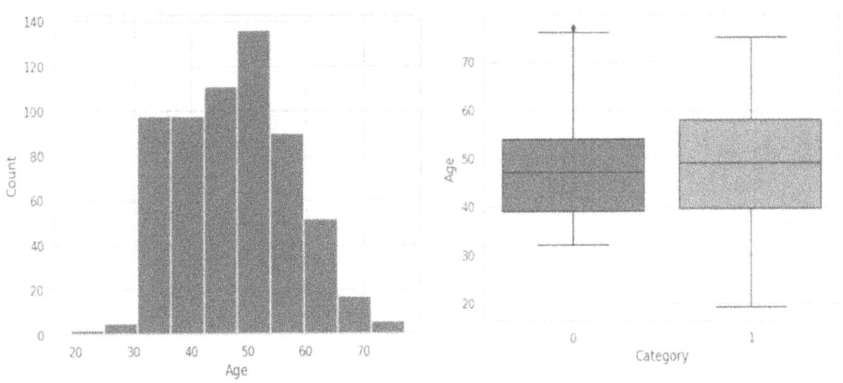

Figure 12.6 Visualization of age.

Figure 12.7 Visualization of age with category.

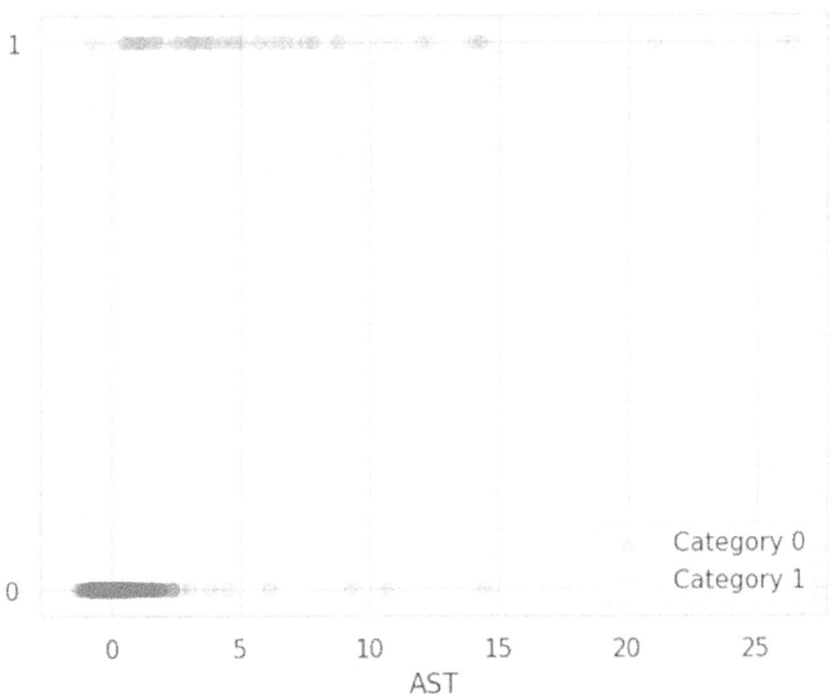

Figure 12.8 Scatterplot of category separately.

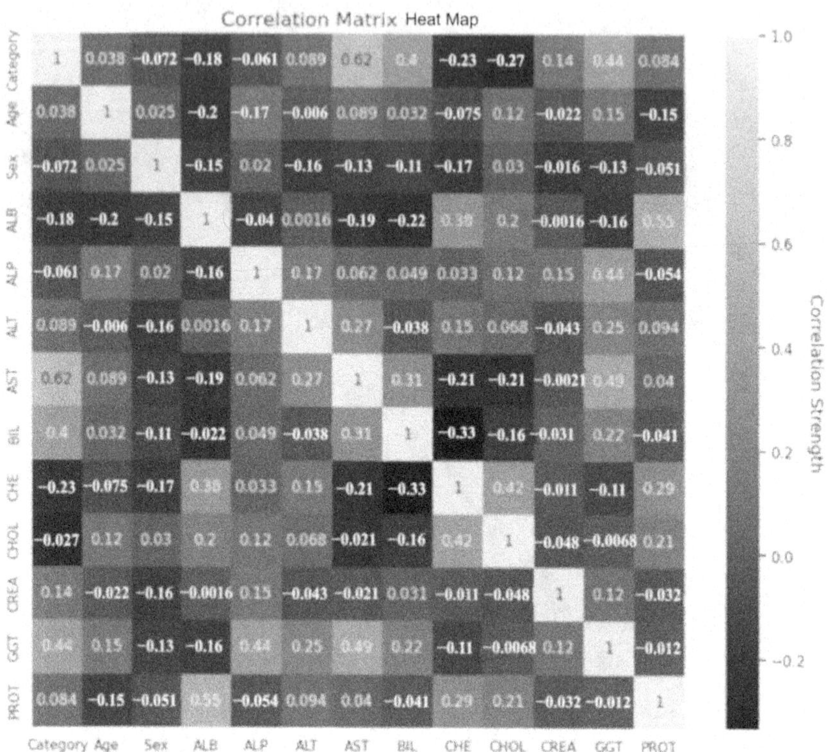

Figure 12.9 Correlation between all columns.

After completing each of the previous procedures, we move on to implement a total of 12 machine learning models. The dataset was split in an 80:20 ratio for training and testing. As shown in Figure 12.3, it is clear that our dataset displays an imbalance. To achieve a balanced dataset, the undersampling technique is employed. Undersampling is a method that reduces the number of samples in the majority class in line with the number of samples in the minority class [39]. This process aims to create a more equal distribution of data for training machine learning models. This approach minimizes potential bias to the dominant class and improves the overall accuracy of classification. Next, we execute a comparative analysis of the results obtained from the machine learning models by evaluating the accuracy, recall, and precision metrics for both the before balancing and after balancing datasets.

12.4 EXPERIMENTAL RESULT

The tables in this research show the results of the machine learning algorithms used to detect hepatitis C.

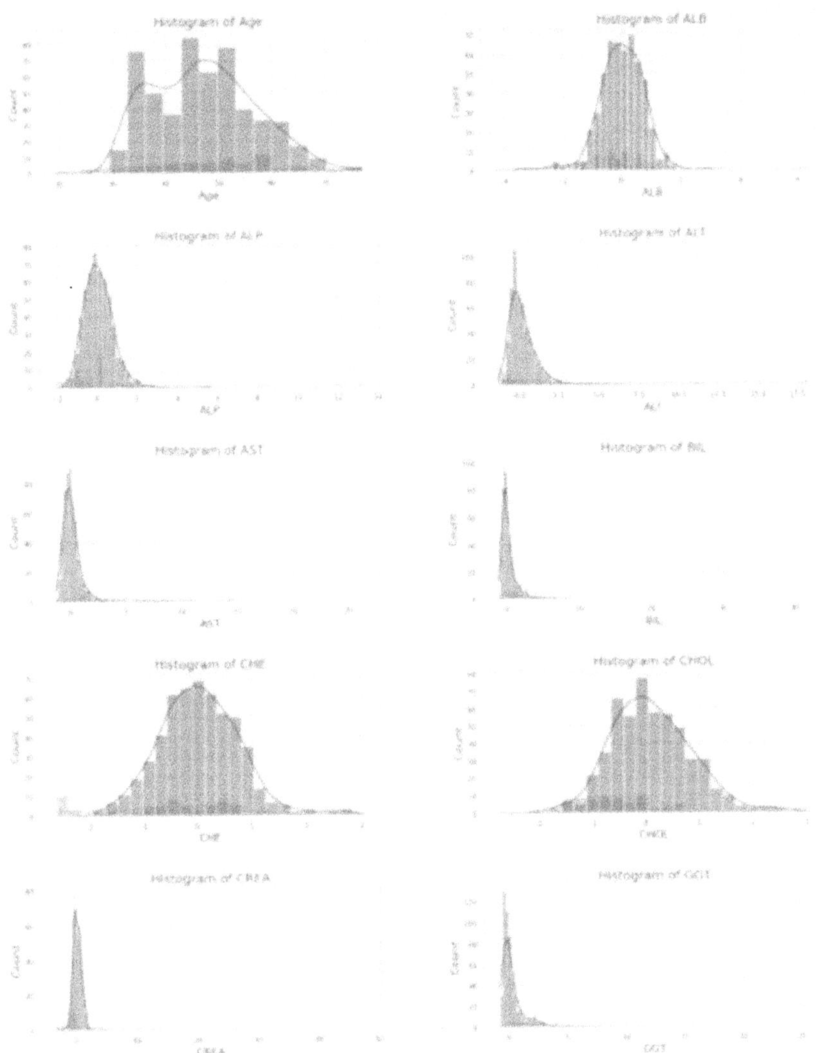

Figure 12.10 Distribution of various blood test results by category.

The results of the machine learning algorithms on the imbalanced dataset are shown in Table 12.3. Before performing the balancing process, the LogitBoost algorithm displayed a level of accuracy reaching 0.98. Additionally, it demonstrated a precision of 0.99 and a recall of 0.92. The extreme gradient boost (XGBoost) algorithm showed outstanding results, achieving an accuracy of 0.97, precision of 0.99, and recall of 0.83. More algorithms, namely, gradient boosting, AdaBoost, and RF displayed important levels of accuracy and precision.

Table 12.3 Before balancing the dataset, the result of machine learning algorithms

Algorithms	*Accuracy*	*Precision*	*Recall*
Random Forest	0.93	0.99	0.62
k-Nearest Neighbors	0.84	0.83	0.21
Gradient Boosting	0.94	0.99	0.71
Extreme Gradient Boost	0.97	0.99	0.83
Extra Trees	0.91	0.93	0.58
AdaBoost	0.95	0.99	0.75
LogitBoost	**0.98**	**0.99**	**0.92**
CatBoost	0.94	0.99	0.71
Support Vector Machine	0.84	0.99	0.17
Naive Bayes	0.88	0.76	0.54
Neural Network (Multilayer Perceptrons)	0.93	0.99	0.67
Gaussian Process	0.85	0.88	0.29

Table 12.4 After balancing the dataset, the result of machine learning algorithms

Algorithms	*Accuracy*	*Precision*	*Recall*
Random Forest	0.97	0.93	0.99
k-Nearest Neighbors	0.90	0.99	0.79
Gradient Boosting	0.90	0.87	0.93
Extreme Gradient Boost	0.90	0.87	0.93
Extra Trees	**0.99**	**0.99**	**0.99**
AdaBoost	0.93	0.88	0.99
LogitBoost	0.90	0.87	0.93
CatBoost	0.93	0.93	0.93
Support Vector Machine	0.83	0.99	0.64
Naive Bayes	0.87	0.86	0.86
Neural Network (Multilayer Perceptrons)	0.97	0.93	0.99
Gaussian Process	0.87	0.92	0.79

Table 12.4 shows what was determined after dataset balancing. The extra trees algorithm got 0.99 precision, recall, and accuracy. The RF approach performed higher with an accuracy of 0.97, precision of 0.93, and recall of 0.99. The SVM algorithm performed worse compared to other algorithms, having an accuracy rate of 0.83 and a recall rate of 0.64.

Figures 12.11, 12.12, and 12.13 graphically illustrate the contrasting evaluation of the algorithms' accuracy, precision, and recall metrics, each before and after the dataset is balanced. The given figures represent the effect of balancing on the performance metrics, highlighting the observed enhancements after the execution of the balancing methods.

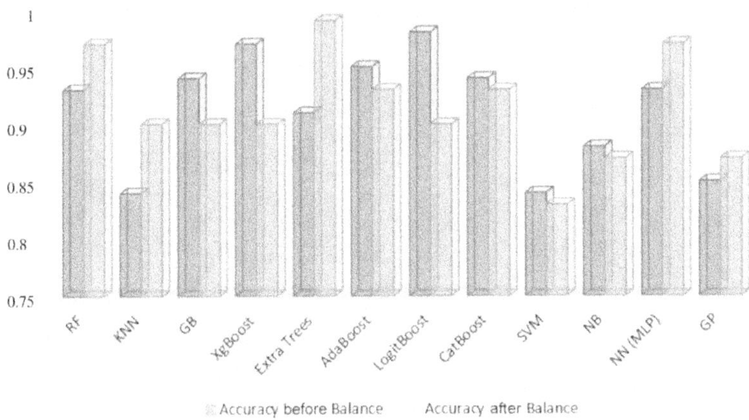

Figure 12.11 Comparison between accuracy before and after balancing the dataset.

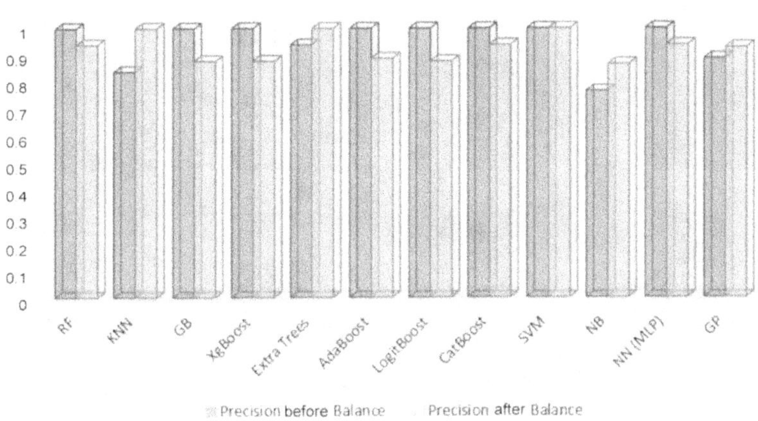

Figure 12.12 Comparison between precision before and after balancing the dataset.

In addition, Figures 12.14 and 12.15 give the confusion matrices both before and after the dataset has been balanced. The matrices provided offer valuable insights into the classification performance of the algorithms, displaying the accurate identification of true positives and true negatives and the incorrect identification of false positives and false negatives.

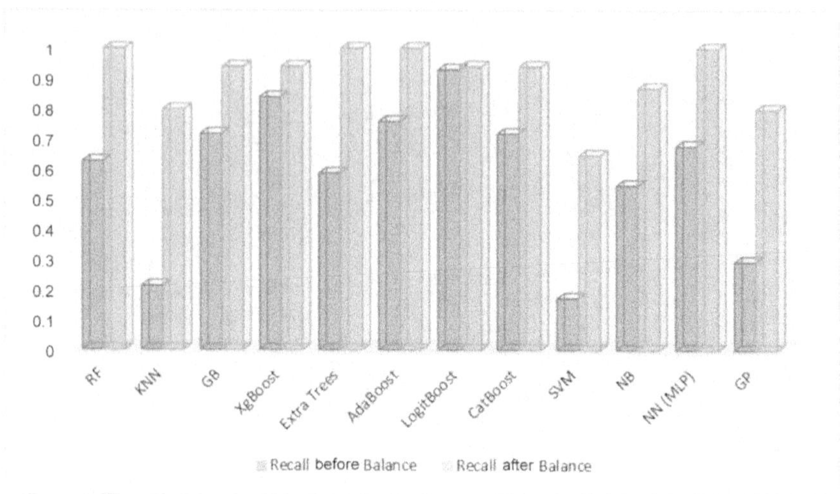

Figure 12.13 Comparison between recall before and after balancing the dataset.

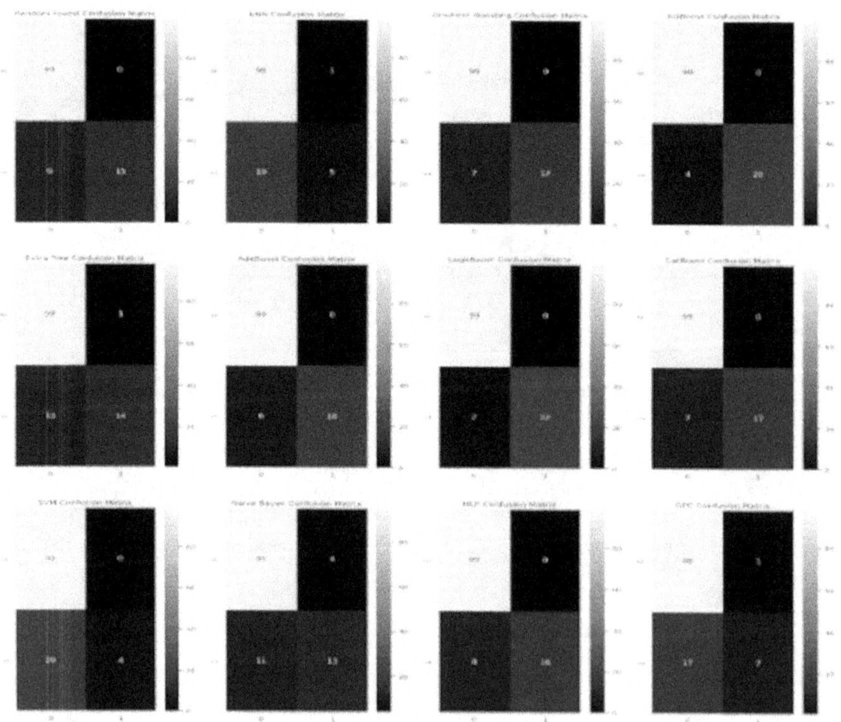

Figure 12.14 Before balancing the confusion matrix.

Figure 12.15 After balancing the confusion matrix.

In general, the experiment results show that the performance of different algorithms used for machine learning was specifically improved by balancing the dataset. The use of a balanced dataset raised accuracy, precision, and recall metrics across multiple algorithms, therefore, showing enhanced abilities in identifying hepatitis C.

12.5 CONCLUSION AND FUTURE WORK

The present study utilized machine learning algorithms to accurately diagnose hepatitis C. We examined and evaluated 12 different models' classification accuracy, precision, and recall. By analyzing relevant scholarly papers and comparing them with other algorithms, we determined the most effective methods for diagnosing hepatitis C. The empirical results demonstrated that machine learning significantly enhances the accuracy of hepatitis C diagnosis. Specifically, the extra trees model achieved an impressive accuracy rate of 99% after implementing undersampling techniques. These remarkable results unequivocally prove that these models can diagnose hepatitis C accurately. It is crucial to note that employing machine learning algorithms in medicine offers several advantages for detecting hepatitis C. The high-accuracy levels exhibited by these models enable early detection, prompt intervention, and ultimately, improve patient outcomes. Moreover, given the global health threat posed by hepatitis C, utilizing advanced computational methods to enhance illness diagnostics becomes imperative. Therefore, this study contributes substantially to the growing research on machine learning in health care. The authors emphasize the significance of advanced computational methods for precise disease diagnosis. Furthermore, these findings underscore the importance of continuous research and development in improving the accuracy and efficacy of diagnostic technologies for hepatitis C. Overall, this study demonstrates that machine learning techniques can effectively improve diagnostic accuracy for hepatitis C. In further study attempts, investigating novel features and data sources, such as genetic data and patient demographics, could yield significant findings for developing more comprehensive diagnostic models. Furthermore, implementing more in-depth study efforts, including diverse and representative datasets, can yield a better understanding of the versatility of models and their practical significance in real-world scenarios.

REFERENCES

[1] A. Hashim, L. Macken, A. Jones, M. McGeer, G. Aithal, and S. Verma, "Community-Based Assessment and Treatment of Hepatitis C Virus-Related Liver Disease, Injecting Drug and Alcohol Use Amongst People Who Are Homeless: A Systematic Review and Meta-Analysis," *International Journal of Drug Policy*, vol. 96, p. 103342, Oct. 2021, doi: 10.1016/J.DRUGPO.2021.103342.

[2] S. Frankova, N. Uzlova, D. Merta, V. Pitova, and J. Sperl, "Predictors of Significant Liver Fibrosis in People with Chronic Hepatitis C Who Inject Drugs in the Czech Republic," *Life*, vol. 13, no. 4, p. 932, Apr. 2023, doi: 10.3390/LIFE13040932.

[3] R. Ahmed *et al.*, "Sofosbuvir/Velpatasvir—A Promising Treatment for Chronic Hepatitis C Virus Infection," *Cureus*, vol. 13, no. 8, Aug. 2021, doi: 10.7759/CUREUS.17237.

[4] M. P. Manns and B. Maasoumy, "Breakthroughs in Hepatitis C Research: From Discovery to Cure," *Nature Reviews Gastroenterology & Hepatology*, vol. 19, no. 8, pp. 533–550, Aug. 2022, doi: 10.1038/s41575-022-00608-8.

[5] J. F. H. Eijsink, M. N. M. T. Al Khayat, C. Boersma, P. G. J. ter Horst, J. C. Wilschut, and M. J. Postma, "Cost-Effectiveness of Hepatitis C Virus Screening, and Subsequent Monitoring or Treatment Among Pregnant Women in the Netherlands," *The European Journal of Health Economics*, vol. 22, no. 1, pp. 75–88, Feb. 2021, doi: 10.1007/s10198-020-01236-2.

[6] S. J. Park and Y. S. Hahn, "Hepatocytes Infected with Hepatitis C Virus Change Immunological Features in the Liver Microenvironment," *Clinical and Molecular Hepatology*, vol. 29, no. 1, p. 65, Jan. 2023, doi: 10.3350/CMH.2022.0032.

[7] I. Gow, N. C. Smith, D. Stark, and J. Ellis, "Laboratory Diagnostics for Human Leishmania Infections: A Polymerase Chain Reaction-Focussed Review of

Detection and Identification Methods," *Parasites & Vectors*, vol. 15, no. 1, pp. 1–24, Nov. 2022, doi: 10.1186/S13071-022-05524-Z.

[8] M. W. Tenforde *et al.*, "Protection of Messenger RNA Vaccines Against Hospitalized Coronavirus Disease 2019 in Adults Over the First Year Following Authorization in the United States," *Clinical Infectious Diseases*, vol. 76, no. 3, pp. e460–e468, Feb. 2023, doi: 10.1093/CID/CIAC381.

[9] L. Parlati, C. Hollande, and S. Pol, "Treatment of Hepatitis C Virus Infection," *Clinics and Research in Hepatology and Gastroenterology*, vol. 45, no. 4, p. 101578, Jul. 2021, doi: 10.1016/J.CLINRE.2020.11.008.

[10] C. R. Balsom, A. Farrell, and D. V. Kelly, "Barriers and Enablers to Testing for Hepatitis C Virus Infection in People Who Inject Drugs—A Scoping Review of the Qualitative Evidence," *BMC Public Health*, vol. 23, no. 1, p. 1038, Dec. 2023, doi: 10.1186/S12889-023-16017-8/TABLES/3.

[11] T. Islam, A. Kundu, N. Islam Khan, C. Chandra Bonik, F. Akter, and M. Jihadul Islam, "Machine Learning Approaches to Predict Breast Cancer: Bangladesh Perspective," *Smart Innovation, Systems and Technologies*, vol. 302, pp. 291–305, 2022, doi: 10.1007/978-981-19-2541-2_23/COVER.

[12] Md. A. Sheakh, Mst. Sazia Tahosin, M. M. Hasan, T. Islam, O. Islam, and M. M. Rana, "Child and Maternal Mortality Risk Factor Analysis Using Machine Learning Approaches," in *2023 11th International Symposium on Digital Forensics and Security (ISDFS)*, IEEE, May 2023, pp. 1–6, doi: 10.1109/ISDFS58141.2023.10131826.

[13] T. A. N. Saputra, K. I. Arizona, M. R. Andrian, F. I. Kurniadi, and B. Juarto, "Random Forest in Detecting Hepatitis C," in *Proceedings—2022 9th International Conference on Information Technology, Computer and Electrical Engineering, ICITACEE 2022*, 2022, pp. 299–302, doi: 10.1109/ICITACEE55701.2022.9924074.

[14] H. Park *et al.*, "Machine Learning Algorithms for Predicting Direct-Acting Antiviral Treatment Failure in Chronic Hepatitis C: An HCV-TARGET Analysis," *Hepatology*, vol. 76, no. 2, pp. 483–491, Aug. 2022, doi: 10.1002/HEP.32347.

[15] S. Setianingsih, M. U. Chasanah, Y. I. Kurniawan, and L. Afuan, "Implementation of Particle Swarm Optimization in k-Nearest Neighbor Algorithm as Optimization Hepatitis C Classification," *Jurnal Teknik Informatika (Jutif)*, vol. 4, no. 2, pp. 457–465, Apr. 2023, doi: 10.52436/1.JUTIF.2023.4.2.980.

[16] N. Ali, D. Srivastava, A. Tiwari, A. Pandey, A. K. Pandey, and A. Sahu, "Predicting Life Expectancy of Hepatitis B Patients using Machine Learning," in *IEEE International Conference on Distributed Computing and Electrical Circuits and Electronics, ICDCECE 2022*, 2022, doi: 10.1109/ICDCECE53908.2022.9793025.

[17] X. Liu, L. Wang, C. H. Liang, Y. P. Lu, T. Yang, and X. Zhang, "An Enhanced Methodology for Predicting Protein-Protein Interactions Between Human and Hepatitis C Virus Via Ensemble Learning Algorithms," *Journal of Biomolecular Structure and Dynamics*, vol. 40, no. 21, pp. 10592–10602, 2022, doi: 10.1080/07391102.2021.1946429/SUPPL_FILE/TBSD_A_1946429_SM8044.DOCX.

[18] D. A. Jadhav, "An Enhanced and Secured Predictive Model of Ada-Boost and Random-Forest Techniques in HCV Detections," *Materials Today: Proceedings*, vol. 51, pp. 186–195, Jan. 2022, doi: 10.1016/J.MATPR.2021.05.071.

[19] C. Geetha and S. Maruthuperumal, "Prediction of Liver Cirrhosis using Ensemble Machine Learning Algorithms," in *2022 1st International Conference on Computer,*

Power and Communications, ICCPC 2022——Proceedings, 2022, pp. 453–457, doi: 10.1109/ICCPC55978.2022.10072150.

[20] J. Jangiti, C. G. Paluri, S. Vadlamani, and S. K. Jindal, "Hepatitis C Severity Prognosis: A Machine Learning Approach," *Journal of Electrical Engineering and Technology*, vol. 18, no. 4, pp. 3253–3264, Feb. 2023, doi: 10.1007/S42835-023-01441-Y/METRICS.

[21] K. Keyvan, M. R. Sohrabi, and F. Motiee, "An Intelligent Method Based on Feed-Forward Artificial Neural Network and Least Square Support Vector Machine for the Simultaneous Spectrophotometric Estimation of Anti Hepatitis C Virus Drugs in Pharmaceutical Formulation and Biological Fluid," *Spectrochimica Acta Part A: Molecular and Biomolecular Spectroscopy*, vol. 263, p. 120190, Dec. 2021, doi: 10.1016/J.SAA.2021.120190.

[22] R. Safdari, A. Deghatipour, M. Gholamzadeh, and K. Maghooli, "Applying Data Mining Techniques to Classify Patients with Suspected Hepatitis C Virus Infection," *Intelligent Medicine*, vol. 2, no. 4, pp. 193–198, Nov. 2022, doi: 10.1016/J.IMED.2021.12.003/ASSET/5C896865-8D61-41F6-A794-CDDEF0C58309/ASSETS/GRAPHIC/2096-9376-02-04-002-F003.PNG.

[23] Ç. Suiçmez, C. Yılmaz, H. T. Kahraman, E. Cengiz, and A. Suiçmez, "Prediction of Hepatitis C Disease with Different Machine Learning and Data Mining Technique," pp. 375–398, 2023, doi: 10.1007/978-3-031-09753-9_27.

[24] L. A. Escamilla, O. Akarsu, E. Di Valentino, and J. A. Vazquez, "Model-Independent Reconstruction of the Interacting Dark Energy Kernel: Binned and Gaussian Process," May 2023, Accessed: Jul. 8, 2023. [Online]. Available: https://arxiv.org/abs/2305.16290v1

[25] Md. T. Islam, T. Ahmed, A. B. M. Raihanur Rashid, T. Islam, Md. S. Rahman, and Md. Tarek Habib, "Convolutional Neural Network Based Partial Face Detection," in *2022 IEEE 7th International conference for Convergence in Technology (I2CT)*, IEEE, Apr. 2022, pp. 1–6, doi: 10.1109/I2CT54291.2022.9825259.

[26] A. Alizargar, Y. L. Chang, and T. H. Tan, "Performance Comparison of Machine Learning Approaches on Hepatitis C Prediction Employing Data Mining Techniques," *Bioengineering*, vol. 10, no. 4, p. 481, Apr. 2023, doi: 10.3390/BIOENGINEERING10040481/S1.

[27] V. Harabor *et al.*, "Machine Learning Approaches for the Prediction of Hepatitis B and C Seropositivity," *International Journal of Environmental Research and Public Health*, vol. 20, no. 3, p. 2380, Jan. 2023, doi: 10.3390/IJERPH20032380.

[28] D. Zhang *et al.*, "Explainable Machine Learning Approach for Hepatitis C Diagnosis Using SFS Feature Selection," *Machines*, vol. 11, no. 3, p. 391, Mar. 2023, doi: 10.3390/MACHINES11030391.

[29] H. Mamdouh Farghaly, M. Y. Shams, and T. Abd El-Hafeez, "Hepatitis C Virus Prediction Based on Machine Learning Framework: A Real-World Case Study in Egypt," *Knowledge and Information Systems*, vol. 65, no. 6, pp. 2595–2617, Jun. 2023, doi: 10.1007/S10115-023-01851-4/TABLES/7.

[30] M. O. Edeh *et al.*, "Artificial Intelligence-Based Ensemble Learning Model for Prediction of Hepatitis C Disease," *Frontiers in Public Health*, vol. 10, p. 892371, Apr. 2022, doi: 10.3389/FPUBH.2022.892371/BIBTEX.

[31] P. L. Septina and J. I. Sihotang, "A Comparative Study on Hepatitis C Predictions Using Machine Learning Algorithms," *8ISC Proceedings: Technology*, pp. 33–42,

Feb. 2022, Accessed: Jul. 4, 2023. [Online]. Available: http://ejournal.unklab.ac.id/index.php/8ISCTE/article/view/684

[32] S. C. Nandipati, C. XinYing, and K. K. Wah, "Hepatitis C Virus (HCV) Prediction by Machine Learning Techniques," *Applications of Modelling and Simulation*, vol. 4, pp. 89–100, Mar. 2020, Accessed: Jul. 4, 2023. [Online]. Available: http://arqiipubl.com/ojs/index.php/AMS_Journal/article/view/122

[33] L. Syafaâ, Z. Zulfatman, I. Pakaya, and M. Lestandy, "Comparison of Machine Learning Classification Methods in Hepatitis C Virus," *Jurnal Online Informatika*, vol. 6, no. 1, pp. 73–78, Jun. 2021, doi: 10.15575/JOIN.V6I1.719.

[34] A. Alotaibi *et al.*, "Explainable Ensemble-Based Machine Learning Models for Detecting the Presence of Cirrhosis in Hepatitis C Patients," *Computation*, vol. 11, no. 6, p. 104, May 2023, doi: 10.3390/COMPUTATION11060104.

[35] S. M. Abd El-Salam *et al.*, "Performance of Machine Learning Approaches on Prediction of Esophageal Varices for Egyptian Chronic Hepatitis C Patients," *Informatics in Medicine Unlocked*, vol. 17, p. 100267, Jan. 2019, doi: 10.1016/J.IMU.2019.100267.

[36] L. Chen, P. Ji, and Y. Ma, "Machine Learning Model for Hepatitis C Diagnosis Customized to Each Patient," *IEEE Access*, vol. 10, pp. 106655–106672, 2022, doi: 10.1109/ACCESS.2022.3210347.

[37] T. Islam *et al.*, *Review Analysis of Ride-Sharing Applications Using Machine Learning Approaches: Bangladesh Perspective*. New York: CRC Press, 2023, doi: 10.1201/9781003253051-7.

[38] "HCV data—UCI Machine Learning Repository," Accessed: Jul. 08, 2023. [Online]. Available: https://archive.ics.uci.edu/dataset/571/hcv+data

[39] A. Guzmán-Ponce, J. S. Sánchez, R. M. Valdovinos, and J. R. Marcial-Romero, "DBIG-US: A Two-Stage Under-Sampling Algorithm to Face the Class Imbalance Problem," *Expert Systems with Applications*, vol. 168, p. 114301, Apr. 2021, doi: 10.1016/J.ESWA.2020.114301.

Chapter 13

A step toward the detection of alzheimer's disease using ensemble learning

Faiza Ayyob, Muhammad Azeem, and Muhammad Imad

13.1 INTRODUCTION

Amyloidosis (Alzheimer's disease) is a progressive neurodegenerative illness that is chronic, irreversible, and incurable. It is characterized clinically by amnesia, cognitive dysfunction, and the progressive loss of numerous other brain functions, as well as changes in everyday living situations. Alzheimer's disease (AD) is a neurological illness that progresses over time [1]. Moreover, with an estimated increase in the number of people with Alzheimer's disease from the current 47 billion to 152 million by 2050, the condition is expected to become more prevalent. There will be significant economic, medical, and societal ramifications for society because of this event. The biology of AD is still incompletely understood, and there is currently no FDA-approved medication that can cure the disease or entirely halt its progression [2]. Amnestic mild cognitive impairment (MCI) is associated with an increased risk of developing AD compared to patients with age-matched healthy cognition. MCI is a common stage in both cognitively normal elderly people and those suffering from AD [3]. It is crucial to diagnose AD early through MCI screening to apply proper management and care methods, as well as explore novel drugs and treatments to slow the progression of the disease. In vivo, experiments on AD-related abnormalities in the brain have been possible because of the development of magnetic resonance imaging (MRI). Many promising applications of machine learning have made use of MRI for the prediction of AD [4]. Furthermore, random forest [5], support vector machine [6], and boosting techniques [7] are some machine learning example models that can be used. In contrast, existing machine learning algorithms often rely on the manual selection of predefined brain areas of interest (ROIs) by recognized MRI markers of AD to do their calculations. Under our current understanding of definitive MRI biomarkers for AD, it looks likely that preselected ROIs will not capture all the information that could be useful in unraveling the intricacies of the illness. Additionally, manual selection can be prone to subjective errors, and it can be time consuming and cumbersome to carry out correctly [8]. Moreover, deep learning (DL) is a more advanced

 DOI: 10.1201/9781003496410-16

approach that incorporates algorithms such as the stacked autoencoder (SAE) [9], deep belief networks (DBNs), and convolutional neural networks (CNNs) [10]. Consequently, for classification purposes, the raw brain image is far too big to be used in its entirety. As a result, to diagnose the disease, it is necessary to preprocess the MRI images and perform feature extraction and classification [11]. DL algorithms have gained in popularity in recent years, and they are now being used to jointly learn features from images, as well as for discrimination for image classification and computer vision, among other applications [12]. While traditional approaches extract handcrafted characteristics from raw image data using field-specific knowledge, deep learning may create a deep neural network architecture from raw image data to develop feature representation; traditional methods do not have this capability. Deep learning is hence capable of discovering complex patterns because of its capabilities. Whether CNNs could interpret the qualities of MRI brain images for diagnosing AD was investigated [13]. The main objective of this research is to develop an automated process using CNN models that can be used to classify MRI pictures and detect Alzheimer's symptoms before the advanced phase. Three DL models, ResNet-50 [14], DensNet-121 [15], and VGG19 [16], are ensembled to detect AD and treat the patient to reduce possible brain malfunction.

13.2 LITERATURE REVIEW

As described by Hemanth, the usage of biological systems for image classification tasks (ANNs) might be problematic because ANNs require a .lengthier convergence period than other methods. Hemanth addressed this problem by constructing two new neuronal networks. The modified counters propagation human brain and the modified Kohen hidden Markova model are two networks that have been proposed to reduce the number of iterations of the ANN. The effectiveness of these networks was evaluated using segmentation of 40 to 4 abnormal brain MRIs, with encouraging results. Suggested network [17]. According to Pokhrel, Han, and Gardner [18], there were two primary steps: an interesting voxel selection phase that utilized a controlled logic model and a classification phase that utilized low-density tractor-trailer segmentation (LDS) to produce final predictions. A system for automatically assessing cortical thickness was introduced by Pokhrel, Han, and Gardner [18] to improve clinical diagnosis of suspected AD. Oyesiku, Somuyiwa, and Oduwole [19] used morphometric data for tracking by utilizing numerous fundamental alterations against respective spatial frequency components. Because of the increased analysis of information and computational resources, researchers in the medical field have begun to rely on deep diagnostics training to aid in their research. Medical data is still in short supply as contrasted to the requirements of deep neural networks; however, this was a disadvantage due to the use of different approaches, such as translation learning or training, which was overcome by research networks in the field of deep neural networks.

Moreover, 2D-CNN and 3D-CNN are used in Yu, Yang, Wang, Leader, Wilson, and Pu [20] to extract function layers by layer for layer-by-layer performance extraction. Payan and colleagues, in addition, used a two-step process in which they first trained on randomly picked 3D patches from MRI brain scans before moving on to investigate the filters for convolutional operations after being taught about those patches. Furthermore, Liu, Chow, Xu, Jiang, Dou, and Zhou [21] create a three-dimensional CNN that makes use of automatic encoder filters in the initial layer. Similarly, Zhou, He, and Jia [22] proposed a system for Alzheimer's disease diagnosis. This system retrieved MRI image features using an independent Convolutional Autoencoder (CAE) 2D-trained model and then conducted task-specific classification for target domains with an adaptive 3D Convolutional Neural Network (CNN). Zhang, Shinomiya, and Yoshida [23] have used the 3D brain MRI model obtained from the ADNI database to construct classification models featuring network architectures like VGG16 and residual neural networks, as well as other techniques. Additionally, Yin et al. [24] present a transfer learning method for selecting multi-domain transmission (MDTFS) to determine the most feature representations and a multi-domain transfer identification (MDTC) to classify the features using a multi-domain transfer learning framework. Goldman's transfer learning method was used to refine the AlexNet-driven data volume on ImageNet and to extend the data, such as the data transformation from the database. Data from ADNI will be used to increase the volume of training data. Billons and colleagues [25] performed a 16-layer form of CNN VGG16, and it has been modified to be more efficient. It has been the defining quality of their employment that they have completed their range of 20 MRI sections that were selected randomly from each subject and defeated other various classifiers. Marcia and colleagues performed a similar assignment, but they used an intelligent strategy to select pieces, instead of selecting data intuitively, by employing the stochastic process for selecting the data. In the following step, the researchers used transfer learning to analyze how well the architectures VGG1 and Incept IV performed [26]. They were able to reach 96% accuracy because of the Inception V4 architecture of the OASIS dataset. It was suggested by Yanqing that brain MRI collected information be used in conjunction with a deep, Conv human brain for the diagnosis of AD. They conducted extensive demon experiments to test the compatibility of their suggested model with the baseline dataset from the Open Access Series imaging investigations, which they found to be insufficient (OASIS). They developed and refined their structure to classify 3D MRI images, Inception-v4 data, and ResNet data using 416 data samples, and they achieved excellent precision in their classification model [27]. In Zhang et al. [28], the authors used the Rs-fMRI dataset for the classification of AD in six stages. They applied different preprocessing techniques and then trained DL models (ResNet-18/1R, OTS, and FT). The proposed network is promising for clinical decision-making and has the possibility of helping early identification of AD and related phases [29] using an ensemble learning approach. They used DenseNet-121, DensNet-161, and DensNet-169.

13.3 METHODOLOGY

We integrated the probability forecast or numerical prediction of four machine learning models, combining them or utilizing the most common observation between models to arrive at a final prediction. A model ensemble is created by fitting four models, making predictions using each of them, and merging the results. It then uses the test data performance to produce an ideal weighted average of the models, which is referred to as an ensemble. We fit an ensemble model by simply adding algorithms to the existing models.

13.3.1 Dataset

The dataset is separated into two files: training and testing, each of which contains approximately 5,000 photos, each of which is classified according to the severity of Alzheimer's illness. The training and testing dataset consists of four classes that indicate four stages of AD, as depicted in Table 13.1.

13.3.2 Ensemble method

Assembling is a term that is frequently seen in machine learning. This method serves as the foundation for a variety of algorithms. To name a few of the most well-known, random forests utilize a technique where a variable number of decision trees are integrated into one operation. Overall, an ensemble can be thought of as a learning strategy in which numerous models are combined in instruction to solve a problem of interest. Because an ensemble tends to perform better than a single individual, this is complete to recover simplification ability. The variety of ways in which an ensemble can be performed knows no bounds. The same can be said for the number of models evaluated as well as the sorts of models analyzed. Keep in mind that choosing components with low bias and high variance is a good habit to develop. This is achieved by simply selecting models with a diverse structure and format. We demonstrate how the ensemble process is used in the neural network area. For greater clarity, we demonstrate alternative approaches to efficiently combining DL architectures. A good example is the image classification domain, where it is typical to encounter models with extremely deep learning capabilities. The same thinking and approach can be pragmatic to a variety of other applications with relative ease. Figure 13.1 highlights the complete overview of the ensemble method used for this work.

Table 13.1 Dataset description

AD Stages	*Training*	*Testing*
Mild Demented	717	179
Moderate Demented	52	12
Non-demented	2,560	640
Very Mild Demented	1,792	448

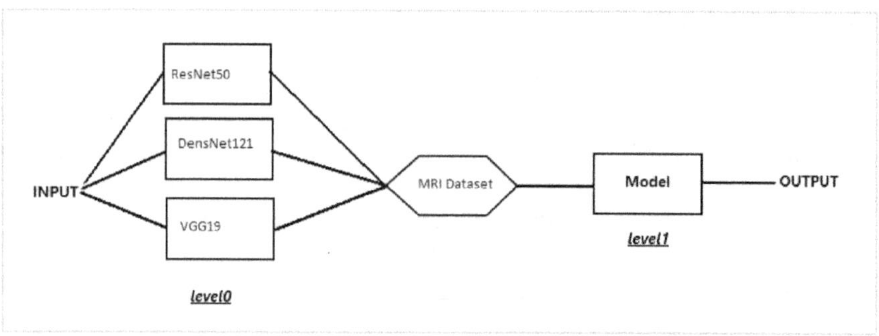

Figure 13.1 Proposed ensemble model's graphic representation.

There are three network classifiers as illustrated in the above diagram, each of which is integrated into a different method at the top of the hierarchy. This example demonstrates how to operate an ensemble of three distinct classifiers. They are pretrained models that have been trained on a dataset, but which we will employ in our task. We can achieve better results with transfer learning precisely because it allows us to initialize our new train in the correct direction by modifying the information we have already acquired. This is feasible by fine-tuning the operation based on our facts. Specifically, we are operating in the ensemble mode, and we are fine-tuning our three ensemble components as we are merging them. In all the approaches discussed, we run the ensemble using the base models VGG19, ResNet-50, and DensNet-121. We fine-tune them, allowing them to retrain on their respective last convolutional blocks because of our operation. The ensemble is only capable of operating in two dimensions. We make the transition from 3D to 2D operation global pooling. Dense layers with fixed dimensions are used to equalize dimensionalities across all the networks that are sending the pooling outputs through them.

13.4 RESULTS AND DISCUSSION

The proposed AD diagnosis model was in comparison to an ensemble model that was tuned to fit 3D brain MRI data, and the results were inconclusive. The results showed that the proposed AD diagnosis model was a successful method of diagnosis. We looked at four criteria for quantitative evaluation and comparison, including accuracy, precision, recall, and F1-score, among other things, and found them to be useful. The dataset for magnetic resonance contains 6,400 data samples. We partitioned the dataset into two parts: a training dataset and a test dataset, with an 80%:20% split between the two portions. It was decided to create a validation dataset using data from the training dataset, which accounted for around 20% of the entire dataset. The training with a mini-batch size of 64 was used as part of

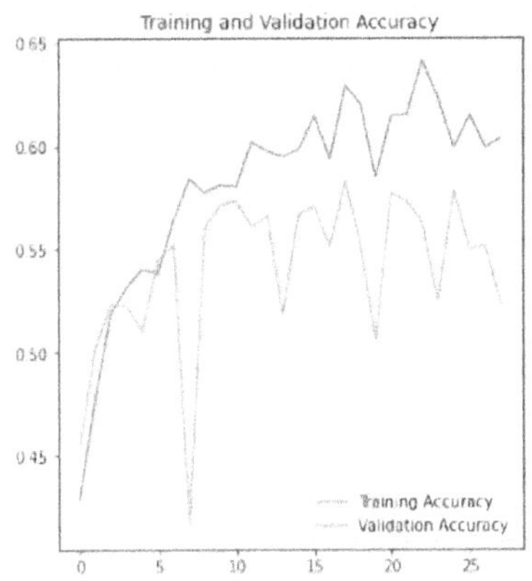

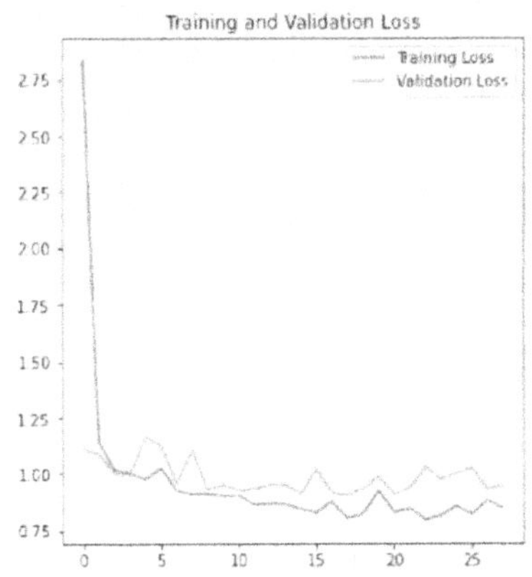

Figure 13.2 Training and validation loss and accuracy.

the optimization process. We used three different basic models. ResNet-50, DensNet-121, and VGG19 are the three networks. Using the stacking approach, we put all these models together in one place. We ran the main model, as well as its majority voting and stacking-based ensemble, in parallel. The precision of the proposed model is 96%, the recall is 95%, the F1-score is 94%, and the total accuracy of the model is 90.15%. Even while the model's performance for the categorization of non-demented patients is adequate, it has much scope to improve when it comes to diagnosing demented individuals. It is feasible that training the proposed model on a dataset that comprises a greater number of samples of demented patients will aid in overcoming this limitation of the dataset. As evidenced by great results in the classification of different stages of AD when compared to existing baseline models, the proposed model offers hope in the fight against AD. Figure 13.2 shows the graphical representation of training and validation accuracy along with training and validation loss.

13.5 CONCLUSION

AD is a progressive neurological brain illness that develops from birth. When AD is detected early, it can assist in avoiding further damage to the brain tissues and aid in the administration of appropriate treatment. A combination of algorithms is used to obtain higher prediction accuracy than any of the algorithms used individually in the ensemble learning process. Because of the increasing popularity of in-depth learning technologies, researchers have begun combining various technologies for a variety of different applications. In this study work, we suggested a technique based on profound learning as a method for integrating algorithms for the categorization of AD, which was later refined and published. We put up three different models. ResNet-50, DensNet-121, and VGG19 are models that have not previously been employed as part of an ensemble model. When merging various base learning models for final MRI image classification, the usefulness of the ensemble approaches, such as majority voting and stacking, is investigated for their efficiency. Regarding correct classification rates for Alzheimer's detection, the ensemble classifier produces the best results, with the precision of the proposed model being 96%, recall being 95%, F1-score being 94%, and overall accuracy of the model being 90.15%. The results also show that the ensemble classifier outperforms the single learning model by a significant margin, suggesting that the proposed approach is both effective and efficient. In the future, we plan to incorporate multimodalities into this ensemble model, as well as improve the overall performance of the model.

REFERENCES

[1] J. E. Galvin *et al.*, "Early stages of Alzheimer's disease: Evolving the care team for optimal patient management," *Frontiers in Neurology*, vol. 11, p. 592302, 2021.

[2] J. Rasmussen and H. Langerman, "Alzheimer's disease–why we need early diagnosis," *Degenerative Neurological and Neuromuscular Disease*, pp. 123–130, 2019.

[3] H. Chen, Y. He, J. Ji, and Y. Shi, "A machine learning method for identifying critical interactions between gene pairs in Alzheimer's disease prediction," *Frontiers in Neurology*, vol. 10, p. 1162, 2019.

[4] P. H. Nguyen *et al.*, "Amyloid oligomers: A joint experimental/computational perspective on Alzheimer's disease, Parkinson's disease, type II diabetes, and amyotrophic lateral sclerosis," *Chemical Reviews*, vol. 121, no. 4, pp. 2545–2647, 2021.

[5] P. Palimkar, R. N. Shaw, and A. Ghosh, "Machine learning technique to prognosis diabetes disease: Random forest classifier approach," in *Advanced Computing and Intelligent Technologies: Proceedings of ICACIT 2021*, 2022: Springer, pp. 219–244.

[6] W. Huang *et al.*, "Railway dangerous goods transportation system risk identification: Comparisons among SVM, PSO-SVM, GA-SVM and GS-SVM," *Applied Soft Computing*, vol. 109, p. 107541, 2021.

[7] A. Mosavi, F. Sajedi Hosseini, B. Choubin, M. Goodarzi, A. A. Dineva, and E. Rafiei Sardooi, "Ensemble boosting and bagging based machine learning models for groundwater potential prediction," *Water Resources Management*, vol. 35, pp. 23–37, 2021.

[8] N. Yamanakkanavar, J. Y. Choi, and B. Lee, "MRI segmentation and classification of human brain using deep learning for diagnosis of Alzheimer's disease: A survey," *Sensors*, vol. 20, no. 11, p. 3243, 2020.

[9] G. Zhang, Y. Liu, and X. Jin, "A survey of autoencoder-based recommender systems," *Frontiers of Computer Science*, vol. 14, pp. 430–450, 2020.

[10] Q. Dong *et al.*, "Modeling hierarchical brain networks via volumetric sparse deep belief network," *IEEE Transactions on Biomedical Engineering*, vol. 67, no. 6, pp. 1739–1748, 2019.

[11] A. Wadhwa, A. Bhardwaj, and V. S. Verma, "A review on brain tumor segmentation of MRI images," *Magnetic Resonance Imaging*, vol. 61, pp. 247–259, 2019.
[12] M. Khojaste-Sarakhsi, S. S. Haghighi, S. F. Ghomi, and E. Marchiori, "Deep learning for Alzheimer's disease diagnosis: A survey," *Artificial Intelligence in Medicine*, vol. 130, p. 102332, 2022.
[13] A. W. Salehi, P. Baglat, B. B. Sharma, G. Gupta, and A. Upadhya, "A CNN model: Earlier diagnosis and classification of Alzheimer disease using MRI," in *2020 International Conference on Smart Electronics and Communication (ICOSEC)*, 2020: IEEE, pp. 156–161.
[14] D. Theckedath and R. Sedamkar, "Detecting affect states using VGG16, ResNet50 and SE-ResNet50 networks," *SN Computer Science*, vol. 1, pp. 1–7, 2020.
[15] M. Chhabra and R. Kumar, "A smart healthcare system based on classifier DenseNet 121 model to detect multiple diseases," in *Mobile Radio Communications and 5G Networks: Proceedings of Second MRCN 2021*, 2022: Springer, pp. 297–312.
[16] N. Dey, Y.-D. Zhang, V. Rajinikanth, R. Pugalenthi, and N. S. M. Raja, "Customized VGG19 architecture for pneumonia detection in chest X-rays," *Pattern Recognition Letters*, vol. 143, pp. 67–74, 2021.
[17] D. A. Otchere, T. O. A. Ganat, R. Gholami, and S. Ridha, "Application of supervised machine learning paradigms in the prediction of petroleum reservoir properties: Comparative analysis of ANN and SVM models," *Journal of Petroleum Science and Engineering*, vol. 200, p. 108182, 2021.
[18] G. Pokhrel, Y. Han, and D. J. Gardner, "Comparative study of the properties of wood flour and wood pellets manufactured from secondary processing mill residues," *Polymers*, vol. 13, no. 15, p. 2487, 2021.
[19] O. O. Oyesiku, A. O. Somuyiwa, and A. O. Oduwole, "Analysis of transport and logistics education regulations and economic development in Nigeria," *Transportation Research Procedia*, vol. 48, pp. 2462–2487, 2020.
[20] J. Yu, B. Yang, J. Wang, J. Leader, D. Wilson, and J. Pu, "2D CNN versus 3D CNN for false-positive reduction in lung cancer screening," *Journal of Medical Imaging*, vol. 7, no. 5, p. 051202, 2020.
[21] Z. Liu, P. Chow, J. Xu, J. Jiang, Y. Dou, and J. Zhou, "A uniform architecture design for accelerating 2D and 3D CNNs on FPGAs," *Electronics*, vol. 8, no. 1, p. 65, 2019.
[22] Z. Zhou, Z. He, and Y. Jia, "AFPNet: A 3D fully convolutional neural network with atrous-convolution feature pyramid for brain tumor segmentation via MRI images," *Neurocomputing*, vol. 402, pp. 235–244, 2020.
[23] H. Zhang, Y. Shinomiya, and S. Yoshida, "3D MRI reconstruction based on 2D generative adversarial network super-resolution," *Sensors*, vol. 21, no. 9, p. 2978, 2021.
[24] Z. Yin *et al.*, "Multi-domain resource multiplexing based secure transmission for satellite-assisted IoT: AO-SCA approach," *IEEE Transactions on Wireless Communications*, vol. 22, no. 11, pp. 7319–7330, 2023.
[25] R. Zambare, R. Deshmukh, C. Awati, S. Shirgave, S. Thorat, and S. Zalte, "Deep learning model for disease identification of cotton plants," *Specialusis Ugdymas*, vol. 1, no. 43, pp. 6684–6695, 2022.
[26] S. Mascarenhas and M. Agarwal, "A comparison between VGG16, VGG19 and ResNet50 architecture frameworks for image classification," in *2021 International*

Conference on Disruptive Technologies for Multi-Disciplinary Research and Applications (CENTCON), 2021, vol. 1: IEEE, pp. 96–99.

[27] L. Wen, X. Li, and L. Gao, "A transfer convolutional neural network for fault diagnosis based on ResNet-50," *Neural Computing and Applications*, vol. 32, pp. 6111–6124, 2020.

[28] C. P. Santana, E. A. de Carvalho, I. D. Rodrigues, G. S. Bastos, A. D. de Souza, and L. L. de Brito, "Rs-fMRI and machine learning for ASD diagnosis: A systematic review and meta-analysis," *Scientific Reports*, vol. 12, no. 1, p. 6030, 2022.

[29] P. P. Dalvi, D. R. Edla, and B. Purushothama, "Diagnosis of coronavirus disease from chest X-Ray images using DenseNet-169 architecture," *SN Computer Science*, vol. 4, no. 3, p. 214, 2023.

Chapter 14

Exploring the use of machine learning algorithms in early detection of liver disease

Rishalatun Jannat Lima, Nafeya Rahman Heya, Md. Musfiqur Rahman Foysal, Md. Muzahidul Islam, Sajidur Rahman Sajid, and Muhammad Usama

14.1 INTRODUCTION

Liver disease is a major public health issue around the world, affecting millions of people and leading to potentially fatal illnesses such as cirrhosis and liver cancer [1]. The early detection and accurate identification of liver illnesses is critical to improving patient outcomes and lowering mortality rates. In recent years, machine learning (ML) algorithms have emerged as powerful tools in the biomedical field, offering promising avenues for predicting and diagnosing liver diseases with enhanced efficiency and cost-effectiveness. The liver, the largest organ in the body, performs around 500 essential processes, including the production of bile, the synthesis of glycogen, the purification of the blood from chemicals and toxins, and the storage of vitamins [2]. With such critical functions, any interruption in liver function can have serious ramifications for general health. However, liver diseases often go undetected in their early stages due to the organ's remarkable regenerative abilities, allowing it to continue functioning despite partial damage. Researchers have resorted to ML approaches to overcome the problems involved with the early diagnosis of liver disorders. ML, a subfield of artificial intelligence, uses algorithms that learn from input-output data pairs to provide computers the ability to learn without explicit programming. In the health care domain, ML has shown great potential in improving diagnostic accuracy, facilitating early intervention, and reducing health care costs [3]. The purpose of this study is to investigate the use of ML algorithms in the early diagnosis of liver disorders. By harnessing the power of classification techniques, we seek to develop a predictive model that can accurately distinguish between patients with liver diseases and those without. Such a model would provide valuable support to health care professionals in making timely and informed decisions, leading to improved patient outcomes. Several ML algorithms will be evaluated in this research [4]. By comparing the performance of these algorithms, we aim to identify the most effective approach for early detection of liver diseases. Accuracy, precision, recall, and F1-score are a

DOI: 10.1201/9781003496410-17

few performance evaluation metrics that will be employed to evaluate how well the models can predict the future. Through analyzing large datasets of medical records, ML algorithms can identify patterns and predict potential outcomes. With this information, health care professionals can make informed decisions on appropriate interventions and treatments for patients. One of the main benefits of ML algorithms in liver disease diagnosis is the ability to detect diseases at earlier stages than traditional methods [5]. Early detection of liver diseases can significantly improve patient outcomes, as it allows for timely intervention and effective treatment before the disease progresses further. Moreover, ML algorithms can help predict the risk of developing liver diseases by analyzing various factors such as lifestyle, genetics, and environmental exposures. This information can be used to create personalized prevention plans for patients. A promising strategy for enhancing patient outcomes and reducing mortality rates is the early detection of liver diseases using ML algorithms. The development of accurate and efficient methods for early detection of liver diseases can empower health care professionals with reliable tools that can enhance diagnosis, improve patient outcomes, and potentially save lives. As the field of ML in health care continues to evolve rapidly, there are several challenges associated with its implementation. These include data privacy concerns, ethical considerations, and the need for specialized training and knowledge among health care professionals. However, if these challenges can be addressed effectively, ML algorithms have the potential to revolutionize liver disease diagnosis and pave the way for more personalized and effective health care solutions.

14.2 LITERATURE REVIEW

Gupta et al. [6] use different ML models for predicting liver disease by using the Indian Liver Patients' Records dataset, including logistic regression, *k*-nearest neighbors (KNN), gradient boosting, decision tree, random forest (RF), extreme gradient boosting, and LightGB. In addition to using base and advanced classifiers, they also performed exploratory data analysis (EDA), data preprocessing, outlier reduction, and SMOTE (synthetic minority oversampling technique). After feature selection, the study indicated that the accuracy of 63% was highest for LightGB and RF. However, the study was limited by its dependence on a single dataset and the exclusion of other potential factors impacting liver disease.

ML techniques were used by Dritsas et al. [7] to predict the prevalence of liver disease. Using the Indian Liver Patients' Records dataset, they assessed various supervised ML models and ensemble methods. The voting classifier outperformed other models after SMOTE with tenfold cross-validation, achieving accuracy, recall, and F-measure of 80.1%, precision of 80.4%, and AUC of 88.4%. The report highlights the value of early liver disease detection without drawing attention to any limitations.

An improved support vector machine classifier (CSA-SVM) was proposed by Devikanniga et al. [8] using the Indian Liver Patients' Records dataset for the accurate diagnosis of liver disease. With a classification accuracy of 99.49%, they excelled at optimizing SVMs using the crow search algorithm.

To identify and predict liver illness, Shaheamlung et al. [9] discuss the use of ML techniques. SVM, KNN, *k*-mean clustering, neural networks, and decision trees are some of the methodologies that are compared in the study. The decision tree's 98.46% testing accuracy was the maximum obtained. The research emphasizes the significance of early prognosis for efficient diagnosis and recovery.

B. Khan et al. [10] aimed to predict liver disease at an early stage using different classification algorithms. They compared RF, logistic regression, and separation algorithm based on the parameters and predictive accuracies. Liver disease datasets from various sources were used. RF performed the best among the algorithms, offering potential for liver disease prediction. The paper contributes to improving health care services and early disease prediction. However, it has the limitation of focusing solely on related works without considering other influencing factors in liver disease prediction.

Singh et al. [11] proposed a software-based approach for predicting liver illness using feature selection and classification approaches. They made use of the Indian Liver Patients' Records dataset from the University of California, Irvine. The accuracy of six different categorization algorithms was examined. Several metrics were used to evaluate the results. The study discusses several software development methodologies, with the waterfall model being suggested for complicated projects and so contributes to the development of intelligent liver disease prediction software.

Nahar et al. [12] conducted a comparison analysis of ensemble algorithms for the prediction of liver disease using the Indian Liver Patients' Records dataset. The study compared AdaBoost, LogitBoost, BeggRep, BeggJ48, and RF models and discovered that LogitBoost was the most accurate. The work contributes by investigating and comparing the use of ensemble methods. The dataset includes 416 records with liver disease and 167 without. For model construction and evaluation, the WEKA toolbox was used.

R. A. Khan et al. [13] conducted a systematic review of computer-aided diagnosis for hepatic lesions, focusing on ultrasonography, computed tomography, and magnetic resonance imaging. The paper analyzes preprocessing, attribute analysis, and classification techniques, highlighting the use of texture properties for attribute analysis and SVMs for classification.

Using ML techniques, Kuzhipallil et al. [14] present a new classifier for the automated diagnosis of liver disease in Indian patients. The study evaluates various classification methods and visualization strategies using the Indian Liver Patients' Records dataset from the UCI repository. Outlier detection is performed using isolation forest, and a feature selection algorithm is used to improve performance. The results show increased accuracy and reduced classification time. The stacking estimator combined with RF achieves the highest accuracy.

14.3 METHODOLOGY

The proposed methodology for exploring the use of ML algorithms in early detection of liver diseases involves several steps. Figure 14.1 illustrates the proposed model workflow of this research.

First, we will collect a dataset of liver disease patients from reliable sources. Once the data is collected, we will check for null values in the dataset and fill them with median values. Then, to convert categorical data to numerical data, we will use label encoding. Next, we will check the correlation between features using a correlation matrix. Any features that are highly correlated or have the same information will be dropped. After this, we will perform data scaling to ensure that all features are on the same scale. Following that, the dataset will be divided into training and testing sets. Then we'll use eight ML methods, including gradient boosting classifier [15], random forest classifier [16], extra tree classifier [17], *k*-nearest neighbors [18], extreme gradient boosting classifier [19], AdaBoost classifier [20], LogitBoost classifier [21], and CatBoost classifier [22] to predict liver illness. To combine the predictions of all the models, we will use a voting classifier to get the final prediction. We will balance the dataset by undersampling the majority class, oversampling the minority class, or employing SMOTE [23] to enhance the model's performance. Following that, the model's performance will be evaluated using a range of metrics, such as accuracy, precision, recall, F1-score, and confusion matrix. We conduct an analysis of the study's findings. Overall, this methodology will provide a systematic strategy to forecasting liver illness using ML algorithms while maintaining the data's quality and dependability.

14.3.1 Data collection and preprocessing

The dataset for this research will be obtained from a combination of sources, including the UCI ML Repository and Kaggle. The dataset contains a total of 1,858 patient records, out of which 1,323 patients have liver disease and 535 patients do not have liver disease. The patients were collected from various locations. The

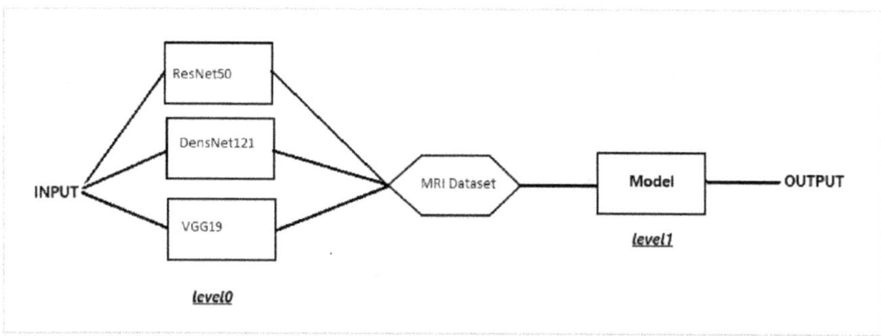

Figure 14.1 Proposed model workflow.

dataset consists of various features, including age, gender, alkaline phosphatase, alanine aminotransferase, aspartate aminotransferase, total bilirubin, direct bilirubin, albumin, total proteins, and albumin and globulin ratio. These features are important for analyzing the patients' liver health. There are 1,402 male patient records and 456 female patient records in the dataset. To maintain anonymity, any patient who is older than 89 is labeled as 90. Before utilizing the dataset for this research, a thorough review will be conducted to ensure its reliability and accuracy, including verifying the data sources and performing data quality checks. Figure 14.2 shows the overall description of the dataset, and Figure 14.3 shows the albumin and globulin ratio feature histogram.

14.4 DATA PREPROCESSING

The first step in the data preprocessing stage involves handling missing values in the dataset. In this case, we found null values in the feature named albumin and globulin ratio. The data in this feature is normally distributed but slightly right-skewed and has been shown in Figure 14.3; we decided to fill up the null values using the median method. We then proceeded to look at two other features named total bilirubin and direct bilirubin and discovered that the datasets contained huge outliers. Next, we performed label encoding on the dataset and looked for correlations between features. We observed that there was a high correlation between total bilirubin and direct bilirubin, as well as alanine aminotransferase and aspartate aminotransferase, and total proteins and albumin. To increase the training speed and accuracy of our model, we decided to eliminate one feature from each correlated pair. Therefore, we removed direct bilirubin, aspartate aminotransferase, and albumin from our model. Furthermore, we noticed that the dataset contained

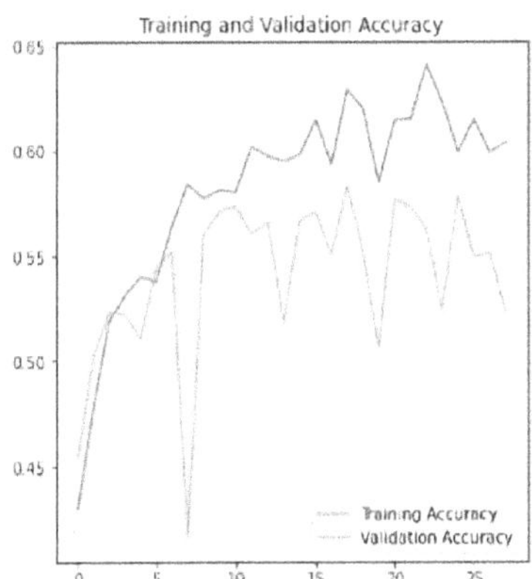

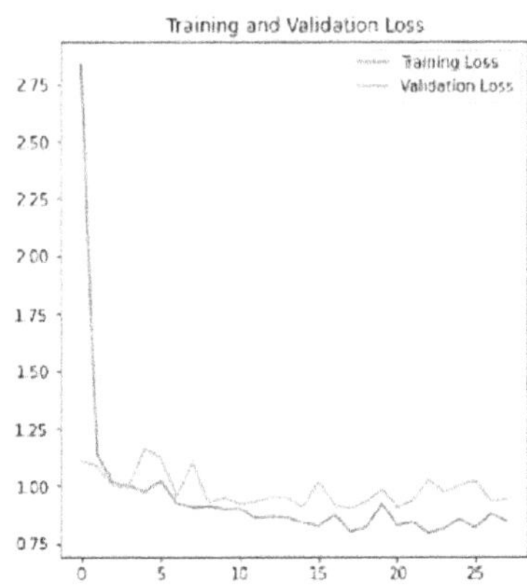

Figure 14.2 Overall description of the dataset.

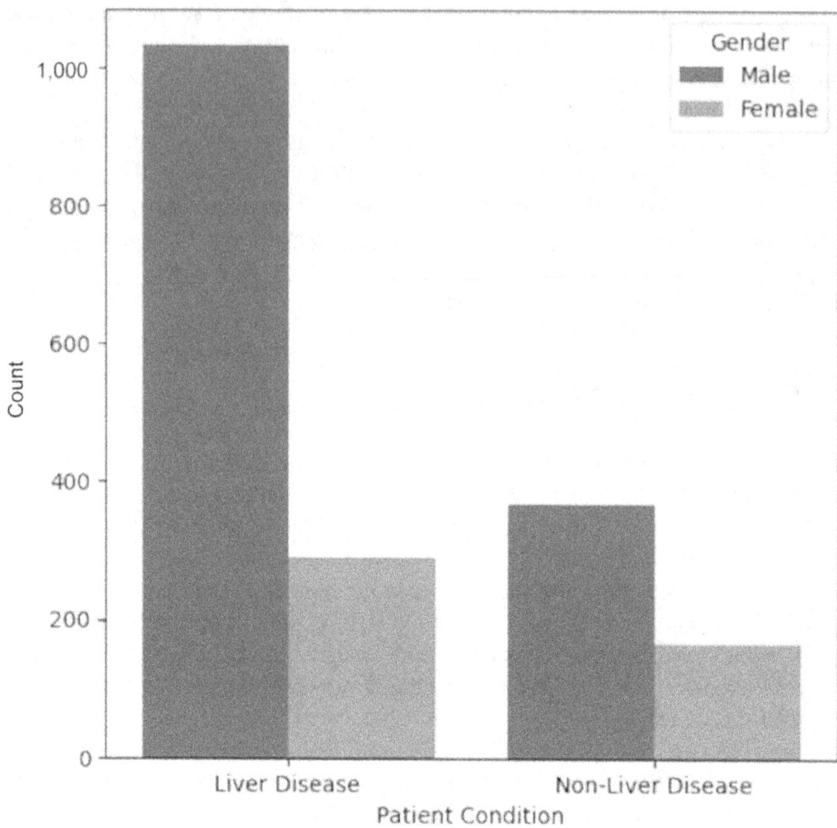

Figure 14.3 Albumin and globulin ratio feature histogram.

unbalanced classes. To solve this problem, we resampled our data to reduce incorrectness. We then scaled the data and used eight ML models to obtain accuracy, precision, recall, and confusion matrix scores. After that, we applied a voting classifier to further improve the accuracy of our model. Finally, to balance the majority and minority classes, we employed undersampling [24], oversampling [25], and SMOTE techniques. The results were analyzed to determine the effectiveness of our approach.

14.5 RESULT AND ANALYSIS

Table 14.1 shows the result before balancing the dataset. The extra tree algorithm had the highest accuracy (AUC) of 85% among all the algorithms on the given dataset, with a precision of 0.75 and recall of 0.6. The CatBoost algorithm ranked second in terms of accuracy with a score of 82%, followed by AdaBoost and

Table 14.1 Comparison of algorithms before balancing the dataset

Algorithms	*AUC*	*Precision*	*Recall*
RF	78%	0.58	0.47
KNN	74%	0.48	0.43
Gradient Boost	74%	0.5	0.43
XGBoost	73%	0.46	0.43
Extra Tree	85%	0.75	0.6
AdaBoost	77%	0.56	0.5
LogitBoost	78%	0.57	0.53
CatBoost	82%	0.7	0.53

Note: KNN, k-nearest neighbors; RF, random forest; XGBoost, extreme gradient boost.

LogitBoost with scores of 77% and 78%, respectively. The KNN and XGBoost algorithms scored 74% on accuracy, while gradient boost scored 0.74% on accuracy with a precision of 0.5 and recall of 0.43. The RF algorithm had an accuracy of 0.78, which was higher than the scores of KNN, XGBoost, and gradient boost but lower than extra tree. Finally, the ensemble model created using AdaBoost, extra tree, and CatBoost achieved a score of 0.82 accuracy, with a precision of 0.7 and recall of 0.53.

Undoubtedly, imbalanced data can have a negative impact on ML techniques. To tackle this issue, the current study adopted undersampling, oversampling, and SMOTE approaches on the dataset under consideration.

Table 14.2 shows the results of various ML algorithms after implementing undersampling. The gradient boost algorithm had the highest accuracy score of 0.58, followed by the voting classifier consisting of AdaBoost, extra tree, and CatBoost algorithms with an accuracy score of 0.57. Among the individual algorithms listed, RF, KNN, extra tree, AdaBoost, and CatBoost had the highest precision scores, all with a precision score of 0.54 or 0.55. The algorithm with the highest recall score was the gradient boost algorithm, with a score of 0.64. Note that precision and recall scores were not given for all algorithms in the table.

Table 14.3 displays the outcomes of various machine learning algorithms after oversampling the dataset. The extra tree algorithm had the highest accuracy score of 0.92, followed by XGBoost with a score of 0.90, RF with a score of 0.89, and CatBoost with a score of 0.86. Among the individual algorithms listed, extra tree had the highest precision score of 0.89, while KNN had the highest recall score of 0.76.

The results of all ML algorithms after applying SMOTE to the dataset are presented in Table 14.4. The extra tree and RF algorithms achieved the highest accuracy scores of 0.81, followed by CatBoost with a score of 0.81 as well. XGBoost obtained an accuracy score of 0.8, while gradient boost and LogitBoost achieved scores of 0.78. KNN demonstrated an accuracy score of 0.78, whereas AdaBoost showed a score of 0.74. In terms of precision, RF and XGBoost exhibited the

Table 14.2 Comparison of algorithms after undersampling the dataset

Algorithms	*AUC*	*Precision*	*Recall*
RF	52%	0.52	0.48
KNN	49%	0.49	0.52
Gradient Boost	58%	0.57	0.64
XGBoost	54%	0.53	0.55
Extra Tree	52%	0.52	0.52
AdaBoost	55%	0.54	0.58
LogitBoost	51%	0.5	0.45
CatBoost	55%	0.54	0.61

Note: KNN, k-nearest neighbors; RF, random forest; XGBoost, extreme gradient boost.

Table 14.3 Comparison of algorithms after oversampling the dataset

Algorithms	*AUC*	*Precision*	*Recall*
RF	89%	0.82	0.96
KNN	62%	0.56	0.76
Gradient Boost	83%	0.76	0.91
XGBoost	90%	0.86	0.95
Extra Tree	92%	0.89	0.95
AdaBoost	80%	0.73	0.87
LogitBoost	81%	0.74	0.88
CatBoost	86%	0.8	0.93

Note: KNN, k-nearest neighbors; RF, random forest; XGBoost, extreme gradient boost.

Table 14.4 Comparison of algorithms after applying SMOTE to the dataset

Algorithms	*AUC*	*Precision*	*Recall*
RF	81%	0.84	0.8
KNN	78%	0.75	0.95
Gradient Boost	78%	0.8	0.82
XGBoost	80%	0.83	0.8
Extra Tree	81%	0.83	0.82
AdaBoost	74%	0.78	0.73
LogitBoost	78%	0.83	0.77
CatBoost	81%	0.83	0.84

Note: KNN, k-nearest neighbors; RF, random forest; SMOTE, synthetic minority oversampling technique; XGBoost, extreme gradient boost.

highest score of 0.83, followed closely by extra tree and CatBoost with scores of 0.83 and 0.84, respectively. Logistic regression displayed a precision score of 0.83, while KNN had the lowest precision score of 0.75. Regarding recall, KNN exhibited the highest score of 0.95, indicating that the algorithm was effective at correctly identifying true positives. CatBoost demonstrated the second-highest recall score of 0.82, followed by gradient boost with a score of 0.82 as well. Meanwhile, AdaBoost exhibited the lowest recall score of 0.73. Finally, the voting classifier consisting of AdaBoost, extra tree, and CatBoost algorithms achieved an accuracy score of 0.8, precision score of 0.82, and recall score of 0.83.

14.6 DISCUSSION

The results of the research are quite interesting, as we have explored the effectiveness of various ML algorithms in a balanced dataset. The use of a voting classifier is commendable, as it allows for an ensemble approach to decision-making, which can improve accuracy. The incorporation of undersampling, oversampling, and SMOTE techniques is also a good practice, as it ensures that the dataset used for training is representative of the actual data distribution. This approach helps to mitigate issues related to class imbalance, which is a common challenge in many ML applications. Notably, this study found that the extra classifier in each sector led to the highest accuracy. This shows that adding more classifiers could increase the performance of ML models in real-world applications. Overall, this study underscores the importance of carefully selecting appropriate ML algorithms and ensuring balanced datasets for achieving optimal model performance. These findings could be useful to researchers and practitioners working in this field.

14.7 CONCLUSION AND FUTURE WORK

Using various supervised ML classifiers, we attempted to provide an accurate diagnosis approach for patients with early detection of liver disease. After careful analysis of patient information parameters and implementing eight different algorithms, we found that the extra tree classifier achieved the highest accuracy score of each section, followed closely by CatBoost algorithms. We also tested a voting classifier consisting of AdaBoost, extra tree, and CatBoost algorithms, which demonstrated its effectiveness in improving overall performance. Our work has significant implications for health care research and development, particularly in low-income countries where access to medical resources is limited. By providing early detection and advice on maintaining liver health, our approach could help prevent chronic illness and contribute toward improving public health. Moving forward, there is still scope for further research in this area, including exploring additional algorithms and refining the prediction model to improve accuracy.

Overall, this work represents an important step toward developing a more reliable and effective diagnostic tool for liver disease.

REFERENCES

[1] S. K. Asrani, H. Devarbhavi, J. Eaton, and P. S. Kamath, "Burden of liver diseases in the world," *J Hepatol*, vol. 70, no. 1, pp. 151–171, Jan. 2019, doi: 10.1016/J.JHEP.2018.09.014.

[2] T. Messelmani *et al.*, "Liver organ-on-chip models for toxicity studies and risk assessment," *Lab Chip*, vol. 22, no. 13, pp. 2423–2450, Jun. 2022, doi: 10.1039/D2LC00307D.

[3] T. Islam, A. Kundu, N. Islam Khan, C. Chandra Bonik, F. Akter, and M. Jihadul Islam, "Machine learning approaches to predict breast cancer: Bangladesh perspective," pp. 291–305, 2022, doi: 10.1007/978-981-19-2541-2_23.

[4] T. Islam *et al.*, "Review analysis of ride-sharing applications using machine learning approaches," in *Computational Statistical Methodologies and Modeling for Artificial Intelligence*, New York: CRC Press, 2023, pp. 99–122, doi: 10.1201/9781003253051-7.

[5] Q. Md, S. Kulkarni, C. J. Joshua, T. Vaichole, S. Mohan, and C. Iwendi, "Enhanced preprocessing approach using ensemble machine learning algorithms for detecting liver disease," *Biomedicines*, vol. 11, no. 2, p. 581, Feb. 2023, doi: 10.3390/BIOMEDICINES11020581.

[6] K. Gupta, N. Jiwani, N. Afreen, and D. Divyarani, "Liver disease prediction using machine learning classification techniques," in *Proceedings—2022 IEEE 11th International Conference on Communication Systems and Network Technologies, CSNT 2022*, Institute of Electrical and Electronics Engineers Inc., 2022, pp. 221–226, doi: 10.1109/CSNT54456.2022.9787574.

[7] E. Dritsas and M. Trigka, "Supervised machine learning models for liver disease risk prediction," *Computers*, vol. 12, no. 1, p. 19, Jan. 2023, doi: 10.3390/computers12010019.

[8] D. Devikanniga, A. Ramu, and A. Haldorai, "Efficient diagnosis of liver disease using support vector machine optimized with crows search algorithm," *EAI Endorsed Trans. Energy Web*, vol. 7, no. 29, 2020, doi: 10.4108/EAI.13-7-2018.164177.

[9] G. Shaheamlung, H. Kaur, and M. Kaur, "A Survey on machine learning techniques for the diagnosis of liver disease," in *Proceedings of International Conference on Intelligent Engineering and Management, ICIEM 2020*, Jun. 2020, pp. 337–341, doi: 10.1109/ICIEM48762.2020.9160097.

[10] B. Khan, P. K. Shukla, M. K. Ahirwar, and M. Mishra, "Strategic analysis in prediction of liver disease using different classification algorithms," pp. 437–449, Jan. 1 AD, https://services.igi-global.com/resolvedoi/resolve.aspx?doi=10.4018/978-1-7998-2742-9.ch022, doi: 10.4018/978-1-7998-2742-9.CH022.

[11] J. Singh, S. Bagga, and R. Kaur, "Software-based prediction of liver disease with feature selection and classification techniques," *Procedia Comput Sci*, vol. 167, pp. 1970–1980, Jan. 2020, doi: 10.1016/J.PROCS.2020.03.226.

[12] N. Nahar, F. Ara, M. A. I. Neloy, V. Barua, M. S. Hossain, and K. Andersson, "A comparative analysis of the ensemble method for liver disease prediction," in

ICIET 2019–2nd International Conference on Innovation in Engineering and Technology, Dec. 2019, doi: 10.1109/ICIET48527.2019.9290507.

[13] R. A. Khan, Y. Luo, and F. X. Wu, "Machine learning based liver disease diagnosis: A systematic review," *Neurocomputing*, vol. 468, pp. 492–509, Jan. 2022, doi: 10.1016/J.NEUCOM.2021.08.138.

[14] M. A. Kuzhippallil, C. Joseph, and A. Kannan, "Comparative analysis of machine learning techniques for Indian liver disease patients," in *2020 6th International Conference on Advanced Computing and Communication Systems, ICACCS 2020*, Mar. 2020, pp. 778–782, doi: 10.1109/ICACCS48705.2020.9074368.

[15] R. Blagus and L. Lusa, "Gradient boosting for high-dimensional prediction of rare events," *Comput Stat Data Anal*, vol. 113, pp. 19–37, Sep. 2017, doi: 10.1016/J.CSDA.2016.07.016.

[16] M. Pal, "Random forest classifier for remote sensing classification," *Int J Remote Sens*, vol. 26, no. 1, pp. 217–222, Jan. 2007, doi: 10.1080/01431160412331269698.

[17] A. Sharaff and H. Gupta, "Extra-tree classifier with metaheuristics approach for email classification," *Adv Intell Syst Comput*, vol. 924, pp. 189–197, 2019, doi: 10.1007/978-981-13-6861-5_17/COVER.

[18] Z. Zhang, "Introduction to machine learning: k-nearest neighbors," *Ann Transl Med*, vol. 4, no. 11, Jun. 2016, doi: 10.21037/ATM.2016.03.37.

[19] Y. C. Chang, K. H. Chang, and G. J. Wu, "Application of eXtreme gradient boosting trees in the construction of credit risk assessment models for financial institutions," *Appl Soft Comput*, vol. 73, pp. 914–920, Dec. 2018, doi: 10.1016/J.ASOC.2018.09.029.

[20] T. K. An and M. H. Kim, "A new diverse AdaBoost classifier," in *Proceedings—International Conference on Artificial Intelligence and Computational Intelligence, AICI 2010*, vol. 1, 2010, pp. 359–363, doi: 10.1109/AICI.2010.82.

[21] M. S. Tehrany, S. Jones, F. Shabani, F. Martínez-Álvarez, and D. Tien Bui, "A novel ensemble modeling approach for the spatial prediction of tropical forest fire susceptibility using LogitBoost machine learning classifier and multi-source geospatial data," *Theor Appl Climatol*, vol. 137, no. 1–2, pp. 637–653, Jul. 2019, doi: 10.1007/S00704-018-2628-9/METRICS.

[22] E. B. Postnikov, D. A. Esmedljaeva, and A. I. Lavrova, "A CatBoost machine learning for prognosis of pathogen's drug resistance in pulmonary tuberculosis," in *LifeTech 2020–2020 IEEE 2nd Global Conference on Life Sciences and Technologies*, Mar. 2020, pp. 86–87, doi: 10.1109/LIFETECH48969.2020.1570619054.

[23] T. Islam, M. A. Hosen, A. Mony, M. T. Hasan, I. Jahan, and A. Kundu, "A proposed Bi-LSTM method to fake news detection," in *2022 International Conference for Advancement in Technology (ICONAT)*, IEEE, Jan. 2022, pp. 1–5.

[24] T. Islam, A. Vuyia, M. Hasan, and M. M. Rana, "Cardiovascular disease prediction using machine learning approaches," in *2023 International Conference on Computational Intelligence and Sustainable Engineering Solutions (CISES)*, IEEE, Apr. 2023, pp. 813–819.

[25] M. T. Islam, T. Ahmed, A. R. Rashid, T. Islam, M. S. Rahman, and M. T. Habib, "Convolutional neural network based partial face detection," in *2022 IEEE 7th International Conference for Convergence in Technology (I2CT)*, IEEE, Apr. 2022, pp. 1–6.

Chapter 15

Intelligent transportation channels for smart cities

Ubaid Ullah, Muhammad Usama, Zaid Muhammad, Abdullah Akbar, Shahzad Latif, and Rahat Ullah

15.1 INTRODUCTION

Due to the massive urbanization in recent years, more than half of the world's population lives in urban areas. Numerous issues have been created by this fast urbanization, such as worsening traffic jams, higher pollution levels, and pressure on the transportation system [1]. To address these issues and create sustainable urban environments, the concept of "smart cities" has emerged [2]. The foundation of a smart city is the integration of cutting-edge technology and data-driven systems to enhance all aspects of urban life. Intelligent transportation systems (ITS) are necessary for these "smart cities" since they provide innovative and practical answers to the issues with urban mobility [3]. The necessity to improve urban dwellers' quality of life while maintaining environmental sustainability has given rise to the concept of "smart cities" [4]. Smart cities aspire to maximize resource use, enhance public services, and promote better connectedness between residents and infrastructure by utilizing technology and data-driven solutions. The transportation system of an intelligent city is essential to its success because it must be adequate, dependable, and flexible enough to meet changing urban environment demands. Traditional transportation networks can't successfully handle these contemporary difficulties because of their constrained capacity and outmoded infrastructure. As a result, the ITS has drawn great interest as a revolutionary method of urban movement [5].

An urban region is considered a "smart city" if it uses cutting-edge technology, data analytics, and creative solutions to improve the quality of life for its citizens, increase the effectiveness of urban operations, and encourage sustainable growth [6]. To improve a variety of areas of city life, including transportation, energy, infrastructure, government, public services, and citizen involvement, it integrates information and communication technology (ICT), internet of things (IoT) gadgets, and data-driven systems [7]. Smart cities use cutting-edge technology to collect real-time data and enable seamless interaction across various urban components, such as sensors, linked devices, and communication networks [8]. This technological integration enables effective resource management, enhanced

 DOI: 10.1201/9781003496410-18

decision-making, and the capacity to react quickly to changing circumstances. The cornerstone of intelligent cities is data-driven decision-making, which uses data analytics, artificial intelligence (AI), and machine learning (ML) algorithms to glean valuable insights from massive volumes of data. These insights enable predictive and proactive approaches to urban management, promote informed decision-making, and simplify providing inhabitants with specialized and individualized services. In smart cities, efficiency and sustainability are top goals [9]. Smart cities strive to reduce their environmental impact while increasing the overall effectiveness of urban systems by optimizing resource use, cutting energy use, managing trash properly, and encouraging eco-friendly behaviors [10].

Furthermore, intelligent cities adopt a citizen-centric approach, actively engaging citizens in the decision-making process and leveraging technology to enhance citizen participation and engagement [11]. Through open data initiatives, online platforms, and collaborative governance models, smart cities empower citizens to contribute to urban environments' planning, development, and improvement [12]. This chapter aims to delve into the world of intelligent transportation channels and their role in shaping smart cities. This chapter will explore the key concepts, technologies, and challenges associated with ITS. Furthermore, we will investigate the benefits of implementing these systems and their potential impact on transforming the urban landscape. The chapter aims to contribute to the existing body of knowledge by presenting a comprehensive analysis of the current state of intelligent transportation channels and offering valuable insights into their future development.

15.2 LITERATURE

Rodolfo et al. present recent advances in ITS within the context of smart cities. The authors explore various components of ITS, such as traffic management, connected vehicles, real-time data analytics, and intelligent infrastructure. The paper highlights the potential benefits of integrating ITS into smart cities, including reduced congestion, enhanced safety, and improved overall transportation efficiency. It also discusses the challenges, such as data privacy and cybersecurity, interoperability, and the need for robust infrastructure, which must be addressed for successful implementation [13].

Xiaochenn et al. emphasize that deep learning holds significant potential for revolutionizing traffic sensing and prediction in intelligent cities. However, it also points out the importance of addressing the existing challenges to fully realize the benefits of these techniques in shaping the future of urban mobility and transportation management [14]. Umer Majeed et al. highlight blockchain technology's potential in empowering IoT-based smart cities with secure and efficient data management, transaction processing, and governance mechanisms. However, the paper also acknowledges the existing challenges that must be addressed to fully harness the benefits of blockchain in shaping the future of smart cities [15]. Ali Gohar et al.

advocate for adopting 5G technologies in smart cities, particularly in the domain of ITS. By leveraging 5G's capabilities, cities can unlock transformative transportation services that enhance traffic management, optimize mobility, and pave the way for a more connected and intelligent urban future [16]. Table 15.1 provides a concise overview of the standard techniques employed in intelligent transportation channels for smart cities, along with a description and limitations.

Table 15.1 Exploring intelligent transportation channels for smart cities, a comprehensive overview of techniques and limitations

References	*Technique Used*	*Description*	*Limitations*
[17, 18]	Big data analysis	Big data analytics to optimize traffic flow and predict congestion patterns in intelligent cities	Data privacy concerns, data quality, and computational resources are required
[19–21]	IoT and V2X communication	Internet of things (IoT) and vehicle-to-everything (V2X) communication for smart mobility	Dependence on robust communication infrastructure and potential security vulnerabilities
[22–24]	AI-based traffic management	Artificial intelligence (AI) for dynamic traffic management and adaptive signal control	Algorithm complexity, real-time data acquisition, and system calibration challenges
[25, 26]	Autonomous vehicles integration	Integration of autonomous vehicles into existing transportation systems	Regulatory and safety concerns, human-machine interaction, and high initial costs
[27]	Multimodal transportation integration	Seamless integration of various transportation modes for enhanced mobility	Coordination between different stakeholders, interoperability issues, and user behavior
[28]	Environmental impact assessment	Environmental impact of ITCs on urban areas and emissions reduction	Limited long-term data, challenges in attributing changes solely to ITCs, and varied environmental factors affecting results
[29]	Data privacy and security	Data privacy and security issues associated with ITCs and user information	Ensuring compliance with privacy regulations, safeguarding against potential breaches, and building public trust in data handling practices
[30]	Cloud computing for traffic management	Cloud-based solutions for managing traffic data and coordinating ITCs	Reliance on stable internet connectivity, potential data access, and security concerns

(Continued)

Table 15.1 (Continued)

References	*Technique Used*	*Description*	*Limitations*
[31]	Edge computing in vehicular networks	Edge computing for real-time data processing in vehicular networks	Edge infrastructure reliability, latency challenges, and network scalability
[32]	Blockchain for secure transactions	Blockchain technology to ensure secure and transparent transactions in ITCs	Blockchain scalability, energy consumption, and integration complexity

Note: ITC, intelligent transportation channel.

15.3 TRANSPORTATION IN SMART CITY DEVELOPMENT

15.3.1 Importance of transportation in smart city development

Transportation plays a crucial role in the development and functioning of smart cities. It is a fundamental aspect that affects various facets of urban life and contributes significantly to a city's overall efficiency, sustainability, and livability. Here are some key reasons highlighting the importance of transportation in innovative city development [33, 34].

15.3.1.1 Connectivity and mobility

Efficient transportation systems provide seamless connectivity and mobility options within a city. Smart cities focus on developing integrated and multimodal transportation networks that encompass public transit, cycling lanes, pedestrian-friendly infrastructure, and intelligent parking solutions. This connectivity allows residents to access jobs, education, health care, and other essential services [35].

15.3.1.2 Traffic management and congestion reduction

Innovative transportation solutions employ technologies like intelligent traffic management systems (ITMS) to monitor and manage traffic flow in real time. This includes adaptive traffic signal control, dynamic route guidance, and incident management. By optimizing traffic flow and reducing congestion, smart cities enhance travel times, decrease fuel consumption, and improve air quality [36].

15.3.1.3 Sustainability and environmental impact

Transportation significantly contributes to greenhouse gas emissions and air pollution in urban areas. Intelligent city transportation initiatives emphasize sustainable mobility options, such as electric vehicles (EVs), bike-sharing programs, and carpooling services. These initiatives help reduce carbon footprint, improve air quality, and promote eco-friendly transportation alternatives [37].

15.3.1.4 Data-driven decision-making

Transportation systems generate vast amounts of data, including traffic volumes, travel patterns, and user preferences. Smart cities leverage this data to gain insights into transportation behavior, plan infrastructure improvements, optimize route planning, and enhance transportation services. Data analytics and real-time monitoring enable evidence-based decision-making, ensuring that transportation systems are efficient and responsive to evolving needs [30].

15.3.1.5 Safety and security

The ITS contributes to enhancing road safety and security. Smart cities deploy surveillance technologies, advanced driver assistance systems, and ITMS to promptly detect and respond to incidents. This proactive approach helps prevent accidents, improves emergency response times, and enhances overall safety for residents and commuters [38].

15.3.1.6 Economic growth and productivity

Efficient transportation networks are vital for economic growth and productivity in cities. Well-connected and accessible transportation infrastructure attracts businesses, encourages investment, and facilitates the movement of goods and services. Smart cities focus on developing efficient logistics systems, intelligent freight management, and last-mile delivery solutions to optimize urban logistics and support economic activities [39].

15.3.2 Key goals and objectives of intelligent cities

The goals and objectives of intelligent cities revolve around creating sustainable, efficient, inclusive, and livable urban environments that enhance the quality of life for residents [30]. While specific goals may vary based on local priorities and challenges, the following are vital goals commonly associated with innovative city initiatives.

15.3.2.1 Sustainability and environmental stewardship

Smart cities aim to reduce their environmental footprint and promote sustainable practices. This includes optimizing energy consumption, integrating renewable energy sources, implementing eco-friendly transportation options, managing water resources efficiently, and adopting green building standards [40].

15.3.2.2 Efficient urban infrastructure and services

Smart cities focus on optimizing urban infrastructure and resources to improve efficiency. This involves deploying smart grids for energy distribution, implementing ITS, optimizing waste management processes, and utilizing technology for efficient water and resource management [41].

15.3.2.3 Enhanced quality of life

Smart cities seek to enhance people's quality of life by facilitating easier access to necessary services, including public safety, transportation, health care, and education. They leverage technology and data to deliver personalized and citizen-centric services, enhance public spaces, promote cultural initiatives, and create inclusive communities [42, 43].

15.3.2.4 Economic growth and innovation

Smart cities strive to foster economic growth and promote innovation. They create an ecosystem that attracts businesses, start-ups, and research institutions, offering opportunities for entrepreneurship, collaboration, and investment. Smart cities focus on developing digital infrastructure, supporting innovation hubs, and fostering a culture of entrepreneurship and creativity [44].

15.3.2.5 Citizen engagement and participation

Smart cities actively involve and promote citizen participation in decision-making processes. They leverage technology platforms, open data initiatives, and participatory governance models to empower residents to contribute ideas, provide feedback, and actively shape the urban environment. This citizen-centric approach enhances transparency, accountability, and social inclusion [44].

15.3.2.6 Resilience and safety

Smart cities prioritize resilience to address potential risks and ensure the safety of residents. They implement robust emergency response systems, employ predictive analytics for disaster management, enhance public safety through surveillance systems and intelligent sensors, and develop infrastructure that can withstand natural and manufactured hazards [45].

15.3.2.7 Data-driven decision-making

Data analytics, AI, and ML are used in smart cities to acquire insights and guide decision-making. Smart cities enhance urban planning, resource allocation, and service delivery by leveraging data from various sources, including sensors,

Figure 15.1 Importance of transportation in smart city development.

gadgets, and citizen input. This helps them to serve their citizens' changing demands better [46]. Figure 15.1 shows the importance and goals of smart cities.

15.4 INTELLIGENT TRANSPORTATION CHANNELS

Intelligent transportation channels are a transformative aspect of modern mobility, leveraging cutting-edge technologies to optimize transportation systems and address the challenges of growing urbanization [47]. These systems continually monitor traffic conditions by collecting real-time data from several sources, including sensors, cameras, and global positioning system (GPS) devices. By analyzing this data, transportation authorities can make informed decisions, dynamically adjust signal timings, reroute vehicles during congestion, and proactively manage traffic flow. Connected vehicle technology is another vital intelligent transportation channel. Automobiles with communication capabilities may share information with infrastructure and other vehicles. This enables them to share their positions, speeds, and potential hazards on the road, fostering improved safety and better vehicle coordination [48]. Additionally, this connectivity can pave the way for autonomous driving, further enhancing road safety and efficiency.

Intelligent public transportation is geared toward optimizing traditional public transportation systems. By employing innovative technologies, authorities can analyze data on commuter patterns, travel demands, and operational performance. This information allows for more efficient route planning, dynamic scheduling, and effective fleet management, providing passengers with real-time updates and enhancing the overall public transportation experience [49, 50]. Furthermore, integrating EVs is vital in creating sustainable and environmentally friendly transportation systems. Intelligent transportation channels include the development of an extensive network of EV charging infrastructure, which allows for smart charging solutions that manage energy demand and ensure efficient use of resources. This paves the way for wider EV adoption and reduced carbon emissions in the transportation sector [51, 52]. Figure 15.2 shows various transportation channels.

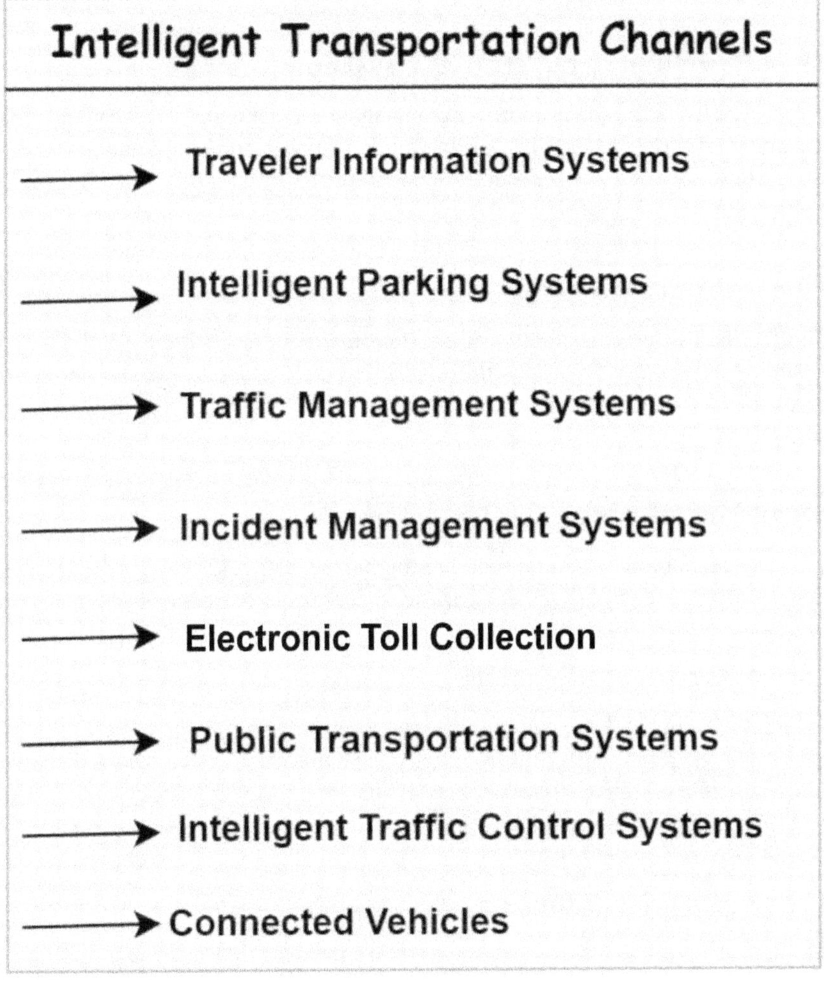

Figure 15.2 Intelligent transportation channels.

15.4.1 Traffic management systems

These systems utilize sensors, cameras, and other data collecting techniques to continuously monitor the state of the traffic. The information is then used to optimize signal timings, manage congestion, and provide real-time traffic updates to drivers.

15.4.2 Intelligent traffic control systems

These systems employ adaptive traffic signal control algorithms to dynamically alter signal timings in accordance with traffic flow. They optimize traffic flow, reduce congestion, and improve overall travel time.

15.4.3 Incident management systems

These systems help in identifying and managing incidents such as accidents, road hazards, and breakdowns. They enable quick response and coordination between emergency services and transportation authorities, minimizing the impact on traffic.

15.4.4 Public transportation systems

The ITS is also used in public transportation networks, including for real-time bus/train arrival and departure information, ticketing systems, passenger counting, and fleet management, making public transportation more efficient and convenient.

15.4.5 Intelligent parking systems

These systems reduce drivers' time looking for parking and urban traffic congestion by directing them to available parking spaces using sensors and data analytics.

15.4.6 Connected vehicles

Vehicles used in ITS have communication systems that allow for vehicle-to- vehicle (V2V) and vehicle-to-infrastructure (V2I) connections. As a result, driving is safer and more effective since cars may exchange information about traffic conditions and possible risks and improve routes.

15.4.7 Traveler information systems

To help drivers and passengers make well-informed decisions regarding their travel plans, these systems offer real-time travel information, such as traffic situations, alternate routes, public transit timetables, and road/weather conditions.

15.4.8 How intelligent transportation channels contribute to smart city goals

Intelligent transportation channels play a significant role in achieving the goals of smart cities.

15.4.8.1 Efficient traffic management

Intelligent transportation networks aid in streamlining traffic, easing congestion, and cutting down on travel time. Traffic management systems may dynamically change signal timings, redirect traffic to other routes, and provide drivers with correct information by employing real-time data from sensors, cameras, and other sources. This effectiveness makes cities more sustainable and livable by easing traffic congestion, enabling less fuel consumption, and improving air quality [53, 54].

15.4.8.2 Improved public transportation

Public transit networks are more effective and convenient thanks to intelligent transportation technologies. Commuters can better plan their trips when given real-time information on bus/train timetables, arrival times, and service disruptions. Amenities like passenger counts and automated ticketing systems facilitate effective resource allocation and revenue collection. These systems promote a modal shift from private automobiles to public transit, lowering traffic congestion and pollution by making public transportation more alluring and user friendly [55].

15.4.8.3 Enhanced safety

Roadway safety has been enhanced thanks to intelligent transportation routes. Incident management systems make quick reactions to accidents, failures, and other occurrences possible, which reduces their negative effects on traffic flow and helps avoid further accidents. Connected vehicle technologies facilitate V2V communication, allowing vehicles to exchange information about hazards, traffic conditions, and pedestrians, thus reducing the risk of collisions. By improving safety, these systems create a secure transportation environment for both drivers and pedestrians [55].

15.4.8.4 Sustainable mobility

Smart cities aim to promote sustainable modes of transportation. Intelligent transportation channels support this goal by integrating various sustainable transportation options into the urban fabric. By providing accurate and timely information about bike-sharing stations, EV charging stations, and multimodal transportation routes, these systems encourage the use of eco-friendly modes of transportation.

They also facilitate better management of transportation demand and help identify opportunities for alternative transportation initiatives like carpooling or ride-sharing [56].

15.4.8.5 Data-driven decision-making

The ITS generates vast amounts of data related to traffic patterns, travel behavior, and transportation infrastructure. This data can be analyzed to gain valuable insights into transportation trends, identify bottlenecks, and inform urban planning decisions. By leveraging data analytics and predictive modeling, city authorities can optimize transportation infrastructure investments, plan for future transportation needs, and make informed policy decisions [57].

15.4.8.6 Improved quality of life

Intelligent transportation channels contribute to an overall improved quality of life in smart cities. By reducing traffic congestion and travel time, these systems enhance mobility and reduce stress levels for residents. They also promote equitable access to transportation by providing information about accessible routes and public transportation options. Additionally, by minimizing environmental impacts such as air pollution and greenhouse gas emissions, these systems contribute to healthier and more sustainable living environments [58].

15.4.9 Benefits and advantages of implementing intelligent transportation channels

Implementing innovative transportation channels offers numerous benefits and advantages, including the following.

15.4.9.1 Improved traffic management

Intelligent transportation channels optimize traffic flow using real-time data, sensors, and advanced algorithms. This reduces congestion, smoothens traffic operations, and improves commuter travel times. Efficient traffic management enhances productivity, reduces fuel consumption, and lowers greenhouse gas emissions [59].

15.4.9.2 Enhanced safety

The ITS contributes to safer roadways, enabling incident detection and management systems to quickly identify and respond to accidents, hazards, and emergencies. Additionally, connected vehicle technologies facilitate vehicle communication, alerting drivers about potential risks and improving overall road safety [60].

15.4.9.3 Reduced environmental impact

Intelligent transportation channels help minimize vehicle idle time and unnecessary fuel consumption by optimizing traffic flow and reducing congestion. This leads to reduced greenhouse gas emissions and improved air quality. Encouraging the use of sustainable transportation modes, such as public transit and cycling, further contributes to a greener and more environmentally friendly transportation system [61].

15.4.9.4 Improved public transportation efficiency

The effectiveness and dependability of public transportation networks are improved through ITS. Passengers can better plan their trips when they have access to real-time information on bus/train timetables, delays, and service interruptions. Automated fare collection and electronic ticketing systems improve operations and cut down on wait times, making public transportation more comfortable and appealing [61].

15.4.9.5 Enhanced mobility options

Intelligent transportation channels integrate various modes of transportation, including public transit, cycling, and car-sharing services, into a seamless network. This promotes multimodal transportation and provides travelers with more choices, flexibility, and convenience. It enables users to plan and navigate their journeys using a combination of transportation modes based on their preferences and needs [62].

15.4.9.6 Data-driven decision-making

The ITS generates vast amounts of data related to traffic patterns, travel behavior, and transportation infrastructure. This data can be analyzed to gain valuable insights into transportation trends, identify bottlenecks, and inform urban planning decisions. By leveraging data analytics and predictive modeling, transportation authorities can optimize infrastructure investments, plan for future transportation needs, and make informed policy decisions [63].

15.4.9.7 Improved quality of life

Intelligent transportation channels contribute to an overall improved quality of life for city residents. By reducing traffic congestion, travel times, and stress levels, these systems enhance mobility and make commuting more enjoyable. They also promote equitable access to transportation, providing real-time information about accessible routes and public transportation options for individuals with disabilities or special needs [64].

15.4.9.8 Economic benefits

Efficient transportation systems have positive economic impacts. Intelligent transportation channels enhance productivity and economic efficiency by reducing travel times and congestion. They also attract businesses and investment by providing a well-connected and accessible transportation infrastructure [65].

15.4.10 Technologies enabling intelligent transportation channels

15.4.10.1 Internet of things in transportation

The transportation sector has radically transformed due to the IoT, which has changed how infrastructure, cars, and passengers are linked and communicated. The idea of linked automobiles is a crucial IoT application in transportation. Vehicles may interact with one another and the local infrastructure by being outfitted with connections, sensors, and actuators. This makes it possible for communication between vehicles (V2V) and between vehicles and infrastructure (V2I), which enhances driving safety, efficiency, and convenience. Real-time data on traffic, accidents, and road dangers may be shared between connected cars, enabling drivers to make educated judgments and take preventive measures [66]. Intelligent traffic management is a different use of IoT. Real-time information on traffic flow, congestion, and road conditions is gathered through IoT-based sensors and devices on roads, traffic signals, and infrastructure. The timing of traffic signals is then optimized, traffic lanes are managed dynamically, and drivers are given real-time information thanks to the analysis of this data. The ITMS helps more efficient transportation operations and with shorter travel times by easing traffic congestion and enhancing traffic flow [67].

Smart parking system deployment is also made possible by IoT. Parking spot occupancy may be tracked in real-time using sensors and cameras placed in parking lots and on the streets. Drivers receive this information via smartphone applications, which direct them to open parking spaces and shorten the time spent looking for parking. Urban parking congestion is minimized thanks to smart parking technologies, which also improve overall efficiency and reduce traffic. IoT is crucial in logistics and supply chain management for streamlining processes. Vehicles, containers, and shipments can have IoT sensors added to them so that their position, temperature, and other details can be monitored the whole time they are being transported. Reducing delays, losses, and inventory management expenses through timely delivery of products is made possible by effective route planning and real-time monitoring [67].

Public transportation systems are being transformed by the IoT. The passenger experience is improved through connected buses and trains that deliver real-time updates on departure schedules, delays, and service interruptions [68]. Smart ticketing systems enable contactless payments and seamless transfers between different modes of public transportation [69]. Additionally, IoT-based passenger counting systems optimize capacity management, ensuring efficient utilization of resources. Safety and security are enhanced through IoT in transportation. Sensors and cameras can detect hazards, accidents, or abnormal driving behavior, triggering alerts to relevant authorities or emergency services for quick response. IoT-based surveillance systems also contribute to preventing theft and vandalism and ensuring the security of transportation assets and infrastructure. Furthermore, IoT generates vast amounts of data related to transportation, which can be analyzed to derive valuable insights. Advanced analytics and ML algorithms can optimize operations, inform decision-making, and plan for future infrastructure developments. IoT data helps transportation authorities and companies identify areas for improvement, allocate resources effectively, and enhance overall transportation efficiency [70].

15.4.10.2 Big data and analytics

15.4.10.2.1 Utilizing big data for transportation planning and optimization

Utilizing big data for transportation planning and optimization has become increasingly important in addressing the challenges of modern transportation systems. Here's how big data can be beneficial in this context: First, big data provides a wealth of information about transportation patterns, traffic flows, and travel behaviors. By analyzing large-scale datasets from various sources, such as GPS data, traffic sensors, social media, and mobile applications, transportation planners can gain valuable insights into travel demand, congestion hot spots, and transportation trends. This data-driven approach allows for a more accurate understanding of transportation needs and helps in designing effective transportation plans and strategies [71].

Second, big data analytics can assist in optimizing transportation operations and infrastructure. By analyzing real-time data, transportation agencies can identify traffic bottlenecks, monitor traffic conditions, and dynamically adjust signal timings and traffic management strategies. This increases the effectiveness of transportation networks by enhancing overall traffic flow and reducing congestion. Additionally, big data can aid in optimizing public transit routes and schedules, leading to improved service reliability and reduced travel times for commuters [72]. Moreover, big data enables predictive analytics, which can be valuable for transportation planning. By analyzing historical data combined with real-time information, predictive models can forecast future travel patterns, congestion levels, and demand fluctuations. This allows transportation planners to proactively anticipate and address potential issues, allocate resources efficiently, and optimize transportation networks based on anticipated needs. Big data also plays a crucial role in multimodal transportation planning. By integrating data from various transportation modes, such as public transit, cycling, and ride-sharing services, transportation planners can develop comprehensive transportation solutions that encourage modal shifts and promote sustainable travel options. This approach helps in reducing congestion, enhancing connectivity, and providing seamless mobility experiences for travelers [73].

Furthermore, big data analytics can support transportation infrastructure planning and investment decisions. By analyzing data on population growth, land use patterns, and transportation demand, planners can identify areas that require new infrastructure or improvements to existing facilities. This data-driven approach ensures that transportation investments are strategically allocated to meet current and future transportation needs. However, it's essential to consider privacy and ethical considerations when utilizing big data for transportation planning. Safeguarding personal data and ensuring compliance with privacy regulations are crucial aspects of responsible data usage [4].

15.4.10.2.2 Data collection methods and sources

When it comes to collecting data for transportation planning and optimization, various methods and sources can be utilized. Here are some common data collection methods and sources.

15.4.10.2.3 Traffic sensors

Traffic sensors, such as loop detectors, radar, or infrared sensors, are commonly deployed on roads to collect data on traffic flow, speed, and occupancy. These sensors provide real-time information about vehicle movements and can be used to analyze traffic patterns and congestion levels [74].

15.4.10.2.4 GPS and mobile applications

GPS technology and mobile applications provide a rich source of data on travel behavior and routes. GPS devices or smartphone applications can track the movement of vehicles, providing information on travel times, routes chosen, and traffic conditions. This data can be aggregated and analyzed to understand travel patterns and optimize transportation systems [75].

15.4.10.2.5 Surveys

Surveys, including household surveys, travel diaries, and origin-destination surveys, gather information directly from individuals about their travel behavior, preferences, and characteristics. Surveys can provide insights into commuting patterns, trip purposes, mode choice, and factors influencing travel decisions [75].

15.4.10.2.6 Intelligent transportation systems

ITS technologies, such as electronic toll collection systems, automated license plate readers, and CCTV cameras, generate valuable data for transportation planning. These systems capture data on traffic volumes, travel times, and incident management, enabling real-time monitoring and analysis of transportation networks [76].

15.4.10.2.7 *Social media and web data*

Social media platforms and web-based applications can provide valuable insights into transportation-related information. Analysis of social media posts, geo-tagged photos, and check-ins can provide data on popular destinations, travel experiences, and public sentiment related to transportation [77].

15.4.10.2.8 *Smart card data*

Smart card systems used in public transit, such as contactless fare cards, generate data on ridership, entry-exit points, and trip patterns. Analysis of smart card data can help optimize public transportation services, identify demand patterns, and evaluate fare structures [78].

15.4.10.2.9 *Remote sensing technologies*

Data on land use, transportation infrastructure, and geographical patterns are provided through remote sensing technologies such as satellite images and aerial surveys. These data sources are particularly useful for analyzing urban development, transportation network connectivity, and environmental factors influencing transportation planning [79].

15.4.10.2.10 *Crowdsourcing*

Crowdsourcing platforms and applications allow users to contribute data voluntarily. For transportation planning, crowdsourcing can involve collecting data on road conditions, traffic incidents, or cyclist/pedestrian routes from individuals who share their experiences through mobile apps or dedicated platforms [80].

15.4.10.3 *Analytical techniques for extracting insights from transportation data*

Analytical techniques play a crucial role in extracting valuable insights from transportation data. Here are some commonly used techniques for analyzing transportation data.

15.4.10.3.1 *Descriptive analytics*

Descriptive analytics focuses on summarizing and visualizing transportation data to understand its characteristics and patterns. It involves techniques such as data aggregation, data profiling, and data visualization. Descriptive analytics can help identify trends, patterns, and anomalies in transportation data, providing a foundation for further analysis [81].

15.4.10.3.2 Geospatial analysis

Geospatial analysis involves analyzing transportation data in a spatial context. It leverages geographic information system (GIS) technology to understand the relationships between transportation infrastructure, land use, and travel patterns. Geospatial techniques can be used to identify congestion hot spots, visualize traffic flows, analyze accessibility to transportation services, and optimize route planning [82]. Figure 15.3 serves as a visual aid to help readers comprehend the practical application of geospatial analysis in resource management and showcases the intricate details and information that can be extracted from geographic data.

15.4.10.3.3 Network analysis

Network analysis techniques focus on analyzing transportation networks and their characteristics. This involves examining network connectivity, linkages, travel times, and flow patterns. Network analysis techniques can help identify bottlenecks and critical nodes and evaluate the efficiency and resilience of transportation systems [83].

15.4.10.3.4 Predictive analytics

Using statistical modeling and ML techniques, predictive analytics foresees future transportation patterns and consequences. It involves analyzing historical data to identify ways and relationships that can then be used to make predictions. Predictive analytics can assist in predicting traffic congestion, travel demand, and transportation infrastructure needs, enabling proactive decision-making and planning [84, 85].

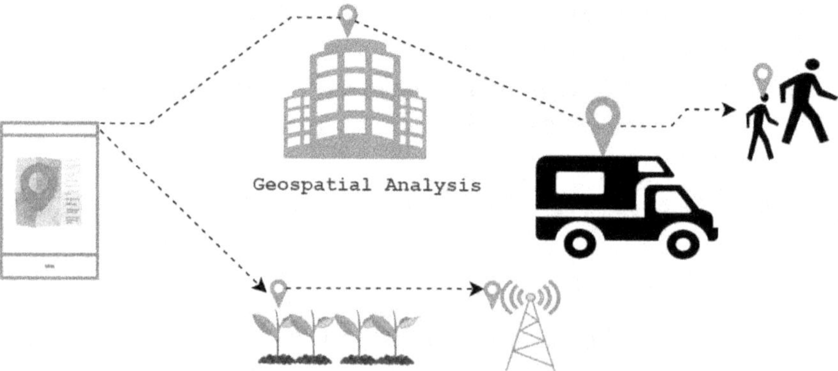

Figure 15.3 Geospatial analysis in action.

15.4.10.3.5 Optimization modeling

Optimization modeling techniques aim to find the best solutions to transportation problems by optimizing various factors. These models consider multiple variables, constraints, and objectives to optimize transportation operations and resource allocation. Optimization modeling can be applied to route optimization, vehicle scheduling, and transportation network design, among other applications [86].

15.4.10.3.6 Data mining

Large datasets are analyzed using data mining techniques to find hidden patterns and correlations. To find connections, clusters, or prediction models in the transportation data, ML methods are used. Data mining techniques can uncover insights related to travel behavior, mode choice, and factors influencing transportation decisions [87]. Figure 15.4 provides an overview of the data mining process, illustrating the key steps and components involved in extracting valuable insights from large datasets.

15.4.10.3.7 Simulation and modeling

Simulation and modeling techniques use mathematical and computational models to simulate transportation scenarios and evaluate their outcomes. These techniques can help assess the impact of different transportation interventions, test hypotheses, and forecast the effects of policy changes on transportation systems [88].

15.4.10.3.8 Social network analysis

Transportation system social networks are studied using social network analysis. It focuses on comprehending the connections between people, groups, or other entities within the context of transportation. Social network analysis may be used to examine collaboration and decision-making processes in transportation systems, identify key people, comprehend information flow, and more [89].

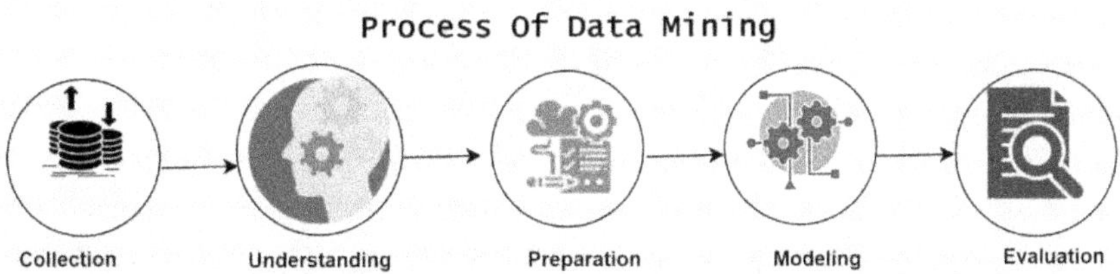

Figure 15.4 Process of data mining.

15.4.11 Artificial intelligence and machine learning in transportation

15.4.11.1 Introduction to AI and ML in transportation

The transportation sector is one of several sectors that are being quickly transformed by AI and ML. Efficiency, safety, and sustainability in transportation systems have all significantly improved as a result of the integration of AI and ML technology. The way we move people and things is being transformed by AI and ML, from driverless cars to ITMS [90]. Autonomous vehicles represent another breakthrough enabled by AI and ML. These vehicles utilize ML algorithms to learn from massive datasets collected by sensors and cameras, enabling them to recognize objects, interpret road conditions, and make real-time decisions [91]. AI systems power the navigation, perception, and interaction capabilities of autonomous vehicles, paving the way for safer and more efficient transportation. Predictive maintenance is another vital application of AI and ML in transportation. By analyzing data on vehicle performance, sensor readings, and historical maintenance records, ML algorithms can predict when components are likely to fail or require maintenance [92]. This predictive approach allows for proactive maintenance scheduling, reducing downtime, improving reliability, and optimizing maintenance costs, and ML also contributes to demand prediction and route optimization. By analyzing historical travel data, weather patterns, and other variables, these technologies can predict travel demand and optimize route planning. This facilitates efficient route selection, reduces travel times, and improves overall transportation efficiency. Furthermore, AI and ML enable enhanced customer experiences and personalized services. Through the analysis of individual travel patterns, preferences, and feedback, transportation providers can offer tailored recommendations, real-time updates, and optimized travel options to passengers, creating a more convenient and personalized journey for them [93].

15.4.12 Applications of AI and ML in traffic prediction, congestion management, and autonomous vehicles

AI and ML technologies have made significant contributions to traffic prediction, congestion management, and the development of autonomous vehicles. Let's explore their applications in each of these areas.

15.4.12.1 Traffic prediction

To forecast traffic patterns and circumstances, AI and ML systems can evaluate enormous volumes of both historical and real-time traffic data. By considering factors such as historical traffic flows, weather conditions, special events, and time of day, these models can generate accurate traffic predictions. This information is valuable for transportation authorities, drivers, and navigation systems, enabling them to plan routes, estimate travel times, and make informed decisions [94].

15.4.12.2 Congestion management

AI and ML techniques aid in managing traffic congestion more efficiently. Real-time traffic data from various sources, including sensors, cameras, and GPS devices, is analyzed to identify congestion hot spots and predict traffic bottlenecks. Using this information, traffic management systems can dynamically adjust signal timings, optimize lane configurations, and suggest alternative routes to alleviate congestion. These AI-powered systems improve traffic flow, reduce travel times, and enhance the overall efficiency of transportation networks [95].

15.4.12.3 Autonomous vehicles

AI and ML are at the core of the development of autonomous vehicles. ML algorithms learn from massive amounts of data collected by sensors, cameras, and other devices mounted on autonomous vehicles. They enable vehicles to perceive their surroundings, recognize objects, and make real-time decisions based on the data [96]. AI systems provide the intelligence and decision-making capabilities necessary for autonomous vehicles to navigate, respond to traffic situations, and interact with other vehicles and infrastructure. The performance and safety of autonomous cars are further improved by the ongoing learning and adaptation of ML algorithms [97]. By connecting and coordinating with other cars, these technologies allow autonomous vehicles to function effectively, lower the chance of accidents brought on by human mistakes, and improve traffic flow. Advanced driving assistance systems (ADAS), which support human drivers by offering features like adaptive cruise control, lane-keeping aid, and collision avoidance, also benefit from AI and ML [98]. Transportation systems can increase efficiency, decrease traffic, improve safety, and provide a smoother travel experience by utilizing AI and ML in traffic prediction, congestion management, and autonomous cars. The potential exists to significantly revolutionize how we view and interact with transportation via continued study and improvements in these fields [99].

15.4.13 Challenges and ethical considerations in AI implementation for transportation

The implementation of AI in transportation brings various challenges and ethical considerations that need to be addressed. Here are some key aspects to consider.

15.4.13.1 Data quality and privacy

For training and decision-making, AI systems significantly rely on data. To avoid biased or incorrect outcomes, it is essential to ensure the quality, correctness, and representativeness of the data. Concerns regarding privacy and data protection are also raised by the usage of personal data. Proper data anonymization, consent management, and compliance with privacy regulations are essential to maintain user trust and safeguard individual privacy [100].

15.4.13.2 Algorithmic bias and fairness

AI systems are susceptible to bias, which can result in discriminatory outcomes. Biased training data or biased algorithms can lead to unfair treatment or inequitable outcomes for certain groups of individuals. It is crucial to carefully evaluate and address biases throughout the AI development process, ensuring fairness and equal treatment for all users [101].

15.4.13.3 Safety and reliability

Safety is paramount in transportation systems, especially in the context of autonomous vehicles. AI systems need to be thoroughly tested and validated to ensure their reliability and ability to handle unexpected scenarios. Robust fail-safe mechanisms and backup systems should be in place to mitigate risks and ensure the safety of passengers, drivers, and pedestrians [102].

15.4.13.4 Human-machine interaction

As AI technologies are integrated into transportation systems, effective human-machine interaction becomes crucial. Ensuring clear communication, understandable interfaces, and appropriate levels of control and autonomy are vital to maintaining user confidence and safety. User acceptance and trust in AI- powered transportation systems will heavily depend on seamless and intuitive human-machine interaction [103].

15.4.13.5 Liability and accountability

The introduction of AI raises questions regarding liability and accountability. Determining responsibility in the event of accidents or failures involving autonomous vehicles or AI systems can be challenging. Establishing legal frameworks and regulations that clarify liability and accountability is essential to address these concerns and ensure a fair and responsible deployment of AI in transportation [103].

15.4.13.6 Workforce displacement and reskilling

The implementation of AI technologies, particularly in autonomous vehicles, may have implications for the workforce. The automation of certain tasks could lead to job displacement in certain sectors of the transportation industry. Adequate planning and measures for reskilling and reemployment should be in place to support affected workers and ensure a just transition [104].

15.4.13.7 Ethical decision-making and transparency

AI systems in transportation may face situations where ethical decisions need to be made, such as prioritizing certain actions in potential accidents. Ensuring transparency in the decision-making process, providing clear explanations for AI-generated decisions, and involving stakeholders in the development of ethical frameworks are important to address ethical concerns and promote accountability [105].

15.5 INTELLIGENT TRAFFIC MANAGEMENT SYSTEMS

15.5.1 Components of intelligent traffic management systems

Systems for intelligent traffic management are intended to improve the sustainability, safety, and efficiency of traffic flow in metropolitan environments. They utilize advanced technologies, data analysis, and communication systems to optimize traffic control and management [106]. The components of ITMS can vary depending on the specific implementation and requirements, but here are some common elements.

15.5.1.1 Traffic monitoring

The ITMS frequently uses a variety of sensors and detectors to keep track of traffic conditions. These may include cameras, radar equipment, infrared sensors, and loop detectors buried in the surface of the road. These sensors collect data on traffic volume, speed, density, and other parameters [107].

15.5.1.2 Data collection and processing

Real-time data is gathered by the ITMS from monitoring sensors and additional sources, including weather reports, traffic incident reports, and vehicle tracking systems. This data is collected, aggregated, and processed to generate insights into traffic patterns and conditions [108].

15.5.1.3 Traffic control centers

These centralized control centers serve as the nerve center of the ITMS. They receive and analyze the collected data, monitor traffic conditions, and make informed decisions for optimizing traffic flow. Traffic engineers and operators use specialized software and interfaces to control traffic signals, manage dynamic message signs (DMS), and respond to incidents or congestion [109].

15.5.1.4 Traffic signal control

Advanced traffic signal management systems that dynamically change signal timings based on real-time traffic circumstances are frequently included in the ITMS. These systems employ algorithms to enhance the timing and phasing of traffic signals to decrease congestion, shorten delays, and increase traffic flow [110].

15.5.1.5 Incident detection and management

The ITMS incorporates systems for detecting and managing traffic incidents, such as accidents, breakdowns, or road hazards. These systems can automatically identify incidents through data analysis or through reports from surveillance cameras or roadside assistance services. They facilitate prompt responses to incidents, including emergency services coordination, incident clearance, and rerouting [111].

15.5.1.6 Traveler information systems

The ITMS provides real-time information to drivers and travelers to assist them in making informed decisions. This can include dynamic message signs (DMS) along roads, smartphone applications, websites, or even in-vehicle navigation systems. Information may include traffic conditions, travel times, alternate routes, public transportation options, and parking availability [112].

15.5.1.7 Intelligent transportation systems integration

The ITMS often integrates with a broader system for intelligent transportation, such as public transportation systems, traffic management for freight and logistics, and connected vehicle technologies. This integration facilitates the exchange of data and coordination between different transportation modes and systems [113].

15.5.1.8 Communication infrastructure

To transfer data between sensors, control hubs, and linked devices, the ITMS depends on reliable communication networks. Fiber optic cables, wireless networks, or specialized communication systems like cellular vehicle-to-everything (C-V2X) or dedicated short-range communications (DSRC) technologies can all be a part of this infrastructure [114].

15.5.1.9 Analytical tools and algorithms

The ITMS leverages advanced analytics and algorithms to process large volumes of data and extract meaningful insights. ML and AI techniques may be employed to predict traffic patterns, optimize signal timings, detect anomalies, and generate proactive management strategies [115].

15.5.2 Traffic signal control and optimization

Traffic signal control and optimization is a crucial component of the ITMS. This aspect focuses on utilizing advanced algorithms and real-time data to optimize the timing and phasing of traffic signals. Traffic signal control systems in ITMS analyze the traffic data collected from sensors, such as cameras, loop detectors, or radar devices, to understand the current traffic conditions at intersections. This data includes information on traffic volume, speed, and congestion levels. By processing this information, the system can make informed decisions to adjust the signal timings dynamically. The optimization algorithms employed in traffic signal control aim to minimize congestion, reduce delays, and improve traffic flow efficiency [116]. These algorithms consider various factors, including traffic volume, historical data, time of day, and even special events. They calculate the optimal green signal duration for each traffic movement at an intersection based on the current and predicted traffic conditions. With real-time data and advanced algorithms, the traffic signal control system can respond quickly to changing traffic patterns. If congestion is detected at a particular intersection, the system may allocate more green time to the congested approach or adjust the signal timings to provide green waves, allowing a smoother progression of traffic [117].

Moreover, traffic signal control and optimization can be coordinated across multiple intersections, creating a network-wide approach known as arterial coordination. In this scenario, the system analyzes the traffic flow across a series of intersections and adjusts the signal timings collectively to create green waves, allowing vehicles to travel with minimal stops and reduced delays along a corridor. By optimizing traffic signal control, the ITMS can significantly improve traffic efficiency, reduce travel times, and enhance overall traffic safety. These systems are designed to adapt to changing traffic conditions and provide real-time adjustments, ensuring that traffic signals are responsive to the immediate needs of the road network [118].

15.5.3 Incident detection and management

Incident detection and management is a critical component of the ITMS that focuses on promptly detecting and effectively managing traffic incidents on roadways. This component aims to minimize the impact of incidents on traffic flow, enhance safety, and facilitate quick incident clearance [119]. The ITMS incorporates various technologies and systems to detect incidents in real time. These can include surveillance cameras, vehicle detection sensors, incident reporting from drivers or witnesses, and integration with emergency services. The collected data is analyzed to identify incidents such as accidents, breakdowns, road hazards, or any abnormal traffic patterns [120].

Once an incident is detected, the ITMS initiates a series of actions for efficient management. These actions can include the following.

15.5.3.1 Incident verification and classification

The system verifies the incident through multiple data sources and classifies it based on its severity, type, and impact on traffic. This step helps in determining the appropriate response and resource allocation [121].

15.5.3.2 Incident notification

The ITMS alerts the relevant authorities, including emergency services, traffic management centers, and maintenance crews, about the incident. This ensures timely response and coordination among different agencies [122].

15.5.3.3 Traffic control measures

The system may implement temporary traffic control measures to manage the incident scene effectively. This can include adjusting traffic signal timings, activating DMS to inform drivers about the incident, and deploying temporary barriers or lane closures to ensure safety [123].

15.5.3.4 Incident response and coordination

The ITMS facilitates communication and coordination between different response teams involved in incident management. It enables the exchange of information between emergency services, tow trucks, maintenance crews, and other relevant stakeholders, ensuring a coordinated and efficient response [124].

15.5.3.5 Incident clearance

The ITMS plays a crucial role in expediting the incident clearance process. It assists in coordinating the removal of damaged vehicles, debris, or any other obstructions from the roadway. By optimizing incident clearance, traffic disruptions are minimized, and normal traffic flow is restored as quickly as possible [125].

15.5.3.6 Traveler information dissemination

The ITMS provides real-time information to drivers and travelers about any incident and its impact on traffic conditions. This can be done through DMS, mobile applications, websites, or in-vehicle navigation systems. Timely and accurate information allows drivers to make informed decisions, choose alternative routes, or adjust their travel plans accordingly [126].

15.6 SMART MOBILITY SOLUTIONS

15.6.1 Connected and autonomous vehicles

The automobile industry has undergone a revolutionary change thanks to connected and autonomous vehicles (CAVs), which combine cutting-edge communication and sensor technology to let cars operate to varied degrees of automation and connection. CAVs have the potential to transform transportation by increasing security, effectiveness, and mobility in general [127].

15.6.1.1 Connected vehicles

The ability to communicate wirelessly gives connected cars the ability to share information with other cars, infrastructure, and other systems. For this communication, technology such as DSRC, C-V2X, or Wi-Fi can be employed. Some essential attributes and advantages of linked automobiles [128] are as follows.

15.6.1.2 Vehicle-to-vehicle communication

Connected cars are capable of exchanging real-time data about their location, velocity, direction, and other pertinent information with one another. Due to the availability of cooperative applications like junction management, platooning, and collision avoidance, traffic safety as a whole has increased [129].

15.6.1.3 Vehicle-to-infrastructure communication

Roadside infrastructure, such as traffic lights, road signs, and toll booths, may be communicated with by connected automobiles. This enables easy interaction with transportation management systems, real-time traffic statistics, and optimal traffic light timings [130].

15.6.1.4 Vehicle-to-pedestrian communication

Connected vehicles can also communicate with pedestrians' devices, such as smartphones or wearables, to enhance safety at crosswalks or crowded areas. These interactions can provide warnings or alerts to both drivers and pedestrians, reducing the risk of accidents [130]. Figure 15.5 shows various modes of communication and connectivity for vehicles.

15.6.1.5 Real-time traffic and navigation information

Connected vehicles can receive up-to-date traffic and navigation information from centralized systems or crowdsourced data. This enables drivers to make informed decisions about route planning, avoiding congestion, and optimizing travel times [130].

15.6.2 Autonomous vehicles

Advanced sensing, perception, and decision-making systems in autonomous vehicles, commonly referred to as self-driving cars or driverless automobiles, enable them to function in certain driving circumstances without human involvement [131]. They rely on various technologies, such as the following.

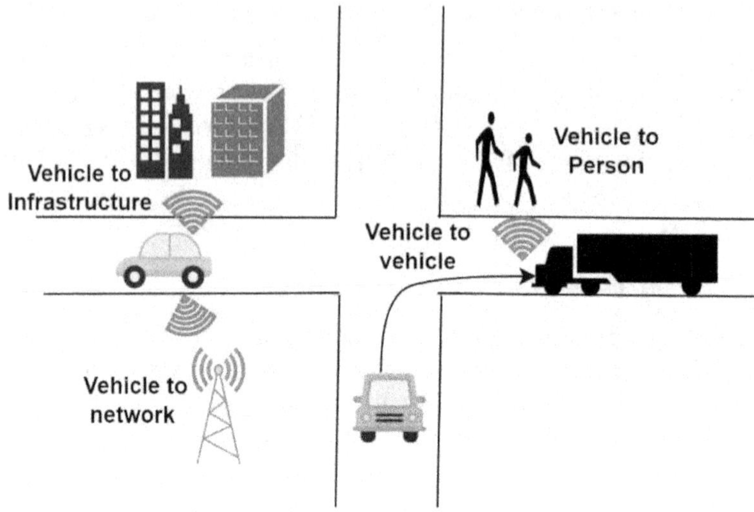

Figure 15.5 Various modes of communication and connectivity for vehicles, including vehicle to infrastructure, vehicle to road and building, and vehicle to person.

15.6.2.1 Sensors

To understand their surroundings, autonomous cars employ a variety of sensors, including cameras, radar, lidar, and ultrasonic sensors. These sensors provide a 360-degree view of the vehicle's surroundings, detecting objects, pedestrians, and other vehicles [132].

15.6.2.2 Artificial intelligence and machine learning

Autonomous vehicles employ AI algorithms and ML techniques to process sensor data, interpret complex driving scenarios, and make real-time decisions. These systems learn from experience and improve their driving capabilities over time [94].

15.6.2.3 Control systems

Autonomous vehicles have sophisticated control systems that regulate acceleration, braking, and steering based on input from the perception and decision-making systems. These systems ensure precise and safe vehicle movements [133].

15.6.2.4 Safety and redundancy

Autonomous vehicles prioritize safety and incorporate redundant systems to mitigate risks. This includes redundant sensors, redundant processing units, and fail-safe mechanisms to handle critical situations or system failures [134].

15.7 SUSTAINABLE TRANSPORTATION SOLUTIONS

15.7.1 Electric mobility

The usage of EVs as a sustainable and low-emission substitute for conventional cars powered by internal combustion engines (ICEs) is referred to as "electric mobility." Due to its advantages for the environment and developments in battery and charging technology, electric mobility has significantly increased in popularity in recent years [135]. The ability of electric mobility to reduce greenhouse gas emissions is one of its main benefits. As they run on energy stored in rechargeable batteries rather than burning fossil fuels, EVs have no exhaust emissions. We can drastically reduce air pollution and stop climate change by switching from ICEs to electric automobiles. EVs have other advantages besides being environmentally friendly. Compared to ICE cars, they have reduced running expenses since electricity is often less expensive than gasoline or diesel fuel. Additionally, EVs might save owners money since they require less maintenance and have fewer moving components [136]. Its environmental credentials are further strengthened by the incorporation of renewable energy sources and electric transportation. Using electricity produced from renewable resources, such as solar or wind power, to charge electric cars guarantees that overall energy usage is clean and sustainable [137]. This synergy between electric mobility and renewable energy facilitates a greener transportation system. To support the adoption of electric mobility, governments and organizations worldwide have implemented various incentives and policies. These include financial incentives like tax credits, grants for charging infrastructure, and preferential parking or access to restricted areas. Such measures incentivize consumers to embrace EVs and foster the development of a comprehensive charging infrastructure network [138].

The growth of electric mobility also drives technological advancements in the automotive industry. Battery technology continues to improve, enabling longer driving ranges and shorter charging times. Charging infrastructure is expanding with the deployment of public charging stations and fast-charging networks. Additionally, advancements in in-vehicle connectivity and smart grid systems facilitate efficient energy management and optimize charging processes [139]. Though electric mobility continues to evolve, challenges remain. A larger charging infrastructure is essential to reduce range driver anxiety and guarantee easy access to charging stations. Battery technology needs further advancement to improve energy storage capacity and reduce costs. Standardization of charging protocols and interoperability of charging stations are necessary for seamless integration into the transportation network [140].

15.7.2 Smart parking systems

Smart parking systems utilize advanced technologies and data integration to improve the efficiency and convenience of parking. These systems aim to address parking challenges in urban areas, such as limited parking spaces, congestion, and time wasted searching for available parking spots [141]. Here are some key aspects of smart parking systems.

15.7.2.1 Real-time parking availability

Smart parking systems use a variety of sensors, such as cameras or ultrasonic sensors, to determine the state of parking spots' occupancy in real time. This data is then transmitted to a central control system or mobile application, allowing drivers to access up-to-date information about available parking spaces in their vicinity [142].

15.7.2.2 Parking guidance and navigation

Smart parking systems provide drivers with real-time guidance and navigation to available parking spaces. This can be done through mobile apps, in-vehicle navigation systems, or digital signage that displays the number of available spaces and guides drivers to the nearest open spot. This reduces the time and frustration associated with finding parking and optimizes the utilization of parking facilities [143].

15.7.2.3 Reservation and prebooking

Some smart parking systems enable drivers to reserve parking spaces in advance, either through mobile apps or online platforms. This feature is particularly useful for high-demand areas or during special events, ensuring that drivers have a guaranteed parking spot upon arrival [143].

15.7.2.4 Payment and integration

Smart parking systems facilitate convenient and seamless payment methods. They can integrate with mobile payment solutions, allowing drivers to pay for parking digitally without the need for physical payment at a meter or ticket machine. Integration with parking apps or e-wallets simplifies the payment process and enhances the overall user experience [143].

15.7.2.5 Data analytics and management

Smart parking systems collect and analyze data on parking occupancy, duration, and patterns. This data can provide valuable insights for urban planning, such as identifying high-demand areas, optimizing parking pricing, and predicting future parking needs. It enables authorities and parking operators to make informed decisions for managing parking resources effectively [143].

15.7.2.6 Sustainability and efficiency

By lessening traffic congestion and emissions brought on by cars circling in search of parking, smart parking systems support sustainability. These devices assist in optimizing traffic flow, reduce fuel consumption, and enhance air quality in metropolitan areas by directing cars to open spaces [144].

15.7.2.7 Enforcement and security

Smart parking systems can enhance enforcement capabilities by utilizing license plate recognition or digital permits to monitor parking compliance. They can also incorporate security features such as CCTV cameras or remote monitoring to ensure the safety of parked vehicles and deter criminal activities [145].

15.7.2.8 Integration with urban mobility solutions

Smart parking systems can integrate with broader urban mobility solutions, such as public transportation systems or ride-sharing platforms. This integration allows for seamless connections between different modes of transportation and encourages the use of sustainable and multimodal transportation options.

Smart parking systems offer significant benefits in terms of convenience, efficiency, and sustainability. By leveraging technology and data, these systems improve the overall parking experience for drivers while optimizing the utilization of parking spaces. As urban areas continue to face parking challenges, smart parking systems play a crucial role in shaping smarter and more connected cities [146].

15.7.3 Bike-sharing and micro-mobility solutions

Bike-sharing and micro-mobility solutions have emerged as popular alternatives to traditional transportation modes, providing convenient and sustainable options for short-distance travel in urban areas. These systems promote the use of bicycles, electric scooters, and other small-scale modes of transportation [147]. Here are some key aspects of bike-sharing and micro-mobility solutions.

15.7.3.1 Bike-sharing programs

Bike-sharing programs allow individuals to rent bicycles for short periods, typically on a self-service basis. These programs provide docking stations where users can pick up and drop off bikes. Users can access bikes through mobile apps or membership cards, making it easy and convenient to access bicycles for short trips [148].

15.7.3.2 Electric scooters and micro-electric vehicles

In addition to bicycles, micro-mobility solutions often include electric scooters and other micro-electric vehicles. These small EVs offer a flexible and efficient way to cover short distances, providing an alternative to walking or using traditional transportation modes. Electric scooters are typically available for rent through mobile apps, allowing users to unlock and use them for a specific duration [148].

15.7.3.3 Last-mile connectivity

Bike-sharing and micro-mobility solutions are particularly valuable for last-mile connectivity, bridging the gap between public transportation and final destinations. Users can easily access bikes or electric scooters at transportation hubs or public transit stations, allowing them to complete the last leg of their journey conveniently and quickly [149].

15.7.4 Flexibility and convenience

Bike-sharing and micro-mobility systems offer users the flexibility to travel at their own pace and choose the most convenient routes. Users can pick up and drop off bikes or electric scooters at designated stations or approved parking areas, providing for a hassle-free experience. The availability of these options at various locations allows users to find a nearby vehicle easily [150].

15.7.4.1 Sustainable and environmentally friendly

Bike-sharing and micro-mobility solutions promote sustainable transportation by reducing carbon emissions and decreasing the reliance on fossil fuels. By encouraging cycling and the use of electric scooters, these systems contribute to cleaner air quality, lower noise pollution, and a more sustainable urban environment [151].

15.7.4.2 Integration with technology

Bike-sharing and micro-mobility systems often integrate with mobile apps, allowing users to find available bikes or electric scooters, unlock them, and make payments seamlessly. These apps provide real-time information about vehicle availability and nearby stations and trip planning features, enhancing the overall user experience [152].

15.8 CHALLENGES AND FUTURE DIRECTIONS

15.8.1 Data privacy and security

In the digital era, data privacy and security have become paramount concerns across various domains, including with the ITS. Safeguarding personal information and ensuring secure data practices are essential for maintaining individuals' privacy rights. To secure sensitive data, organizations must adhere to data protection laws and put in place strong security measures. This includes employing secure communication protocols for data transmission, implementing access controls and authorization mechanisms, and anonymizing or aggregating data to minimize privacy risks. Obtaining informed consent and being transparent about data practices are crucial, as is securely retaining and deleting data as per regulations. Cybersecurity measures like firewalls and regular security audits help prevent unauthorized access and data breaches. Privacy impact assessments and data governance ensure that privacy considerations are integrated into system design and operations. Maintaining data privacy and security is critical to fostering public trust and confidence in the ITS.

15.8.2 Integration and interoperability

Integration and interoperability are critical factors in successfully implementing the ITS and innovative city initiatives. These concepts involve seamless connectivity and interaction between various components, procedures, and modes of transportation. Integration refers to a harmonious combination of different ITS technologies, data sources, and functionalities. It involves aggregating and coordinating data from multiple sensors, devices, and systems, such as traffic management systems, public transportation networks, and data analytics platforms. Through integration, disparate components can work together to provide a comprehensive and unified view of the transportation ecosystem. Interoperability ensures that different systems, technologies, and stakeholders can communicate and exchange data effectively. It involves developing and implementing standardized protocols, formats, and interfaces that enable seamless data sharing and collaboration across various platforms and devices. Interoperability allows different transportation modes and subsystems to interact and operate together, enhancing efficiency, safety, and user experience.

By promoting integration and interoperability, the ITS can achieve several benefits. First, it enables the aggregation and analysis of data from multiple sources, providing a holistic understanding of transportation patterns and facilitating informed decision-making. Second, it allows for better coordination and optimization of traffic management, transit operations, and other transportation services. This leads to improved efficiency, reduced congestion, and enhanced overall mobility. Moreover, integration and interoperability support the development of innovative and user-centric services. For example, integrated multimodal trip planning and payment systems enable seamless journeys across different modes of transportation. It also allows the integration of emerging technologies like CAVs, electric mobility, and intelligent parking systems into the broader transportation ecosystem. To achieve effective integration and interoperability, collaboration among various stakeholders is crucial. This includes government agencies, transportation authorities, technology providers, and standards organizations. Cooperation and adherence to common standards and specifications facilitate compatibility and exchange of data and services across different systems and platforms.

15.8.3 Future trends and emerging technologies

The future of transportation and urban mobility will be shaped by several exciting trends and emerging technologies. Electric and autonomous vehicles are expected to gain momentum with the increasing adoption of EVs and the development of self-driving capabilities. These advancements will revolutionize transportation by improving sustainability, safety, and efficiency [153]. Connected and intelligent transportation systems will play a vital role, integrating communication technologies and data analytics to create a seamless and efficient network. Real-time data sharing, optimized traffic management, and personalized traveler information systems will enhance the overall transportation experience. The concept of mobility-as-a-service (MaaS) will gain prominence, offering integrated platforms that combine various transportation modes and provide convenient, on-demand mobility options [154]. This approach will transform the way people plan and access transportation, simplifying payment systems and providing flexible choices for users. Shared mobility and on-demand services will continue to grow, with car-sharing, bike-sharing, and ride-hailing services becoming more prevalent. These options provide convenient alternatives to traditional modes of transportation, reducing congestion and improving resource utilization [154].

Sustainable and smart infrastructure will be a priority, focusing on the expansion of EV charging infrastructure, integration of renewable energy sources, and implementation of smart traffic management systems. Infrastructure development will also cater to alternative modes of transportation, such as dedicated bike lanes and pedestrian-friendly spaces [155]. The utilization of big data and AI will increase, enabling better decision-making and optimization of transportation systems [156]. AI algorithms will process vast amounts of data to predict traffic patterns, optimize routes, and improve overall efficiency and safety. Emerging technologies like hyperloop and next-generation high-speed rail systems hold the potential to transform long-distance transportation, offering ultrafast and energy-efficient travel options [157]. Additionally, blockchain technology may enhance security, privacy, and transparency in transportation systems, enabling secure transactions, vehicle ownership verification, and peer-to-peer transactions in mobility services. Last, the emergence of urban air mobility (UAM) with electric vertical takeoff and landing aircraft could introduce aerial transportation options, reducing congestion and improving connectivity in densely populated areas. While these future trends and emerging technologies hold great promise, their successful implementation will depend on collaborative efforts among stakeholders, supportive policies and regulations, and addressing infrastructure and societal challenges. Nonetheless, they offer exciting possibilities for creating more sustainable, efficient, and connected transportation systems in the future [158].

15.9 CONCLUSION

This chapter has offered a thorough investigation of the revolutionary possibilities of ITS in the context of smart cities. We have examined the fundamental ideas, innovations, and advantages of implementing intelligent transportation channels, which will revolutionize urban mobility and influence the cities of the future. Intelligent transportation networks are a game-changing response to the problems that urbanization and rising traffic congestion are creating. Smart cities may build integrated, effective, and sustainable transportation networks that serve their residents' requirements by combining cutting-edge technology like IoT, AI, big data, and autonomous cars.

City officials can improve traffic flow, lessen congestion, and boost overall safety using real-time data from a variety of sources, making travel easier and more pleasurable for both locals and visitors. Additionally, the incorporation of autonomous vehicles and connected infrastructure offers a glimpse into a future where transportation is not only efficient but also environmentally friendly, due to a reduction in carbon emissions and the promotion of eco-friendly mobility.

The enormous potential of data-driven decision-making in transportation planning is one of the important lessons from this chapter. Through the collection and analysis of options, intelligent transportation channels contribute to the broader goal of building sustainable and livable smart cities. The importance of CAVs is emphasized, highlighting their potential to transform transportation by leveraging advanced communication and sensing technologies. The book explores the benefits of CAVs, such as improved safety, increased efficiency, and enhanced connectivity within the transportation network. Multimodal transportation solutions are discussed as a means to integrate various modes of transportation, including public transit, cycling, walking, and ride-sharing. These solutions aim to create a seamless and efficient transportation network that offers flexible and convenient options for travelers.

Electric mobility emerges as a sustainable alternative to traditional vehicles, focusing on EVs. The book highlights the benefits of EVs, including reduced emissions, lower operating costs, and improved energy efficiency, contributing to a cleaner and more sustainable transportation system. Data privacy and security are recognized as crucial considerations in ITS. The chapter emphasizes the need for protecting personal data, secure data transmission, access controls, consent transparency, and compliance with data protection regulations. Ensuring data privacy and security is vital to maintaining public trust and confidence in ITS. The chapter concludes by discussing future trends and emerging technologies that will shape transportation and urban mobility. These include electric and autonomous vehicles, connected and intelligent transportation systems, MaaS, sustainable infrastructure, big data analytics, AI, and UAM. These trends and technologies have the potential to revolutionize transportation, enhance efficiency, and create more sustainable and connected cities. The journey toward creating intelligent transportation channels for smart cities requires collaborative efforts from governments, city planners, technology developers, and citizens alike. By aligning visions, leveraging technological advancements, and fostering an inclusive approach, we can build transportation ecosystems that empower our cities and pave the way for a more connected, accessible, and sustainable urban future.

REFERENCES

[1] C. S. Lai *et al.*, "A review of technical standards for smart cities," *Clean Technol.*, vol. 2, no. 3, pp. 290–310, 2020.

[2] T. M. Ghazal *et al.*, "IoT for smart cities: Machine learning approaches in smart healthcare—A review," *Futur. Internet*, vol. 13, no. 8, p. 218, 2021.
[3] M. A. Ahad, S. Paiva, G. Tripathi, and N. Feroz, "Enabling technologies and sustainable smart cities," *Sustain. Cities Soc.*, vol. 61, p. 102301, 2020.
[4] A. R. Javed *et al.*, "Future smart cities: Requirements, emerging technologies, applications, challenges, and future aspects," *Cities*, vol. 129, p. 103794, 2022.
[5] M. A. Khan, B. A. Alvi, A. Safi, and I. U. Khan, "Drones for good in smart cities: A review," in *Proceedings of the International Conference on Electrical, Electronics, Computers, Communication, Mechanical and Computing*, 2018, pp. 1–6.
[6] T. Yigitcanlar, N. Kankanamge, and K. Vella, "How are smart city concepts and technologies perceived and utilized? A systematic geo-Twitter analysis of smart cities in Australia," *J. Urban Technol.*, vol. 28, no. 1–2, pp. 135–154, 2021.
[7] D. Chen, P. Wawrzynski, and Z. Lv, "Cyber security in smart cities: A review of deep learning- based applications and case studies," *Sustain. Cities Soc.*, vol. 66, p. 102655, 2021.
[8] E. Ismagilova, L. Hughes, Y. K. Dwivedi, and K. R. Raman, "Smart cities: Advances in research—An information systems perspective," *Int. J. Inf. Manage.*, vol. 47, pp. 88–100, 2019.
[9] P. Bellini, P. Nesi, and G. Pantaleo, "IoT-enabled smart cities: A review of concepts, frameworks and key technologies," *Appl. Sci.*, vol. 12, no. 3, p. 1607, 2022.
[10] R. Sánchez-Corcuera *et al.*, "Smart cities survey: Technologies, application domains and challenges for the cities of the future," *Int. J. Distrib. Sens. Networks*, vol. 15, no. 6, p. 1550147719853984, 2019.
[11] I. U. Khan, A. Abdollahi, A. Jamil, B. Baig, M. A. Aziz, and F. Subhan, "A novel design of FANET routing protocol aided 5G communication using IoT," *J. Mob. Multimed.*, pp. 1333–1354, 2022.
[12] C. K. Toh, J. A. Sanguesa, J. C. Cano, and F. J. Martinez, "Advances in smart roads for future smart cities," *Proc. R. Soc. A*, vol. 476, no. 2233, p. 20190439, 2020.
[13] R. I. Meneguette, R. De Grande, and A. A. Loureiro, *Intelligent transport system in smart cities*, Springer International Publisher, 2018.
[14] X. Fan *et al.*, "Deep learning for intelligent traffic sensing and prediction: Recent advances and future challenges," *CCF Trans. Pervasive Comput. Interact.*, vol. 2, pp. 240–260, 2020.
[15] U. Majeed, L. U. Khan, I. Yaqoob, S. M. A. Kazmi, K. Salah, and C. S. Hong, "Blockchain for IoT-based smart cities: Recent advances, requirements, and future challenges," *J. Netw. Comput. Appl.*, vol. 181, p. 103007, 2021.
[16] A. Gohar and G. Nencioni, "The role of 5G technologies in a smart city: The case for intelligent transportation system," *Sustainability*, vol. 13, no. 9, p. 5188, 2021.
[17] S. M. Abdullah *et al.*, "Optimizing traffic flow in smart cities: Soft GRU-based recurrent neural networks for enhanced congestion prediction using deep learning," *Sustainability*, vol. 15, no. 7, p. 5949, Mar. 2023, doi: 10.3390/su15075949.
[18] J. Chang, S. Nimer Kadry, and S. Krishnamoorthy, "Review and synthesis of Big Data analytics and computing for smart sustainable cities," *IET Intell. Transp. Syst.*, vol. 14, no. 11, pp. 1363–1370, 2020.
[19] H. Hejazi and L. Bokor, "A survey on the use-cases and deployment efforts toward converged internet of things (IoT) and vehicle-to-everything (V2X) environments," *Acta Tech. Jaurinensis*, vol. 15, no. 2, pp. 58–73, 2022.

[20] H. U. Mustakim, "5G vehicular network for smart vehicles in smart city: A review," *J. Comput. Electron. Telecommun.*, vol. 1, no. 1, 2020.
[21] M. Muhammad and G. A. Safdar, "Survey on existing authentication issues for cellular-assisted V2X communication," *Veh. Commun.*, vol. 12, pp. 50–65, 2018.
[22] A. Sharma, Y. Awasthi, and S. Kumar, "The role of blockchain, AI and IoT for smart road traffic management system," in *2020 IEEE India Council International Subsections Conference (INDISCON)*, IEEE, 2020, pp. 289–296.
[23] S. Konduri, K. Raju, and G. Verma, "Dynamic road traffic signal control system using artificial intelligence," *Int. J. All Res. Educ. Sci. Methods*, vol. 11, no. 7, pp. 1158–1168, 2023.
[24] J. Bharadiya, "Artificial intelligence in transportation systems: A critical review," *Am. J. Comput. Eng.*, vol. 6, no. 1, pp. 34–45, Jun. 2023, doi: 10.47672/ajce.1487.
[25] R. Etminani-Ghasrodashti, R. Ketankumar Patel, S. Kermanshachi, J. Michael Rosenberger, D. Weinreich, and A. Foss, "Integration of shared autonomous vehicles (SAVs) into existing transportation services: A focus group study," *Transp. Res. Interdiscip. Perspect.*, vol. 12, p. 100481, Dec. 2021, doi: 10.1016/j.trip.2021.100481.
[26] R. K. Patel, R. Etminani-Ghasrodashti, S. Kermanshachi, J. M. Rosenberger, and D. Weinreich, "Exploring preferences towards integrating the autonomous vehicles with the current microtransit services: A disability focus group study," in *International Conference on Transportation and Development 2021*, 2021, pp. 355–366.
[27] J. A. Guerrero-Ibanez, S. Zeadally, and J. Contreras-Castillo, "Integration challenges of intelligent transportation systems with connected vehicle, cloud computing, and internet of things technologies," *IEEE Wirel. Commun.*, vol. 22, no. 6, pp. 122–128, Dec. 2015, doi: 10.1109/MWC.2015.7368833.
[28] N. Islam and S. Shin, "Robust deep learning models for OFDM-based image communication systems in intelligent transportation systems (ITS) for smart cities," *Electronics*, vol. 12, no. 11, p. 2425, 2023.
[29] A. Alkhodair, S. P. Mohanty, and E. Kougianos, "FlexiChain 3.0: Distributed ledger technology-based intelligent transportation for vehicular digital asset exchange in smart cities," *Sensors*, vol. 23, no. 8, p. 4114, 2023.
[30] C. Liu and L. Ke, "Cloud assisted Internet of things intelligent transportation system and the traffic control system in the smart city," *J. Control Decis.*, vol. 10, no. 2, pp. 174–187, 2023.
[31] S. Kale and A. Choudhary, "Intelligent transportation system in India," *Int. Res. J. Modern. Eng. Technol. Sci.*, vol. 5, no. 7, pp. 1066–1068, 2023.
[32] X. Deng *et al.*, "A review of 6G autonomous intelligent transportation systems: Mechanisms, applications and challenges," *J. Syst. Archit.*, p. 102929, 2023.
[33] A. Sestino, M. Sestino, L. Ingrang, and G. Guido, "Sustainable smart mobility and intelligent transport systems as enabling factors of smart city transition oriented digital transition strategies: Evidences from Italy," *Electron. J. Knowl. Manag.*, vol. 1, pp. 1–21, 2023.
[34] I. U. Khan *et al.*, "RSSI-controlled long-range communication in secured IoT-enabled unmanned aerial vehicles," *Mob. Inf. Syst.*, vol. 2021, pp. 1–11, 2021.
[35] E. A. Kurniati, R. J. Sumabrata, and S. Maimunah, "Development of MAAS (mobility as a service) implementation framework in Indonesia's city," *Int. J. Eng. Adv. Res.*, vol. 5, no. 2, pp. 11–21, 2023.

[36] V. T. Tran, "Improving traffic management efficiency through reinforcement learning-based traffic signal control and citywide transit simulation," Master Theses and Doctoral Dissertations, 2023. https://scholar.utc.edu/theses/823

[37] Y. Bai, H. Lin, A. M. Abed, A. Deifalla, T. R. Alsenani, and S. Elattar, "A comprehensive investigation of a water and energy-based waste integrated system: Techno-eco-environmental-sustainability aspects," *Chemosphere*, vol. 327, p. 138454, 2023.

[38] X. Xiao *et al.*, "Parking prediction in smart cities: A survey," *IEEE Trans. Intell. Transp. Syst.*, vol. 24, no. 10, 2023.

[39] N. M. C. Galego and R. M. Pascoal, "Cybersecurity in smart cities: Technology and data security in intelligent transport systems," in *Perspectives and Trends in Education and Technology: Selected Papers from ICITED 2021*, Springer, 2022, pp. 17–33.

[40] R. S. Shankaran and L. Rajendran, "Intelligent transport systems and traffic management," *B. Smart Cities Concepts, Pract. Appl.*, 2022.

[41] A. Rehman, K. Haseeb, T. Saba, J. Lloret, and Z. Ahmed, "Towards resilient and secure cooperative behavior of intelligent transportation system using sensor technologies," *IEEE Sens. J.*, vol. 22, no. 7, pp. 7352–7360, 2022.

[42] M. M. Kamruzzaman, "Key technologies, applications and trends of internet of things for energy-efficient 6G wireless communication in smart cities," *Energies*, vol. 15, no. 15, p. 5608, 2022.

[43] I. U. Khan *et al.*, "Intelligent detection system enabled attack probability using Markov chain in aerial networks," *Wirel. Commun. Mob. Comput.*, vol. 2021, pp. 1–9, 2021.

[44] S. Zhou, C. Wei, C. Song, X. Pan, W. Chang, and L. Yang, "Short-term traffic flow prediction of the smart city using 5G internet of vehicles based on edge computing," *IEEE Trans. Intell. Transp. Syst.*, vol. 24, no. 2, pp. 2229–2238, 2022.

[45] X. Xia, R. Yu, and S. Zhang, "Evaluating the impact of smart city policy on carbon emission efficiency," *Land*, vol. 12, no. 7, p. 1292, 2023.

[46] K. N. Qureshi and A. H. Abdullah, "A survey on intelligent transportation systems," *Middle-East J. Sci. Res.*, vol. 15, no. 5, pp. 629–642, 2013.

[47] L. Figueiredo, I. Jesus, J. A. T. Machado, J. R. Ferreira, and J. L. M. De Carvalho, "Towards the development of intelligent transportation systems," in *ITSC 2001. 2001 IEEE Intelligent Transportation Systems. Proceedings (Cat. No. 01TH8585)*, IEEE, 2001, pp. 1206–1211.

[48] L. Qi, "Research on intelligent transportation system technologies and applications," in *2008 Workshop on Power Electronics and Intelligent Transportation System*, IEEE, 2008, pp. 529–531.

[49] S. An, B.-H. Lee, and D.-R. Shin, "A survey of intelligent transportation systems," in *2011 Third International Conference on Computational Intelligence, Communication Systems and Networks*, IEEE, 2011, pp. 332–337.

[50] I. U. Khan, S. B. H. Shah, L. Wang, M. A. Aziz, T. Stephan, and N. Kumar, "Routing protocols & unmanned aerial vehicles autonomous localization in flying networks," *Int. J. Commun. Syst.*, p. e4885, 2021.

[51] J. Guerrero-Ibáñez, S. Zeadally, and J. Contreras-Castillo, "Sensor technologies for intelligent transportation systems," *Sensors*, vol. 18, no. 4, p. 1212, Apr. 2018, doi: 10.3390/s18041212.

[52] I. U. Khan *et al.*, "Reinforce based optimization in wireless communication technologies and routing techniques using internet of flying vehicles," in *The 4th*

International conference on future networks and distributed systems (ICFNDS), 2020, pp. 1–6.

[53] G. Dimitrakopoulos and P. Demestichas, "Intelligent transportation systems," *IEEE Veh. Technol. Mag.*, vol. 5, no. 1, pp. 77–84, 2010.

[54] I. U. Khan, I. M. Qureshi, M. A. Aziz, T. A. Cheema, and S. B. H. Shah, "Smart IoT control-based nature inspired energy efficient routing protocol for flying ad hoc network (FANET)," *IEEE Access*, vol. 8, pp. 56371–56378, 2020.

[55] M. Alam, J. Ferreira, and J. Fonseca, "Introduction to intelligent transportation systems," *Intell. Transp. Syst. Dependable Veh. Commun. Improv. Road Saf.*, pp. 1–17, 2016.

[56] C. N. E. Anagnostopoulos, I. E. Anagnostopoulos, V. Loumos, and E. Kayafas, "A license plate-recognition algorithm for intelligent transportation system applications," *IEEE Trans. Intell. Transp. Syst.*, vol. 7, no. 3, pp. 377–392, 2006.

[57] S. Ezell, "Explaining international IT application leadership: Intelligent transportation systems," 2010. https://trid.trb.org/View/911843

[58] M. Alam, J. Ferreira, and J. Fonseca, "Intelligent transportation systems," *Stud. Syst. Decis. Control*, vol. 52, pp. XIV–270, 2016.

[59] J.-H. Lee and S.-Y. Jung, "SNR analyses of the multi-spectral light channels for optical wireless LED communications in intelligent transportation system," in *2014 IEEE 79th Vehicular Technology Conference (VTC Spring)*, IEEE, 2014, pp. 1–5.

[60] A. Ghazal, C.-X. Wang, B. Ai, D. Yuan, and H. Haas, "A nonstationary wideband MIMO channel model for high-mobility intelligent transportation systems," *IEEE Trans. Intell. Transp. Syst.*, vol. 16, no. 2, pp. 885–897, 2014.

[61] F. Arena, G. Pau, and A. Severino, "A review on IEEE 802.11 p for intelligent transportation systems," *J. Sens. Actuator Networks*, vol. 9, no. 2, p. 22, 2020.

[62] X. Cheng, L. Yang, and X. Shen, "D2D for intelligent transportation systems: A feasibility study," *IEEE Trans. Intell. Transp. Syst.*, vol. 16, no. 4, pp. 1784–1793, 2015.

[63] S. K. Sripathi Venkata Naga, R. Yesuraj, S. Munuswamy, and K. Arputharaj, "A comprehensive survey on certificate-less authentication schemes for vehicular ad hoc networks in intelligent transportation systems," *Sensors*, vol. 23, no. 5, p. 2682, 2023.

[64] O. O. Ajayi, A. B. Bagula, H. C. Maluleke, and I. A. Odun-Ayo, "Transport inequalities and the adoption of intelligent transportation systems in Africa: A research landscape," *Sustainability*, vol. 13, no. 22, p. 12891, 2021.

[65] S. Kaffash, A. T. Nguyen, and J. Zhu, "Big data algorithms and applications in intelligent transportation system: A review and bibliometric analysis," *Int. J. Prod. Econ.*, vol. 231, p. 107868, 2021.

[66] A. A. Laghari, K. Wu, R. A. Laghari, M. Ali, and A. A. Khan, "A review and state of art of Internet of Things (IoT)," *Arch. Comput. Methods Eng.*, pp. 1–19, 2021.

[67] N. Hossein Motlagh, M. Mohammadrezaei, J. Hunt, and B. Zakeri, "Internet of Things (IoT) and the energy sector," *Energies*, vol. 13, no. 2, p. 494, 2020.

[68] F. Meneghello, M. Calore, D. Zucchetto, M. Polese, and A. Zanella, "IoT: Internet of threats? A survey of practical security vulnerabilities in real IoT devices," *IEEE Internet Things J.*, vol. 6, no. 5, pp. 8182–8201, 2019.

[69] J. Wang, M. K. Lim, C. Wang, and M.-L. Tseng, "The evolution of the Internet of Things (IoT) over the past 20 years," *Comput. Ind. Eng.*, vol. 155, p. 107174, 2021.

[70] M. H. Kashani, M. Madanipour, M. Nikravan, P. Asghari, and E. Mahdipour, "A systematic review of IoT in healthcare: Applications, techniques, and trends," *J. Netw. Comput. Appl.*, vol. 192, p. 103164, 2021.

[71] X. Ding, Q. Gan, and M. P. Shaker, "Optimal management of parking lots as a big data for electric vehicles using internet of things and long–short term memory," *Energy*, vol. 268, p. 126613, 2023.

[72] H. Nozari, J. Ghahremani-Nahr, M. Fallah, and A. Szmelter-Jarosz, "Assessment of cyber risks in an IoT-based supply chain using a fuzzy decision-making method," *Int. J. Innov. Manag. Econ. Soc. Sci.*, vol. 2, no. 1, 2022.

[73] H. S. Sommerin, "Big data and its implication on supply chain management," Master Thesis, 2019.

[74] M. Selb, "From success assurance to product development: Data analysis at Gebrüder Weiss in the corporate logistics department," in *The Digitalization of Management Accounting: Use Cases from Theory and Practice*, Springer, 2023, pp. 37–56.

[75] A. Wattanapisit, C. H. Teo, S. Wattanapisit, E. Teoh, W. J. Woo, and C. J. Ng, "Can mobile health apps replace GPs? A scoping review of comparisons between mobile apps and GP tasks," *BMC Med. Inform. Decis. Mak.*, vol. 20, no. 1, pp. 1–11, 2020.

[76] A. Jarašūniene, "Research into intelligent transport systems (ITS) technologies and efficiency," *Transport*, vol. 22, no. 2, pp. 61–67, 2007.

[77] M. F. M. A. Hakeem, N. A. Sulaiman, M. Kassim, and N. M. Isa, "IoT bus monitoring system via mobile application," in *2022 IEEE International Conference on Automatic Control and Intelligent Systems (I2CACIS)*, IEEE, 2022, pp. 125–130.

[78] A. Haydari and Y. Yilmaz, "Real-Time Detection and Mitigation of DDoS Attacks in Intelligent Transportation Systems," in *2018 21st International Conference on Intelligent Transportation Systems (ITSC)*, IEEE, Nov. 2018, pp. 157–163, doi: 10.1109/ITSC.2018.8569698.

[79] N. Huq, R. Vosseler, and M. Swimmer, "Cyberattacks against intelligent transportation systems assessing future threats to ITS," TrendLabs Research Paper, p. 65, 2017.

[80] N. Nurelmadina *et al.*, "A systematic review on cognitive radio in low power wide area network for industrial IoT applications," *Sustainability*, vol. 13, no. 1, p. 338, 2021.

[81] J. Zhang, F.-Y. Wang, K. Wang, W.-H. Lin, X. Xu, and C. Chen, "Data-driven intelligent transportation systems: A survey," *IEEE Trans. Intell. Transp. Syst.*, vol. 12, no. 4, pp. 1624–1639, Dec. 2011, doi: 10.1109/TITS.2011.2158001.

[82] L. Dörrzapf, A. Kovács-Győri, B. Resch, and P. Zeile, "Defining and assessing walkability: A concept for an integrated approach using surveys, biosensors and geospatial analysis," *Urban Dev. Issues*, vol. 62, no. 1, pp. 5–15, 2019.

[83] L. Palk, J. T. Okano, L. Dullie, and S. Blower, "Travel time to health-care facilities, mode of transportation, and HIV elimination in Malawi: A geospatial modelling analysis," *Lancet Glob. Heal.*, vol. 8, no. 12, pp. e1555–e1564, 2020.

[84] D. A. Tedjopurnomo, *Learning-based geospatial data analysis for urban traffic and transportation improvement*, RMIT University, 2022.

[85] M. Usama *et al.*, "Predictive modelling of compression strength of waste GP/FA blended expansive soils using multi-expression programming," *Constr. Build. Mater.*, vol. 392, p. 131956, 2023.

[86] B. B. Dua, S. K. Singh, and I. A. Khan, "Automatic reading task prediction on tablet (PC) under the influence of noise and illumination via optimized LSTM," *Adv. Eng. Softw.*, vol. 180, p. 103391, 2023.
[87] J. A. Delgado, J. M. Bauça, M. I. Pastor, and A. Barceló, "Use of data mining in the establishment of age-adjusted reference intervals for parathyroid hormone," *Clin. Chim. Acta*, vol. 508, pp. 217–220, 2020.
[88] H. Ibrahim and B. H. Far, "Simulation-based benefit analysis of pattern recognition application in intelligent transportation systems," in *2015 IEEE 28th Canadian Conference on Electrical and Computer Engineering (CCECE)*, IEEE, 2015, pp. 507–512.
[89] R. Pajecki, F. Ahmed, X. Qu, X. Zheng, Y. Yang, and S. Easa, "Estimating passenger car equivalent of heavy vehicles at roundabout entry using micro-traffic simulation," *Front. Built Environ.*, vol. 5, no. 6, pp. 97–102, Jun. 2019, doi: 10.3389/fbuil.2019.00077.
[90] S. Filom, A. M. Amiri, and S. Razavi, "Applications of machine learning methods in port operations–A systematic literature review," *Transp. Res. Part E Logist. Transp. Rev.*, vol. 161, p. 102722, 2022.
[91] S.-H. Chung, "Applications of smart technologies in logistics and transport: A review," *Transp. Res. Part E Logist. Transp. Rev.*, vol. 153, p. 102455, 2021.
[92] F. Zantalis, G. Koulouras, S. Karabetsos, and D. Kandris, "A review of machine learning and IoT in smart transportation," *Futur. Internet*, vol. 11, no. 4, p. 94, 2019.
[93] A. Boukerche, Y. Tao, and P. Sun, "Artificial intelligence-based vehicular traffic flow prediction methods for supporting intelligent transportation systems," *Comput. Netw.*, vol. 182, p. 107484, 2020.
[94] R. Abduljabbar, H. Dia, S. Liyanage, and S. A. Bagloee, "Applications of artificial intelligence in transport: An overview," *Sustainability*, vol. 11, no. 1, p. 189, 2019.
[95] D. Hahn, A. Munir, and V. Behzadan, "Security and privacy issues in intelligent transportation systems: Classification and challenges," *IEEE Intell. Transp. Syst. Mag.*, vol. 13, no. 1, pp. 181–196, 2019.
[96] J. Wang, L. Zhang, Y. Huang, J. Zhao, and F. Bella, "Safety of autonomous vehicles," *J. Adv. Transp.*, vol. 2020, pp. 1–13, 2020.
[97] K. Kim, J. S. Kim, S. Jeong, J.-H. Park, and H. K. Kim, "Cybersecurity for autonomous vehicles: Review of attacks and defense," *Comput. Secur.*, vol. 103, p. 102150, 2021.
[98] D. J. Yeong, G. Velasco-Hernandez, J. Barry, and J. Walsh, "Sensor and sensor fusion technology in autonomous vehicles: A review," *Sensors*, vol. 21, no. 6, p. 2140, 2021.
[99] S. Feng *et al.*, "Dense reinforcement learning for safety validation of autonomous vehicles," *Nature*, vol. 615, no. 7953, pp. 620–627, 2023.
[100] B. C. Stahl *et al.*, "A systematic review of artificial intelligence impact assessments," *Artif. Intell. Rev.*, pp. 1–33, 2023.
[101] L. Gaur and B. M. Sahoo, *Explainable Artificial Intelligence for Intelligent Transportation Systems: Ethics and Applications*, Springer Nature, 2022.
[102] H. Herath and M. Mittal, "Adoption of artificial intelligence in smart cities: A comprehensive review," *Int. J. Inf. Manag. Data Insights*, vol. 2, no. 1, p. 100076, 2022.

[103] M. Ryan and B. C. Stahl, "Artificial intelligence ethics guidelines for developers and users: Clarifying their content and normative implications," *J. Information, Commun. Ethics Soc.*, vol. 19, no. 1, pp. 61–86, 2020.
[104] I. P. Pradhan and P. Saxena, "Reskilling workforce for the Artificial Intelligence age: Challenges and the way forward," in *The Adoption and Effect of Artificial Intelligence on Human Resources Management, Part B*, Emerald Publishing Limited, 2023, pp. 181–197.
[105] A. V. S. Madhav and A. K. Tyagi, "Explainable Artificial Intelligence (XAI): Connecting artificial decision-making and human trust in autonomous vehicles," in *Proceedings of Third International Conference on Computing, Communications, and Cyber-Security: IC4S 2021*, Springer, 2022, pp. 123–136.
[106] B. Chen, D. Sun, J. Zhou, W. Wong, and Z. Ding, "A future intelligent traffic system with mixed autonomous vehicles and human-driven vehicles," *Inf. Sci. (NY).*, vol. 529, pp. 59–72, 2020.
[107] A. A. Husain, T. Maity, and R. K. Yadav, "Vehicle detection in intelligent transport system under a hazy environment: A survey," *IET Image Process.*, vol. 14, no. 1, pp. 1–10, 2020.
[108] M. S. Sheikh and J. Liang, "A comprehensive survey on VANET security services in traffic management system," *Wirel. Commun. Mob. Comput.*, vol. 2019, pp. 1–23, 2019.
[109] Y. Luo, G. Feng, S. Wan, S. Zhang, V. Li, and W. Kong, "Charging scheduling strategy for different electric vehicles with optimization for convenience of drivers, performance of transport system and distribution network," *Energy*, vol. 194, p. 116807, 2020.
[110] D. Li, L. Deng, and Z. Cai, "Intelligent vehicle network system and smart city management based on genetic algorithms and image perception," *Mech. Syst. Signal Process.*, vol. 141, p. 106623, 2020.
[111] K. Vats, M. Fani, P. Walters, D. A. Clausi, and J. Zelek, "Event detection in coarsely annotated sports videos via parallel multi-receptive field 1D convolutions," in *Proceedings of the IEEE/CVF Conference on Computer Vision and Pattern Recognition Workshops*, 2020, pp. 882–883.
[112] G. Emmanuel, *Railway Route Optimization System*, Kabale University, 2015.
[113] A. K. Jain, "Sustainable transport," in *Climate Resilient, Green and Low Carbon Built Environment*, Springer, 2023, pp. 65–89.
[114] M. Gupta, J. Benson, F. Patwa, and R. Sandhu, "Secure V2V and V2I communication in intelligent transportation using cloudlets," *IEEE Trans. Serv. Comput.*, vol. 15, no. 4, pp. 1912–1925, 2020.
[115] W. AlAlaween, A. H. AlAlawin, M. Mahfouf, O. H. Abdallah, M. A. Shbool, and M. F. Mustafa, "A new framework for warehouse assessment using a Genetic-Algorithm driven analytic network process," *PLoS One*, vol. 16, no. 9, p. e0256999, 2021.
[116] G. Zheng *et al.*, "Diagnosing reinforcement learning for traffic signal control," arXiv Prepr. arXiv1905.04716, 2019.
[117] H. Wei *et al.*, "Presslight: Learning max pressure control to coordinate traffic signals in arterial network," in *Proceedings of the 25th ACM SIGKDD International Conference on Knowledge Discovery & Data Mining*, 2019, pp. 1290–1298.
[118] X. Zang, H. Yao, G. Zheng, N. Xu, K. Xu, and Z. Li, "Metalight: Value-based meta-reinforcement learning for traffic signal control," in *Proceedings of the AAAI Conference on Artificial Intelligence*, 2020, pp. 1153–1160.

[119] L. Li, Y. Lin, B. Du, F. Yang, and B. Ran, "Real-time traffic incident detection based on a hybrid deep learning model," *Transp. A Transp. Sci.*, vol. 18, no. 1, pp. 78–98, 2022.

[120] M. S. Sheikh, J. Liang, and W. Wang, "An improved automatic traffic incident detection technique using a vehicle to infrastructure communication," *J. Adv. Transp.*, vol. 2020, 2020.

[121] Y. Sun, T. Mallick, P. Balaprakash, and J. Macfarlane, "A data-centric weak supervised learning for highway traffic incident detection," *Accid. Anal. Prev.*, vol. 176, p. 106779, 2022.

[122] M. S. Sheikh and A. Regan, "A complex network analysis approach for estimation and detection of traffic incidents based on independent component analysis," *Phys. A Stat. Mech. its Appl.*, vol. 586, p. 126504, 2022.

[123] Z. Hao and X. Zhang, "Characteristics of urban traffic control measures in China in the background of COVID-19," *J. Adv. Comput. Intell. Intell. Informatics*, vol. 27, no. 3, pp. 352–359, 2023.

[124] C. Pasquale, S. Sacone, S. Siri, and A. Ferrara, "Traffic control for freeway networks with sustainability-related objectives: Review and future challenges," *Annu. Rev. Control*, vol. 48, pp. 312–324, 2019.

[125] R. Safiullin, V. Fedotov, and A. Marusin, "Method to evaluate performance of measurement equipment in automated vehicle traffic control systems," *Transp. Res. Procedia*, vol. 50, pp. 20–27, 2020.

[126] A. Drabicki, R. Kucharski, O. Cats, and A. Szarata, "Modelling the effects of real-time crowding information in urban public transport systems," *Transp. A Transp. Sci.*, vol. 17, no. 4, pp. 675–713, 2021.

[127] M. H. Rahman, M. Abdel-Aty, and Y. Wu, "A multi-vehicle communication system to assess the safety and mobility of connected and automated vehicles," *Transp. Res. Part C Emerg. Technol.*, vol. 124, p. 102887, 2021.

[128] L. Cai, W. Lv, L. Xiao, and Z. Xu, "Total carbon emissions minimization in connected and automated vehicle routing problem with speed variables," *Expert Syst. Appl.*, vol. 165, p. 113910, 2021.

[129] O. Kavas-Torris, S. Y. Gelbal, M. R. Cantas, B. Aksun Guvenc, and L. Guvenc, "V2X communication between connected and automated vehicles (CAVs) and unmanned aerial vehicles (UAVs)," *Sensors*, vol. 22, no. 22, p. 8941, 2022.

[130] G. Thandavarayan, M. Sepulcre, and J. Gozalvez, "Generation of cooperative perception messages for connected and automated vehicles," *IEEE Trans. Veh. Technol.*, vol. 69, no. 12, pp. 16336–16341, 2020.

[131] P. Kopelias, E. Demiridi, K. Vogiatzis, A. Skabardonis, and V. Zafiropoulou, "Connected & autonomous vehicles–Environmental impacts–A review," *Sci. Total Environ.*, vol. 712, p. 135237, 2020.

[132] S. Feng, X. Yan, H. Sun, Y. Feng, and H. X. Liu, "Intelligent driving intelligence test for autonomous vehicles with naturalistic and adversarial environment," *Nat. Commun.*, vol. 12, no. 1, p. 748, 2021.

[133] M. T. Riaz *et al.*, "The Intelligent Transportation Systems with Advanced Technology of Sensor and Network," in *2021 International Conference on Computing, Electronic and Electrical Engineering (ICE Cube)*, 2021, pp. 1–6, doi: 10.1109/ICECube53880.2021.9628331.

[134] R. Fusaro, N. Viola, D. Ferretto, V. Vercella, V. Fernandez Villace, and J. Steelant, "Life cycle cost estimation for high-speed transportation systems," *CEAS Sp. J.*, vol. 12, pp. 213–233, 2020.

[135] N. C. Onat, M. Kucukvar, N. N. M. Aboushaqrah, and R. Jabbar, "How sustainable is electric mobility? A comprehensive sustainability assessment approach for the case of Qatar," *Appl. Energy*, vol. 250, pp. 461–477, 2019.

[136] L. A. Garrow, B. J. German, and C. E. Leonard, "Urban air mobility: A comprehensive review and comparative analysis with autonomous and electric ground transportation for informing future research," *Transp. Res. Part C Emerg. Technol.*, vol. 132, p. 103377, 2021.

[137] B. K. Sovacool, J. Kester, L. Noel, and G. Z. de Rubens, "Energy injustice and Nordic electric mobility: Inequality, elitism, and externalities in the electrification of vehicle-to-grid (V2G) transport," *Ecol. Econ.*, vol. 157, pp. 205–217, 2019.

[138] R. Mounce and J. D. Nelson, "On the potential for one-way electric vehicle carsharing in future mobility systems," *Transp. Res. Part A Policy Pract.*, vol. 120, pp. 17–30, 2019.

[139] T. Skrúcaný, M. Kendra, O. Stopka, S. Milojević, T. Figlus, and C. Csiszár, "Impact of the electric mobility implementation on the greenhouse gases production in central European countries," *Sustainability*, vol. 11, no. 18, p. 4948, 2019.

[140] G. Z. De Rubens, L. Noel, J. Kester, and B. K. Sovacool, "The market case for electric mobility: Investigating electric vehicle business models for mass adoption," *Energy*, vol. 194, p. 116841, 2020.

[141] A. Fahim, M. Hasan, and M. A. Chowdhury, "Smart parking systems: Comprehensive review based on various aspects," *Heliyon*, vol. 7, no. 5, 2021.

[142] V. Duong, "Designing a full stack website with free parking for Kielo Office Solutions visitors," Thesis Report, 2023.

[143] S. Kandiş, *Real Time and Multi-Agent Smart Parking System using Computer Vision for Smart cities*, Altınbaş Üniversitesi, 2020.

[144] A. Saha, M. Abedin, and S. Palei, "Genetic engineering: A tool for sustainable production of biofuel from lignocellulosic biomass," *Genet. Metab. Eng. Improv. Biofuel Prod. from Lignocellul. Biomass*, pp. 97–101, 2020.

[145] K. Mishra and P. S. Aithal, "Planning assessment of transport system: A case from Nepal," *Int. J. Appl. Eng. Manag. Lett.*, vol. 6, no. 1, pp. 280–300, 2022.

[146] V. M. Mareeva, A. M. Ahmad, M. S. Ferwati, and S. B. Garba, "Sustainable urban regeneration of blighted neighborhoods: The case of Al Ghanim neighborhood, Doha, Qatar," *Sustainability*, vol. 14, no. 12, p. 6963, 2022.

[147] H. Yang, J. Huo, Y. Bao, X. Li, L. Yang, and C. R. Cherry, "Impact of e-scooter sharing on bike sharing in Chicago," *Transp. Res. Part A Policy Pract.*, vol. 154, pp. 23–36, 2021.

[148] D. Bylieva, V. Lobatyuk, and I. Shestakova, "Shared micromobility: Between physical and digital reality," *Sustainability*, vol. 14, no. 4, p. 2467, 2022.

[149] I. V Makarova, V. G. Mavrin, K. A. Shubenkova, and A. A. Khalyafiev, "Integration of bike lanes network with public transport routes to solve the last mile problem," in *Информационные технологии и инновации на транспорте*, 2019, pp. 94–103.

[150] H. Bencherif, T. Bendib, M. A. Abdi, and A. Meddour, "Environmental impact of flexible energy devices," in *Smart and Flexible Energy Devices*, CRC Press, 2022, pp. 119–132.

[151] M. Jović, E. Tijan, D. Žgaljić, and S. Aksentijević, "Improving maritime transport sustainability using blockchain-based information exchange," *Sustainability*, vol. 12, no. 21, p. 8866, 2020.

[152] D. Bamwesigye and P. Hlavackova, "Analysis of sustainable transport for smart cities," *Sustainability*, vol. 11, no. 7, p. 2140, 2019.

[153] L. Koh, A. Dolgui, and J. Sarkis, "Blockchain in transport and logistics–paradigms and transitions," *Int. J. Prod. Res.*, vol. 58, no. 7, pp. 2054–2062, 2020.

[154] M. Veres and M. Moussa, "Deep learning for intelligent transportation systems: A survey of emerging trends," *IEEE Trans. Intell. Transp. Syst.*, vol. 21, no. 8, pp. 3152–3168, 2019.

[155] S. Fang, Y. Wang, B. Gou, and Y. Xu, "Toward future green maritime transportation: An overview of seaport microgrids and all-electric ships," *IEEE Trans. Veh. Technol.*, vol. 69, no. 1, pp. 207–219, 2019.

[156] K. Shafique, B. A. Khawaja, F. Sabir, S. Qazi, and M. Mustaqim, "Internet of things (IoT) for next-generation smart systems: A review of current challenges, future trends and prospects for emerging 5G-IoT scenarios," *IEEE Access*, vol. 8, pp. 23022–23040, 2020.

[157] Y. Wang, D. Zhang, Y. Liu, B. Dai, and L. H. Lee, "Enhancing transportation systems via deep learning: A survey," *Transp. Res. Part C Emerg. Technol.*, vol. 99, pp. 144–163, 2019.

[158] P. A. Hancock, I. Nourbakhsh, and J. Stewart, "On the future of transportation in an era of automated and autonomous vehicles," *Proc. Natl. Acad. Sci.*, vol. 116, no. 16, pp. 7684–7691, 2019.

Chapter 16

Digital twins for industry 4.0 and 5.0

Indu Bala, Soumen Mondal, and Dipen Bepari

16.1 INTRODUCTION

The industrial revolution was a time of great change and upheaval, as new technologies and processes transformed the world of manufacturing. But with each new era came new challenges and complexities, as the boundaries between man and machine became increasingly blurred. The early days of Industry 1.0 were marked by the introduction of mechanized production processes, which brought with them a new level of efficiency and productivity. With the passage of time, the machines grew more powerful, and the human workers who operated them began to feel increasingly alienated and disconnected from their work. In Industry 2.0, the introduction of electricity brought with it a new era of automation and innovation. The assembly line revolutionized the manufacturing process, allowing factories to produce goods faster and more efficiently than ever before. Since the machines took on more and more tasks, the role of the human worker became increasingly marginalized [1].

Industry 3.0 brought with it a new wave of technological advancements, as computers and automation took center stage. As the semiconductor industry grew and computer-controlled machines and robots were developed, productivity and efficiency increased dramatically. With the technological advancements and the emergence of Industry 4.0, the once-sturdy human element of manufacturing began to waver in the face of cyber-physical systems and the internet of things (IoT). A new era of connectivity and cognizance has been ushered in by smart factories. Advanced analytics and machine learning enable machines to operate at their peak efficiency. Yet, the integration of artificial intelligence (AI) and robotics brought about a burst of perplexity and complexity. As automation took over an increasing number of tasks, the role of the human worker became increasingly uncertain, leaving many to question their place in the rapidly evolving industry.

In very near future, a new paradigm is emerging in the manufacturing sector known as Industry 5.0. It is anticipated that this new generation will focus on creating a collaborative environment where humans and machines coexist in a state of constant flux. The basic notion of Industry 5.0 is to create a more human-centered

 DOI: 10.1201/9781003496410-19

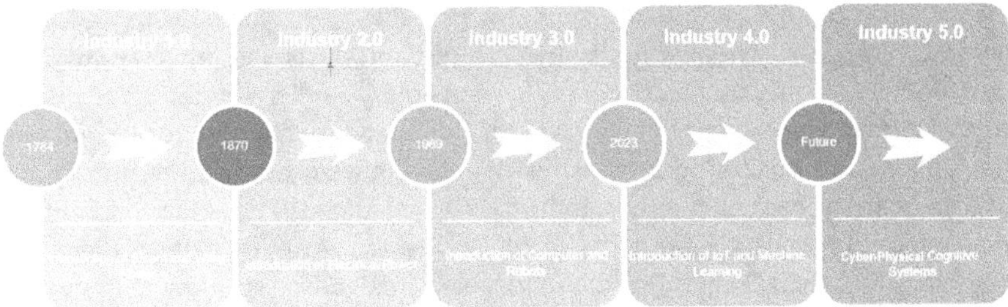

Figure 16.1 Evolution from industry 1.0 to 5.0.

approach to manufacturing to create a balance between automation and the workforce, where humans and machines work together to achieve the best possible outcomes. This approach is designed to create a more sustainable manufacturing process that is better for both workers and the environment [2, 3].

The fusion of virtual reality (VR) and augmented reality (AR) is one of the major components of Industry 5.0. Workers are able to communicate with equipment and procedures in a more natural and engaging way through technologies. Virtual training programs can be developed using AR and VR to give employees the opportunity to learn new skills in an enclosed environment. Another feature of Industry 5.0 is the use of advanced analytics and data visualization tools that enable manufacturers to analyze data in real-time and make informed decisions. They also provide a way to monitor and optimize production processes, leading to increased efficiency and productivity [4]. The evolution of various industrial generations is depicted in Figure 16.1.

16.2 WHAT MAKES INDUSTRY 5.0 DIFFERENT FROM INDUSTRY 4.0?

Industry 4.0 is the integration of digital technologies into the manufacturing sector. It entails utilizing technology like the IoT, AI, robotics, and cloud computing to build smart factories and supply chains that are more effective, adaptable, and customer responsive. However, as the world continues to evolve, so does technology. Industry 4.0 will soon be replaced by the next phase of industrial revolution—that is, Industry 5.0. This new era is focused on merging technology with human skills and creativity to create a sustainable, people-centric manufacturing environment. In Industry 4.0, machines and technology were given priority over human workers. While this was necessary to improve efficiency and productivity, it led to a decline in the importance of human skills and creativity. Technology and people will collaborate to develop a safer and environmentally friendly production process in Industry 5.0 [5].

One of the key differences between Industry 4.0 and Industry 5.0 is the focus on sustainability. Industry 4.0 was focused on maximizing efficiency and reducing waste, but it did not prioritize sustainable practices. Whereas Industry 5.0 is all about creating a sustainable future by designing products and processes that are environmentally friendly and socially responsible. Another key difference is the role of human workers. In Industry 4.0, machines and robots were responsible for most of the manufacturing process, while humans were relegated to monitoring and maintaining the machines. In Industry 5.0, humans will be at the center of the manufacturing process. They will use their skills and creativity to work alongside machines and technology to create unique and innovative products. Industry 5.0 will also require a shift in the way we approach education and training. In the past, education was focused on teaching technical skills that were necessary to operate machines and equipment. However, in Industry 5.0, there will be a greater emphasis on soft skills such as creativity, communication, and problem-solving that will enable workers to collaborate more effectively with machines and technology to create innovative solutions [6, 7]. Table 16.1 shows the key differences between Industry 4.0 and Industry 5.0.

Table 16.1 Differences between industry 4.0 and industry 5.0

Parameter	*Industry 4.0*	*Industry 5.0*
Technology	Connected devices and systems, IoT, AI, and machine learning	Advanced AI, augmented and virtual reality, quantum computing
Data	Big data, real-time data analytics	Contextual data, adaptive and predictive analytics
Automation	Increased automation of processes and systems	Adaptive automation, autonomous decision-making
Human-machine interaction	Humans as operators and maintainers of systems	Humans as collaborators and enablers with machines
Cybersecurity	Network and system security, secure data transmission	Advanced cybersecurity measures, AI-powered threat detection and prevention
Interoperability	Standardized communication protocols, open architecture	Integrated systems, cross-industry collaboration
Sustainability	Resource optimization, energy-efficient processes	Circular economy, sustainable production and consumption
Workforce	Reskilling and upskilling of workers, increased demand for technical skills	Human-centered design, focus on creativity and social skills
Customer experience	Personalized customer experiences, real-time responsiveness	Hyper-personalized experiences, cocreation and collaboration
Business models	Digitally enhanced products and services, data-driven decision-making	Value creation ecosystems, innovation-driven business models

Note: AI, artificial intelligence; IoT, internet of things.

16.3 DT—AN EMERGING TECHNOLOGY FOR INDUSTRY 4.0 AND INDUSTRY 5.0

A bewildering assortment of technologies awaits us as we enter the exciting new world of Industry 4.0 and 5.0; each promises to change the way we live, work, and interact with the world. At the forefront of this technological revolution is digital twin (DT) technology, a cutting-edge approach to creating virtual models of physical objects and systems. The technology enables us to simulate and optimize performance, monitor and predict maintenance needs, and even explore new possibilities that were previously unimaginable [8].

Alongside DT technology, there are many other technologies that are driving the evolution of Industry 4.0 and 5.0. For instance, the IoT facilitates the connection of devices and systems over the internet, enabling the gathering and analysis of enormous volumes of data. AI and machine learning are also critical technologies in Industry 4.0 and 5.0, enabling systems and machines to learn and adapt to new information and situations. Robotics and automation are also transforming the way machines work mechanically, freeing up human workers to focus on more complex and creative tasks. Apart from that, 3D printing, AR, and cloud computing are additional technologies that are driving the evolution of Industry 4.0 and 5.0. These technologies offer new possibilities for design, collaboration, and data storage and analysis [9].

16.3.1 DT technology for manufacturing industry

A new technology called a "DT" makes it possible to build virtual representations of real objects or systems. As shown in Figure 16.2, it is a computerized duplicate created to instantly mirror the system or actual asset. It has been developed to simulate the behavior of a physical asset or system to provide predictive insight, monitor performance, and optimize operations [10].

The necessity to monitor and optimize complex systems led to the development of the idea of DT technology. Dr. Michael Grieves, a professor at the University of Michigan, first proposed the concept in 2002. He suggested using DTs to build simulations of items and processes to enhance their performance. Since then, the technology has evolved to become a critical tool for industries ranging from manufacturing to health care, aerospace, and defense. The need for DT technology arises from the increasing complexity of modern systems. As systems become more complex, they become harder to monitor, maintain, and optimize. In these situations, DT technology offers a mechanism to track the performance of these intricate systems in real time and spot problems before they get out of hand [11].

It provides several benefits to industries that use it such as the following.

1. It enables predictive maintenance, which means that potential issues can be identified and addressed before they cause significant damage. This reduces downtime and improves the efficiency of the system, which can lead to significant cost-savings.

2. It optimizes system performance by creating a cybernetic model of the physical system, for testing different scenarios to operate the system efficiently. This can lead to improved performance, reduced energy consumption, and increased productivity [12].
3. It can be used to improve the design of products by creating a cybernetic model of the product or system to understand potential issues and to try other design alternatives before the product is manufactured. This can lead to improved product quality and reduced manufacturing costs.

The evolution of technology has been driven by advancements in computing, data analytics, and the IoT. The development of high-speed computing systems and powerful data analytics tools has enabled the creation of complex cybernetic models that can simulate the behavior of physical assets or systems. The growth of the IoT has also been a major driver of DT technology. The IoT enables the collection of real-time data from physical assets or systems. This cybernetic model can then be used to monitor the performance of the physical asset or system and identify potential issues [13].

DT technology is being used in a wide range of industries. In manufacturing, it is being used to monitor the performance of production lines, optimize the efficiency of equipment, and improve the quality of products. In health care, the technology is being used to create cybernetic models of patients to monitor their health and predict potential issues. In aerospace and defense, the performance of aircraft can be monitored and potential design issues can be identified and fixed before

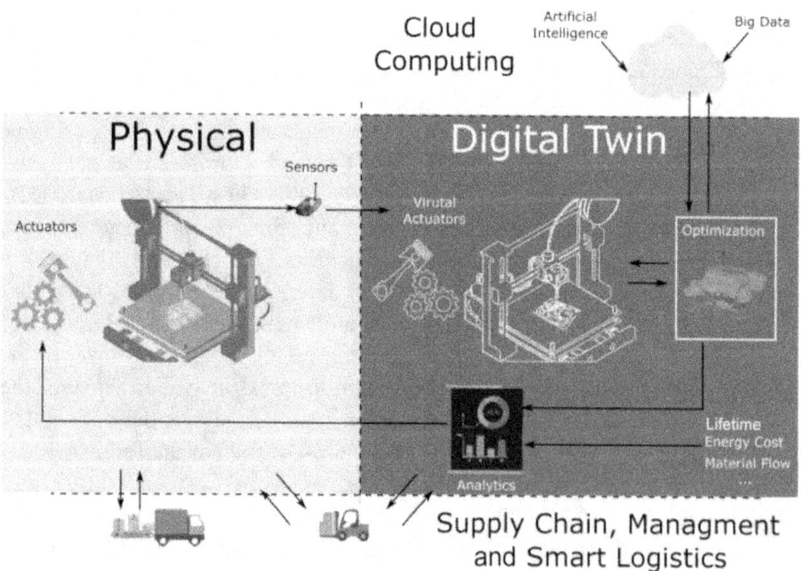

Figure 16.2 DT technology.

they become major problems. As the technology continues to evolve, it is likely to become an essential tool for industries that rely on complex systems and assets [14].

16.3.2 Indispensable role of DT technology in industry 4.0 and 5.0

As discussed earlier, the technology has become an essential tool for industries that rely on complex systems and assets. Its ability to provide a cybernetic model of a physical asset or system, enabling engineers to monitor its performance, predict potential issues, and optimize its operations has made it a critical component of Industry 4.0 and 5.0. In Industry 4.0, DT technology is used to monitor the performance of production lines, optimize the efficiency of equipment, and improve the quality of products. By creating cybernetic models of production lines, engineers can identify potential issues before they cause major problems. This enables predictive maintenance, reduces downtime, and improves efficiency of the production line [15]. Moreover, it could also be beneficial to optimize equipment performance by identifying the most efficient way to operate it with the least energy consumption. By creating cybernetic models of products, designers can identify potential issues and test different design options before the product is manufactured. This reduces the risk of defects and improves product quality.

In Industry 5.0, DT technology is used to create a sustainable, people-centric manufacturing environment. By creating cybernetic models of physical assets or systems, engineers can monitor the performance of the manufacturing process in real time and make adjustments to improve efficiency and productivity. This enables workers to spend more time on tasks that require human skills and creativity, such as problem-solving, critical thinking, and innovation. It can also be used to optimize the environmental impact of manufacturing processes in Industry 5.0. By creating cybernetic models of manufacturing processes, engineers can identify areas for improvement and test different scenarios to reduce waste and minimize the environmental impact of the process. In addition, DT technology can be used to create personalized products in Industry 5.0 [16]. By creating cybernetic models of products and combining them with customer data, manufacturers can create customized products that meet the unique needs and preferences of individual customers. This enables manufacturers to create more personalized products, improving customer satisfaction and loyalty.

16.3.3 Case study on DT for industry 4.0 and 5.0

One example of DT technology being used in Industry 4.0 is Siemens' *"plant data services."* Siemens is a global leader in industrial manufacturing and has developed a DT platform that enables manufacturers to monitor their production lines in real time. Siemens' plant data services platform uses sensors and data analytics to monitor production lines and fault identification before they cause major complications. The platform creates a cybernetic model of the production line to simulate different scenarios and test different operating conditions.

One example of the platform in action is at a Siemens' gas turbine manufacturing plant in Berlin. The plant uses the plant data services platform to monitor the performance of its production line and optimize its operations. The platform enables engineers to minimize downtime and improve the efficiency of its operations. Siemens' plant data services platform has enabled the Berlin plant to reduce its maintenance costs by up to 30% and increase the efficiency of its production line by up to 5%. This has enabled the plant to produce more gas turbines with fewer resources, reducing the environmental impact of its operations.

One more example of DT technology being used in Industry 5.0 is the *"Factory in a Day"* project, which is a European Union–funded research project that aims to develop a modular, reconfigurable, and digitally enabled production system. The project uses DT technology to create a cybernetic model of the production system, enabling engineers to simulate different scenarios and test different operating conditions. The cybernetic model is then used to optimize the production system, reducing waste and improving efficiency. The project has developed a modular production system that can be quickly assembled and disassembled, making it easy to reconfigure the production line to meet changing customer demands. The system is also designed to be environmentally friendly, with a reduced carbon footprint and minimal waste.

16.4 USE CASES OF DT TECHNOLOGY FOR INDUSTRY 4.0 AND 5.0

There are many use cases of DT technology for Industry 4.0 and 5.0. Few examples are as follows.

16.4.1 Industry 4.0

1. *Predictive Maintenance:* DT technology enables manufacturers to perform maintenance tasks before a failure occurs, reducing downtime and improving the efficiency of the production line.
2. *Quality Control:* It can be used to create cybernetic models of products and identify potential issues before they are manufactured. This reduces the risk of defects and improves product quality.
3. *Optimization of Production Processes:* It can be used to optimize the production performance by creating cybernetic models of the production line. This enables engineers to test different operating conditions and identify the most efficient way to operate the production line [17].

16.4.2 Industry 5.0

1. *People-Centric Manufacturing:* DT technology can be used to create a more people-centric manufacturing environment. By creating cybernetic models

of physical assets or systems, engineers can monitor the performance of the manufacturing process in real time and make adjustments to improve efficiency and productivity. This enables workers to spend more time on tasks that require human skills and creativity, such as brainstorming and innovation.

2. *Sustainable Manufacturing:* DT technology can be used to create a more sustainable manufacturing environment by identifying areas for improvement and testing different scenarios to reduce waste and minimize the environmental impact of the process.
3. *Personalized Products:* DT technology can be used to create personalized products by creating cybernetic models of products and combining them with customer data. This enables manufacturers to customize products as per the requirements and preferences of individual customers, improving customer satisfaction and loyalty.
4. *Factory Design and Layout:* It can be used to design and optimize the layout of factories. By creating cybernetic models of factories, engineers can test different layouts and identify potential issues before they are built. This reduces the risk of costly mistakes and improves the efficiency of the manufacturing process.

The use cases of the technology for Industry 4.0 and 5.0 are diverse and can be applied to many different industries and manufacturing processes. By creating cybernetic models of physical assets and systems, manufacturers can optimize their operations, reduce waste, and improve the product quality.

16.5 KEY COMPONENTS OF DT TECHNOLOGY

The key components of DT technology are as follows (refer Figure 16.3):

1. *Physical Object or System:* The physical object or system is the real-world entity that the DT represents.
2. *Sensors and IoT Devices:* DTs depend on sensors and IoT devices for data collection of the physical object or system, including its behavior, performance, and environmental conditions.
3. *Data Processing and Analytics:* The data collected from sensors and IoT devices is managed and examined using algorithms and analytics tools to generate insights and predictions.
4. *Modeling and Simulation:* The technology uses simulation modeling to create a cybernetic representation of the physical object or system, including its geometry, material properties, and behavior.
5. *Visualization and Interaction:* DTs provide interactive and immersive visualization tools to allow users to explore and interact with the cybernetic representation of the physical object or system.

6. *Communication and Connectivity:* DTs are connected to the physical object or system and can communicate with it in real time, enabling remote monitoring, control, and optimization.
7. *Collaboration and Integration:* It facilitates collaboration and integration between stakeholders, including designers, engineers, operators, and maintenance teams, by providing a common digital platform to work on [18].

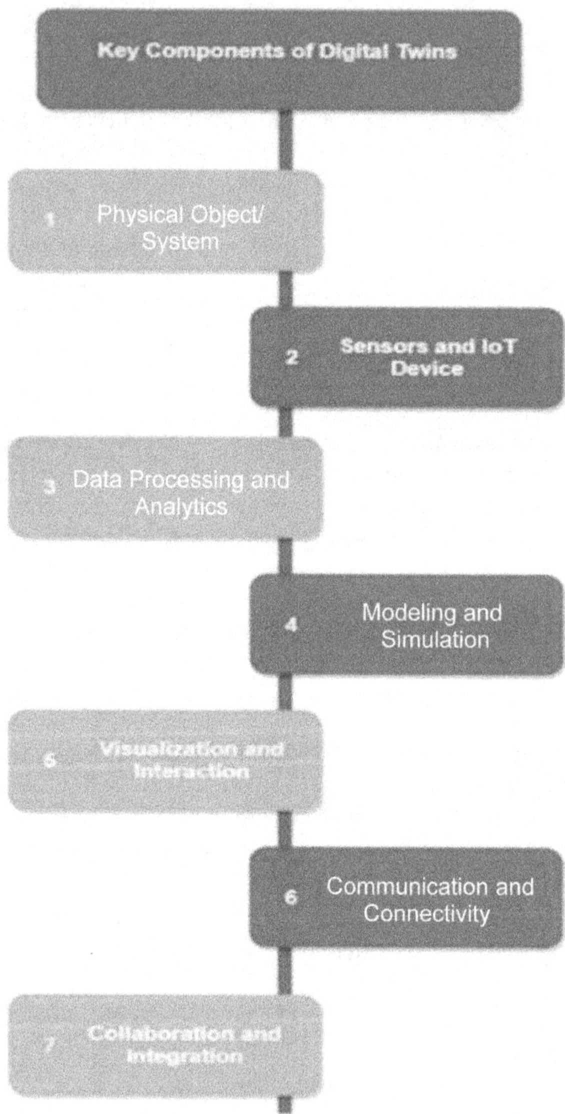

Figure 16.3 Key components of DT technology.

16.6 TYPES OF DTS BEING USED IN INDUSTRIES

There are several types of DTs being used in industries as shown in Figure 16.4, including the following:

1. *Product DT:* It is a cybernetic representation of a physical product, such as a machine, vehicle, or consumer product. It includes the product's design, geometry, materials, and behavior and is used for design, simulation, and optimization.
2. *Process DT:* It is a cybernetic representation of a physical process, such as a manufacturing line or supply chain. It includes the process's components, flow, and behavior and is used for optimization, monitoring, and control.
3. *Performance DT:* It is a cybernetic representation of the performance of a physical object or system, such as a building, power plant, or city. It includes data on the object's behavior, operating conditions, and environmental factors and is used for monitoring, prediction, and optimization.
4. *System DT:* It is a cybernetic representation of an entire complex system, such as a city, transportation network, or ecosystem. It includes data on the system's components, behavior, and interactions and is used for simulation, optimization, and decision-making.
5. *Human DT:* It is a cybernetic representation of an individual person, based on their physiological, behavioral, and environmental data. It is used for personalized health care, wellness, and performance optimization.
6. *Asset DT:* It is a cybernetic representation of a physical asset, such as a building, vehicle, or equipment. It includes data on the asset's performance, maintenance history, and environmental conditions and is used for predictive maintenance, optimization, and risk management.
7. *Network DT:* It is a cybernetic representation of a network of physical objects or systems, such as a power grid, water network, or communication network. It includes data on the network's topology, behavior, and interactions and is used for optimization, monitoring, and control.
8. *Environmental DT*: An environmental DT is a cybernetic representation of a physical environment, such as a city, forest, or ocean. It includes data on the environment's geography, weather, climate, and natural resources and is used for prediction, management, and conservation [19].
9. *Production DT:* It is a cybernetic representation of a production process, such as a manufacturing line or assembly line. It includes data on the process's components, flow, and behavior and is used for optimization, monitoring, and control.
10. *Supply Chain DT:* It is a cybernetic representation of a supply chain, including suppliers, manufacturers, distributors, and retailers. It includes data on the supply chain's components, flow, and behavior and is used for optimization, monitoring, and risk management.

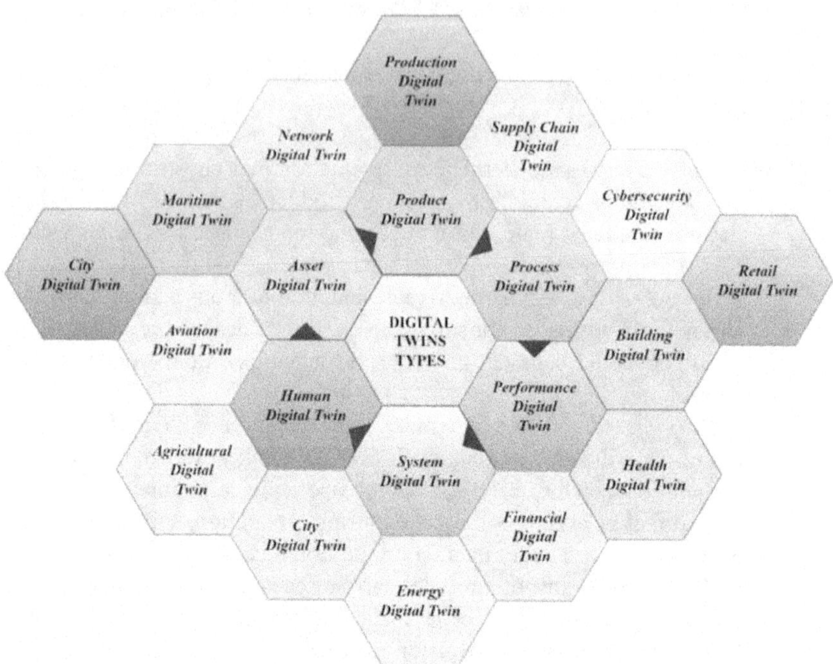

Figure 16.4 DT types.

11. *Cybersecurity DT:* It is a cybernetic representation of a system's security posture. It includes data on the system's vulnerabilities, threats, and risks and is used for testing security measures, identifying vulnerabilities, and simulating cyberattacks.
12. *Building DT:* It is a cybernetic representation of a physical building, including its layout, systems, and performance. It includes data on the building's energy consumption, indoor air quality, and occupancy and is used for energy optimization, maintenance, and occupant comfort.
13. *Health DT:* It is a cybernetic representation of a patient, including their medical history, genetics, and lifestyle factors. It includes data on the patient's physiology, behavior, and environmental factors and is used for personalized health care, disease prevention, and treatment optimization.
14. *Financial DT:* It is a cybernetic representation of a financial system, including markets, assets, and portfolios. It includes data on financial instruments, trading behavior, and market trends and is used for risk management, investment optimization, and financial modeling.
15. *Energy DT:* It is a cybernetic representation of an energy system, including power plants, grids, and storage systems. It includes data on energy production, consumption, and distribution and is used for energy optimization, management, and planning.

16. *City DT:* It is a cybernetic representation of a city, including its infrastructure, buildings, and people. It includes data on the city's transportation systems, energy consumption, air quality, and social behavior and is used for urban planning, resource management, and policymaking.
17. *Agricultural DT:* It is a cybernetic representation of a farm, including its crops, livestock, and environmental factors. It includes data on soil quality, weather conditions, and crop health and is used for precision agriculture, yield prediction, and resource management.
18. *Aviation DT:* It is a cybernetic representation of an aircraft, airport, or air traffic control system. It includes data on the aircraft's performance, maintenance history, and environmental conditions and is used for predictive maintenance, optimization, and risk management.
19. *Maritime DT:* It is a cybernetic representation of a vessel, port, or shipping network. It includes data on the vessel's performance, cargo, and environmental conditions and is used for predictive maintenance, optimization, and risk management.
20. *Retail DT:* It is a cybernetic representation of a retail store or chain, including its layout, inventory, and customer behavior. It includes data on sales trends, customer preferences, and inventory levels and is used for optimization, planning, and marketing.

These are just a few examples of the types of DTs being used in industries, and there are many other applications as well. The use of DTs is growing rapidly, and they are becoming increasingly important for optimizing performance, reducing costs, and improving safety in a wide range of industries.

16.7 BENEFITS OF DT TECHNOLOGY IN MANUFACTURING INDUSTRIES

DTs are not an absolute new concept in the manufacturing sector. However, due to very recent technological developments, small and medium-sized firms have now started using it the most. Accurate digital representations of actual goods or procedures are now possible. Manufacturers can imitate the behavior of a real product, do tests and experiments, and keep track of the machine's efficiency by utilizing DT technology [20]. By lowering prices, raising productivity, and enhancing product quality, this technology has the potential to completely transform the manufacturing sector. Some of the most important advantages of DTs in the manufacturing sector are described here.

1. *Reduced Downtime:* DT technology enables the manufacturing industry to predict and prevent potential equipment failures or malfunctions, thus reducing downtime and increasing operational efficiency.

2. *Improved Quality Control:* DT technology can simulate and analyze the entire production process, enabling manufacturers to identify and fix any quality issues before they occur.
3. *Increased Productivity:* By optimizing production processes and identifying inefficiencies, DT technology can help manufacturers increase productivity and output.
4. *Cost-Savings:* DT technology enables manufacturers to identify and address issues in real time, reducing the need for costly repairs and maintenance.
5. *Enhanced Safety:* DT technology can simulate and analyze potential safety risks, enabling manufacturers to take proactive measures to prevent accidents and injuries.
6. *Better Decision-Making:* With real-time data and analytics, DT technology can provide manufacturers with the insights they need to make informed decisions about their production processes.
7. *Accelerated Innovation:* By simulating and testing new product designs and processes, DT technology can help manufacturers bring new products to market faster and more efficiently.
8. *Improved Supply Chain Management:* DT technology can help manufacturers optimize their supply chain by providing real-time visibility into inventory levels, demand forecasting, and logistics.
9. *Reduced Environmental Impact:* DT technology can help manufacturers identify opportunities to reduce waste, energy consumption, and carbon emissions by optimizing their production processes.
10. *Enhanced Customer Experience:* By improving product quality, reducing lead times, and increasing customization capabilities, DT technology can help manufacturers deliver a better customer experience and build customer loyalty.
11. *Remote Monitoring and Control:* DT technology enables manufacturers to remotely monitor and control their production processes, allowing them to make adjustments and address issues without the need for on-site personnel.
12. *Predictive Maintenance:* Using real-time data and analytics, DT technology can predict when equipment maintenance is required, reducing the risk of unplanned downtime and increasing equipment life span.
13. *Increased Flexibility:* DT technology can help manufacturers quickly adapt to changing market conditions and customer demands by simulating and testing new production processes and product designs.
14. *Improved Collaboration:* DT technology enables manufacturers to collaborate more effectively across departments and with external partners, improving communication and knowledge sharing.
15. *Competitive Advantage:* By leveraging DT technology to improve operational efficiency, product quality, and customer satisfaction, manufacturers can gain a competitive advantage.

16. *Knowledge Management:* DT technology can store and manage vast amounts of data, including historical production data and best practices, enabling manufacturers to make informed decisions and continuously improve their processes.
17. *Training and Education:* DT technology can be used for training and educating employees, allowing them to gain hands-on experience with different production processes and equipment without disrupting operations.
18. *Predictive Analytics:* DT technology can perform advanced analytics on production data, enabling manufacturers to identify trends and patterns and make proactive decisions.
19. *Reduced Risk:* DT technology can help manufacturers reduce risk by simulating and testing different scenarios, allowing them to identify potential issues before they occur.
20. *Regulatory Compliance:* By simulating and analyzing production processes, DT technology can help manufacturers ensure compliance with regulations and standards, reducing the risk of noncompliance penalties and litigation.

16.8 CHALLENGES IN IMPLEMENTATION OF DT TECHNOLOGY

While DT technology offers significant benefits to the manufacturing industry, there are several challenges that need to be addressed to fully exploit its potential for Industry 4.0 and 5.0 (refer Figure 16.5):

1. *Data Management:* DT technology relies on vast amounts of data, which must be collected, stored, and analyzed effectively. This requires robust data management systems and processes to ensure data accuracy, security, and accessibility.
2. *Interoperability:* DT technology often involves multiple systems and platforms, which must be integrated and interoperable to enable seamless data exchange and collaboration.
3. *Complexity:* DT technology can be complex and difficult to implement, requiring specialized skills and expertise. Manufacturers may need to invest in training and development to ensure their workforce can effectively leverage DT technology.
4. *Cost:* Implementing DT technology can be expensive, requiring significant investment in hardware, software, and infrastructure. Manufacturers must carefully evaluate the costs and benefits of DT technology to ensure a positive return on investment.
5. *Culture and Mindset:* Adopting DT technology requires a culture of innovation and a willingness to embrace new ways of working. Manufacturers must be willing to challenge traditional processes and mindsets to fully exploit the potential of DT technology.

6. *Security:* DT technology involves the use of sensitive data, which must be protected from cyber threats and breaches. Manufacturers must implement robust security measures to ensure the confidentiality, integrity, and availability of their data.
7. *Standardization:* There is currently a lack of standardization in DT technology, which can create interoperability issues and limit its scalability. The industry must work together to establish common standards and practices to enable the widespread adoption of DT technology.
8. *Ethics and Privacy:* The use of DT technology raises ethical and privacy concerns, particularly around the use of personal data. Manufacturers must ensure they comply with relevant regulations and ethical standards to protect individuals' rights and privacy.
9. *Scalability:* DT technology must be scalable to enable its use across different industries, applications, and use cases. The industry must develop scalable solutions that can be adapted to different contexts and environments.
10. *Maintenance and Support:* DT technology requires ongoing maintenance and support to ensure its continued effectiveness and value. Manufacturers must establish robust maintenance and support processes to ensure their DT technology remains up-to-date and relevant.
11. *Integration:* DT technology must be integrated with existing production systems and processes to maximize its effectiveness. Manufacturers must ensure that their DT technology is seamlessly integrated into their existing workflows to avoid disruption and minimize the learning curve for employees.
12. *Accuracy:* DT technology must accurately represent the physical system it is modeling to be effective. Manufacturers must ensure that the data used to create the DT is accurate and up-to-date and that the DT is validated against the physical system to ensure its accuracy.
13. *Adoption:* Despite the many benefits of DT technology, there may be a resistance to its adoption within some organizations. Manufacturers must develop effective strategies to communicate the benefits of DT technology and engage stakeholders across the organization to ensure its successful adoption.
14. *Sustainability:* The manufacturing industry is increasingly focused on sustainability, and DT technology can play a role in helping manufacturers reduce their environmental impact. However, manufacturers must ensure that their DT technology is designed and implemented in a sustainable way to avoid creating new environmental challenges.
15. *Intellectual Property:* DT technology may involve the use of proprietary data and intellectual property. Manufacturers must establish robust policies and procedures around data ownership and intellectual property to protect their assets and avoid disputes.

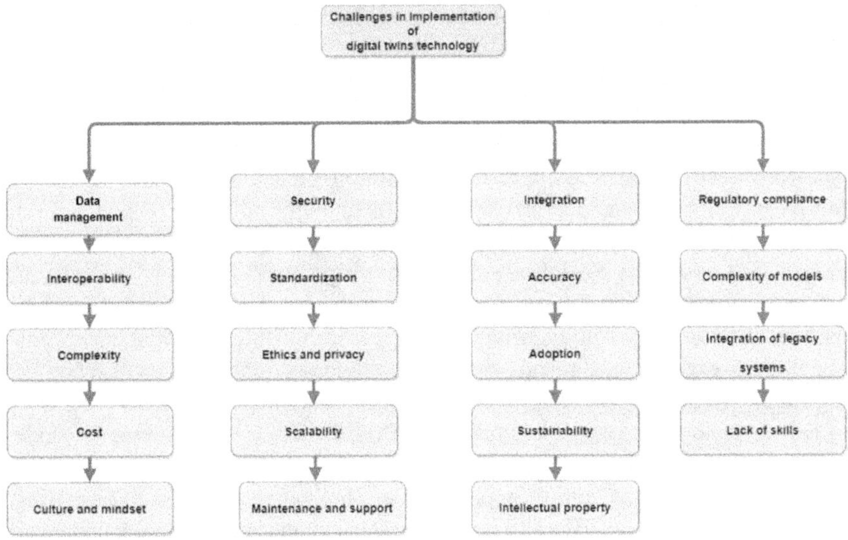

Figure 16.5 Challenges in the implementation of DT technology.

16. *Regulatory Compliance:* The use of DT technology may raise regulatory compliance issues, particularly around data privacy and security. Manufacturers must ensure they comply with relevant regulations and standards to avoid legal and reputational risks.
17. *Complexity of Models:* The complexity of DT models can be a challenge, especially when it comes to creating accurate models for complex systems. Manufacturers must invest in advanced modeling techniques and tools to ensure the accuracy and effectiveness of their DT technology.
18. *Integration of Legacy Systems:* Many manufacturing organizations have legacy systems that are not compatible with DT technology. Manufacturers must develop effective strategies for integrating legacy systems with DT technology to avoid disruption and ensure a smooth transition.
19. *Lack of Skills:* There is a shortage of skilled professionals who can effectively design, develop, and implement DT technology. Manufacturers must invest in training and development programs to ensure they have the necessary skills and expertise to effectively leverage DT technology.

16.9 RECOMMENDATIONS FOR THE SUCCESSFUL ADOPTION OF DT TECHNOLOGY FOR INDUSTRY 4.0 AND 5.0

To successfully adopt DT technology, manufacturers should follow a strategic approach. First, they need to develop a clear DT strategy encompassing goals, objectives, and an implementation road map. This ensures a well-defined direction for utilizing DT technology within the organization. Second, identifying the appropriate use cases tailored to their specific needs and challenges is crucial to maximize the value of DT technology deployment. Third, investing in robust data management systems and processes guarantees accurate, secure, and accessible data for effective utilization. Creating a culture of innovation and collaboration fosters an environment conducive to DT technology adoption. Manufacturers should address organizational change by preparing employees for adoption and emphasize its benefits. Scalability should be a key consideration when designing DT technology solutions, ensuring adaptability to various contexts. Establishing partnerships with technology providers and research institutions enables leveraging external expertise and resources. Developing cross-functional teams comprising experts from different areas of the organization facilitates seamless integration of DT technology into existing workflows. Prioritizing security and privacy safeguards data against cyber threats and breaches. Last, investing in training and development programs equips employees with the necessary skills to effectively leverage DT technology. By following these steps, manufacturers can navigate the DT journey successfully and unlock its transformative potential.

To maximize the benefits of DT technology, manufacturers should undertake several key initiatives. First, they need to engage with stakeholders across the organization, including employees, customers, and partners. This ensures that all stakeholders understand the advantages of DT technology and are supportive of its adoption. Second, manufacturers should measure and evaluate the success of their DT technology solutions regularly. By doing so, they can ensure that these solutions create value, meet goals, and fulfill objectives. Staying up-to-date with the latest technology trends and developments in DT technology is crucial. This allows manufacturers to leverage the most advanced and effective solutions available. Furthermore, fostering a continuous improvement mindset is essential. Manufacturers must consistently seek ways to enhance and refine their DT technology solutions over time. Collaboration with industry peers is another significant aspect. By sharing best practices, learning from one another's experiences, and driving innovation in DT technology, manufacturers can thrive. This collaboration can take the form of participating in industry associations, attending conferences and events, and actively networking with other manufacturers. By implementing these initiatives, manufacturers can optimize their utilization of DT technology and remain at the forefront of innovation within their industry.

16.10 CONCLUSION

In conclusion, DT technology has significant potential to revolutionize the manufacturing industry, particularly in the context of Industry 4.0 and 5.0. By simulating and analyzing production processes, DT technology can help manufacturers improve efficiency, quality, safety, and sustainability while reducing costs and downtime. However, the successful adoption of DT technology requires addressing several challenges, such as data management, interoperability, complexity, cost, culture and mindset, security, standardization, ethics and privacy, scalability, integration, accuracy, adoption, regulatory compliance, complexity of models, integration of legacy systems, lack of skills, and more. To fully exploit the potential of DT technology, manufacturers must develop a clear strategy, identify the right use cases, invest in data management, foster a culture of innovation, address organizational change, focus on scalability, establish partnerships, prioritize security and privacy, emphasize training and development, engage with stakeholders, measure and evaluate success, stay up-to-date with technology trends, foster a continuous improvement mindset, and collaborate with industry peers. By following these recommendations, manufacturers can successfully leverage DT technology to drive innovation, improve productivity, and gain a competitive advantage in the rapidly evolving manufacturing landscape.

REFERENCES

[1] Jones, David, Chris Snider, Aydin Nassehi, Jason Yon, and Ben Hicks. "Characterising the digital twin: A systematic literature review." *CIRP Journal of Manufacturing Science and Technology* 29 (2020): 36–52.
[2] Fuller, Aidan, Zhong Fan, Charles Day, and Chris Barlow. "Digital twin: Enabling technologies, challenges and open research." *IEEE Access* 8 (2020): 108952–108971.
[3] Liu, Mengnan, Shuiliang Fang, Huiyue Dong, and Cunzhi Xu. "Review of digital twin about concepts, technologies, and industrial applications." *Journal of Manufacturing Systems* 58 (2021): 346–361.
[4] Tao, Fei, He Zhang, Ang Liu, and Andrew YC Nee. "Digital twin in industry: State-of-the-art." *IEEE Transactions on Industrial Informatics* 15, no. 4 (2018): 2405–2415.
[5] Haag, Sebastian, and Reiner Anderl. "Digital twin–proof of concept." *Manufacturing Letters* 15 (2018): 64–66.
[6] Rasheed, Adil, Omer San, and Trond Kvamsdal. "Digital twin: Values, challenges and enablers from a modeling perspective." *IEEE Access* 8 (2020): 21980–22012.
[7] Boschert, Stefan, and Roland Rosen. "Digital twin—the simulation aspect." *Mechatronic Futures: Challenges and Solutions for Mechatronic Systems and Their Designers* (2016): 59–74.
[8] Schleich, Benjamin, Nabil Anwer, Luc Mathieu, and Sandro Wartzack. "Shaping the digital twin for design and production engineering." *CIRP Annals* 66, no. 1 (2017): 141–144.
[9] Qi, Qinglin, Fei Tao, Tianliang Hu, Nabil Anwer, Ang Liu, Yongli Wei, Lihui Wang, and A. Y. C. Nee. "Enabling technologies and tools for digital twin." *Journal of Manufacturing Systems* 58 (2021): 3–21.
[10] Wright, Louise, and Stuart Davidson. "How to tell the difference between a model and a digital twin." *Advanced Modeling and Simulation in Engineering Sciences* 7, no. 1 (2020): 1–13.

[11] Singh, Maulshree, Evert Fuenmayor, Eoin P. Hinchy, Yuansong Qiao, Niall Murray, and Declan Devine. "Digital twin: Origin to future." *Applied System Innovation* 4, no. 2 (2021): 36.

[12] Errandonea, Itxaro, Sergio Beltrán, and Saioa Arrizabalaga. "Digital twin for maintenance: A literature review." *Computers in Industry* 123 (2020): 103316.

[13] VanDerHorn, E., and S. Mahadevan. "Digital twin: Generalization, characterization and implementation." *Decision Support Systems* 145 (2021): 113524.

[14] Tao, Fei, Bin Xiao, Qinglin Qi, Jiangfeng Cheng, and Ping Ji. "Digital twin modeling." *Journal of Manufacturing Systems* 64 (2022): 372–389.

[15] Boschert, Stefan, Christoph Heinrich, and Roland Rosen. "Next generation digital twin." In *Proceedings of the TMCE conference*, Las Palmas de Gran Canaria, 2011, vol. 2018, pp. 7–11.

[16] Wu, Yiwen, Ke Zhang, and Yan Zhang. "Digital twin networks: A survey." *IEEE Internet of Things Journal* 8, no. 18 (2021): 13789–13804.

[17] Cimino, Chiara, Elisa Negri, and Luca Fumagalli. "Review of digital twin applications in manufacturing." *Computers in Industry* 113 (2019): 103130.

[18] Tao, Fei, Fangyuan Sui, Ang Liu, Qinglin Qi, Meng Zhang, Boyang Song, Zirong Guo, Stephen C-Y Lu, and Andrew YC Nee. "Digital twin-driven product design framework." *International Journal of Production Research* 57, no. 12 (2019): 3935–3953.

[19] Khajavi, Siavash H, Naser Hossein Motlagh, Alireza Jaribion, Liss C Werner, and Jan Holmström. "Digital twin: Vision, benefits, boundaries, and creation for buildings." *IEEE Access* 7 (2019): 147406–147419.

[20] Vachálek, Ján, Lukás Bartalský, Oliver Rovný, Dana Šišmišová, Martin Morháč, and Milan Lokšík. "The digital twin of an industrial production line within the industry 4.0 concept." In *2017 21st international conference on process control (PC)*, 2017, pp. 258–262. IEEE.

Index

Q

S

T

For Product Safety Concerns and Information please contact our EU representative GPSR@taylorandfrancis.com
Taylor & Francis Verlag GmbH, Kaufingerstraße 24, 80331 München, Germany

www.ingramcontent.com/pod-product-compliance
Lightning Source LLC
LaVergne TN
LVHW020607110826
845149LV00002B/396

* 9 7 8 1 0 3 2 8 0 3 1 8 0 *